01245 KU-626-449

WITHDRAWN FROM STOCK
The University of Liverpool

996

Carbon Partitioning
within and between organisms

ENVIRONMENTAL PLANT BIOLOGY series

Editor: W.J. Davies
Institute of Environmental and Biological Sciences, Division of Biological Sciences,
University of Lancaster, Lancaster LA1 4YQ, UK

Abscisic Acid: physiology and biochemistry

Carbon Partitioning: within and between organisms

Pests and Pathogens: plant responses to foliar attack

Forthcoming titles include:

Water Deficits: plant responses from cell to community

Carbon Partitioning
within and between organisms

C.J. Pollock
AFRC Institute of Grassland and Environmental Research, Welsh Plant Breeding Station, Plas Gogerddan, Aberystwyth, Dyfed SY23 3EB, UK

J.F. Farrar
School of Biological Sciences, University College of North Wales, Bangor, Gwynedd, LL57 2UW, UK

A.J. Gordon
Cell Biology Department, AFRC Institute of Grassland and Environmental Research, Plas Gogerddan, Aberystwyth, Dyfed SY23 3EB, UK

LIVERPOOL
UNIVERSITY

βIOS
SCIENTIFIC
PUBLISHERS

© BIOS Scientific Publishers Limited, 1992

All rights reserved. No part of this book may be reproduced or transmitted, in any form or by any means, without permission.

First published in the United Kingdom 1992 by
BIOS Scientific Publishers Limited,
St Thomas House, Becket Street, Oxford OX1 1SJ, UK.

A CIP catalogue record for this book is available from the British Library.

ISBN 1 872748 95 3

Typeset by Unicus Graphics Ltd, Horsham, UK.
Printed by Information Press Ltd, Oxford, UK.

Preface

Within higher plants, the separation of metabolic function between different organs or organelles, together with the periodic nature of carbon assimilation and energy capture, impose a requirement for the movement and storage of assimilates. In recent years there have been substantial advances in our understanding of the regulatory mechanisms associated with partitioning within and between sources and sinks. In parallel, the potential roles of the transport pathways as an integrating mechanism within the plant have become more widely discussed. The partitioning of assimilates involves crossing of and sequestration behind cell and organelle membranes. Under defined conditions these can also separate different species between which there is a flow of material. It was with the idea that there might be common themes associated with the partitioning and transport of assimilates within and between organisms which led us to formulate this volume. It was based upon a session organized by the Environmental Physiology Group of the Society for Experimental Biology in Lancaster in April 1992. The session covered many aspects of assimilate partitioning but, in order to remain of readable length, the written contributions concentrate upon the partitioning of carbon.

In order to reflect our feeling that common patterns exist, we have tried to group the contributions by the elements of metabolism upon which they concentrate. Photosynthetic carbon metabolism, sucrose synthesis and phloem loading are dealt with in the first four chapters. Phloem unloading and sink metabolism of starch and sucrose are covered in Chapters 5–7, and the remainder of the book is concerned with the integration of partitioning between distant organs. In all of these sections, the problems are also discussed from the standpoint of symbiotic or parasitic associations. Although common elements are revealed, there are also many gaps in our knowledge where extrapolation of current theories of metabolic control to more complex associations are not feasible. Where appropriate, contributors have tried to highlight areas of ignorance in the hope that interest in them will be stimulated.

Finally, we must express our gratitude to the contributors for producing their chapters to such a high standard and within the allotted time-span. We also wish to acknowledge the helpful advice of the Series Editor, Bill Davies, and of Jonathan Ray of BIOS Scientific Publishers Ltd. We are also grateful to Sandra Koura and Vickie Wragg from the SEB Office for shouldering so much of the burden of organising the session.

Chris Pollock (*Aberystwyth*)
John Farrar (*Bangor*)
Tony Gordon (*Aberystwyth*)

Contents

Contributors

Aked, J. The Department of Biology, Biomedical Sciences Building, The University of Southampton, Southampton SO9 3TU, UK

ap Rees, T. Department of Plant Sciences, University of Cambridge, Downing Street, Cambridge CB2 3EA, UK

van Bel, A.J.E. Transport Physiology Research Group, Department of Plant Ecology and Evolutionary Biology, University of Utrecht, Sorbonnelaan 16, 3584 CA Utrecht, The Netherlands

Cechin, I. Department of Environmental Biology, School of Biological Sciences, The University, Manchester M13 PL, UK

Cobb, A.H. Department of Life Sciences, Faculty of Science, Nottingham Polytechnic, Nottingham NG11 8HS, UK

Farrar, J.F. School of Biological Sciences, University College of North Wales, Bangor, Gwynedd, LL57 2UW, UK

Gordon, A.J. Cell Biology Department, AFRC Institute of Grassland and Environmental Research, Plas Gogerddan, Aberystwyth, Dyfed SY23 3EB, UK

Gregory, A.J. The Department of Biology, Biomedical Sciences Building, The University of Southampton, Southampton SO9 3TU, UK

Hall, J.L. The Department of Biology, Biomedical Sciences Building, The University of Southampton, Southampton SO9 3TU, UK

Huber, J.L.A. Departments of Crop Science and Botany, North Carolina State University, Raleigh NC 27695-7631, USA

Huber, S.C. United States Department of Agriculture, Agricultural Research Service, North Carolina State University, Raleigh NC 27695-7631, USA

Leigh, R.A. Rothamsted Experimental Station, Harpenden, Hertfordshire AL5 2JQ, UK

McMichael Jr, R.W. United States Department of Agriculture, Agricultural Research Service, North Carolina State University, Raleigh NC 27695-7631, USA

Minchin, P.E.H. DSIR Fruit and Trees, PO Box 31313, Lower Hutt, New Zealand

Oparka, K.J. Department of Cellular and Environmental Physiology, Scottish Crop Research Institute, Invergowrie, Dundee DD2 5DA, UK

Palta, J.A. Ysgol Gwyddorau Bioleg, Coleg y Brifysgol, Bangor, Gwynedd LL57 2UW, UK

Press, M.C. Department of Environmental Biology, School of Biological Sciences, The University, Manchester M13 9PL, UK

Prior, D.A.M. Department of Cellular and Environmental Physiology, Scottish Crop Research Institute, Invergowrie, Dundee DD2 5DA, UK

Seel, W.E. Department of Environmental Biology, School of Biological Sciences, The University, Manchester M13 9PL, UK

Storr, T. The Department of Biology, Biomedical Sciences Building, The University of Southampton, Southampton SO9 3TU, UK

Thorpe, M.R. DSIR Fruit and Trees, PO Box 31313, Lower Hutt, New Zealand

Tomos, A.D. Ysgol Gwyddorau Bioleg, Coleg y Brifysgol, Bangor, Gwynedd LL57 2UW, UK

Vincent, C.A. Department of Environmental Biology, School of Biological Sciences, The University, Manchester M13 9PL, UK

Viola, R. Department of Cellular and Environmental Physiology, Scottish Crop Research Institute, Invergowrie, Dundee DD2 5DA, UK

Williams, J.H.H. Ysgol Gwyddorau Bioleg, Coleg y Brifysgol, Bangor, Gwynedd LL57 2UW, UK

Williams, M.L. School of Biological Sciences, University College of North Wales, Bangor, Gwynedd LL57 2UW, UK

Wright, K.M. Department of Cellular and Environmental Physiology, Scottish Crop Research Institute, Invergowrie, Dundee DD2 5DA, UK

Abbreviations

ABA	abscisic acid
ADPGlc	adenosine diphosphate glucose
ATPase	adenosine triphosphatase
ATP$_\gamma$S	adenosine 5'-O-(3-thiotriphosphate)
CCCP	carbonyl cyanide m-chlorophenyl hydrazone
CHX	cycloheximide
EDTA	ethylene diaminotetra-acetic acid
EHM	extrahaustorial membrane
EM	electron microscopy
ER	endoplasmic reticulum
ERB	erythrosin B
Fru	fructose
Fru1,6P$_2$	fructose-1,6-bisphosphate
Fru1,6Pase	cytosolic fructose-1,6-phosphatase
Fru2,6P$_2$	fructose-2,6-bisphosphate
Fru6P	fructose-6-phosphate
FSBA	5'-p-fluorosulphonylbenzoyladenosine
FW	fresh weight
Glc	glucose
Glc1P	glucose-1-phosphate
Glc6P	glucose-6-phosphate
GS	glutamine synthetase
INV	invertase
IWF	intercellular washing fluid
Lb	leghaemoglobin
LYCH	Lucifer yellow CH
MDH	malate dehydrogenase
OAA	oxaloacetate
3-OMG	3-O-methylglucose
PCMBS	p-chloromercuribenzene sulphonic acid
PCR	photosynthetic carbon reduction
PEP	phosphoenolpyruvate
PEPC	phosphoenolpyruvate carboxylase
PFK(PPi)	pyrophosphate:fructose-6-phosphate:phosphotransferase
3PGA	3-phosphoglycerate
PGR	plant growth regulator
Pi	inorganic phosphate
PK	pyruvate kinase
PNPG	p-nitro-α-glucoside
PP2A	type 2A protein phosphatase
PPFD	photosynthetic photon flux density
PPi	inorganic pyrophosphate

ABBREVIATIONS

Rubisco	ribulose-1,5-bisphosphate carboxylase/oxygenase
SE	sieve element
SE–CC	sieve element–companion cell
SPPase	sucrose–phosphate phosphatase
SPS	sucrose–phosphate synthase
S:R	shoot:root ratio
SS	sucrose synthase
TCA	tricarboxylic acid
TLE/TLC	thin layer electrophoresis/chromatography
UDP	uridine diphosphate
UDPGlc	uridine diphosphate glucose
UDPGlcPP	uridine diphosphate glucose pyrophosphorylase
UTP	uridine triphosphate

The regulation of sucrose synthesis in leaves

S.C. Huber, J.L.A. Huber and R.W. McMichael Jr

1.1 The cytoplasmic sucrose formation pathway

About 80% of the carbon fixed by a leaf during photosynthesis is used in the synthesis of carbohydrates (starch and sucrose, in most higher plants; Gordon, 1986). The remainder of the fixed carbon is used for biosynthesis of amino acids, organic acids, etc. The products formed during photosynthesis reflect the major function of the mature leaf, which is to assimilate inorganic carbon and nitrogen and produce forms (e.g. sucrose and certain amino acids) which can be trans-located and utilized in other plant parts. Starch is another important end-product of photosynthesis that serves as a temporary reserve form of reduced carbon. Starch is formed in the chloroplast whereas sucrose is synthesized in the cytosol. During photosynthesis, triose phosphates are released from the chloroplasts in exchange for inorganic phosphate (Pi) on the phosphate translocator (Flügge and Heldt, 1991). Conversion of triose–P to sucrose results in the release of Pi which must return to the chloroplast to support continued photosynthesis. Sucrose synthesis and CO_2 assimilation are, therefore, highly interdependent processes. The generalization has emerged that the primary regulation of cellular carbon partitioning resides in the cytosol (Stitt and Quick, 1989; Stitt et al., 1987a). Research over the past decade has identified a hierarchy of mechanisms to control the flux of carbon into sucrose. Some recent developments in these control mechanisms will be discussed.

1.1.1 Regulatory components

Extensive measurements of cytoplasmic metabolite levels have indicated that there are four reactions in the pathway that are displaced from equilibrium: (i) cytosolic fructose-1,6-bisphosphatase (Fru1,6Pase); (ii) sucrose-phosphate synthase (SPS); (iii) sucrose-phosphate phosphatase (SPPase); and (iv) the hydrolysis of inorganic pyrophosphate (PPi) (Gerhardt et al., 1987; Quick et al., 1989a; Wiener et al., 1987). It is generally accepted that the cytoplasmic Fru1,6Pase and SPS are important control points in the pathway (Figure 1.1). The cytoplasmic Fru1,6Pase catalyses the first irreversible reaction of the

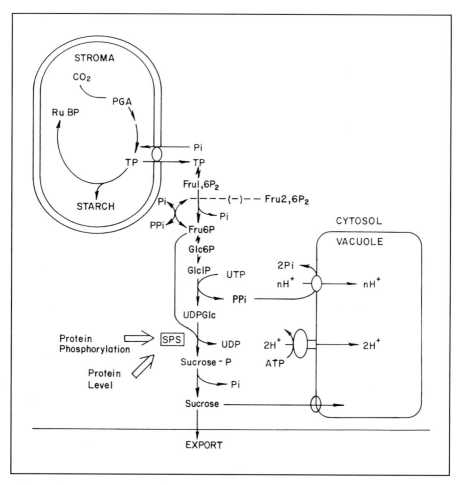

Figure 1.1. *Simplified scheme of the pathways of starch and sucrose synthesis in the mesophyll cell of a C-3 plant. Note regulation of the cytoplasmic Fru1,6Pase by the regulatory metabolite Fru2,6P_2, the potential for control of SPS by protein level and covalent modification (protein phosphorylation), and options for the hydrolysis of cytosolic PPi in the absence of inorganic pyrophosphatase.*

pathway, and is, therefore, in an ideal position to regulate the flow of inter-mediates between the chloroplast and cytosol. The regulatory properties of the cytoplasmic Fru1,6Pase are complex. The activity of this pivotal enzyme can involve interactions among several metabolites and ions, including fructose-1,6-bisphosphate (Fru1,6P_2, substrate), fructose-2,6-bisphosphate (Fru2,6P_2, high affinity inhibitor), AMP (weak inhibitor), Mg^{2+} (necessary for catalysis) and K^+ (alters Fru2,6P_2 sensitivity) (Stitt *et al.*, 1987b; for review see Stitt, 1990). Of particular physiological significance to the regulation of sucrose formation is inhibition of the cytosolic Fru1,6Pase by Fru2,6P_2, which will be discussed below in relation to the 'feedforward' and 'feedback' control of sucrose biosynthesis.

SPS is also an important enzyme of the pathway and is regulated via a hierarchy of mechanisms. First, there is developmental regulation of the steady-state level of SPS enzyme protein. As leaves mature and become 'sources', the increase in the V_{max} activity of SPS correlates closely with an increase in SPS protein (measured immunochemically; Walker and Huber, 1989b). It is also known that the steady-state level of SPS protein is regulated in response to other factors, such as certain stress conditions (see Section 1.3). In mature unstressed leaves of spinach (Walker and Huber, 1989b) and maize (Bruneau et al., 1991) during a normal day/night cycle, the level of SPS protein remains essentially constant. However, at least two mechanisms exist to control the enzymatic activity of the SPS protein molecule. First, SPS is an allosteric enzyme activated by glucose-6-phosphate (Glc6P) and inhibited by Pi (Doehlert and Huber, 1985), and secondly, the activity is controlled by covalent modification (Walker and Huber, 1989b). Changes in SPS activity with light/dark transitions were first observed in barley (Sicher and Kremer, 1984) and Lolium (Pollock and Housley, 1985), and subsequently many other species (S.C. Huber et al., 1989). The biochemical mechanism involved appears to be protein phosphorylation, and is discussed in more detail below (see Section 1.2). The SPS reaction is itself reversible, but is rendered essentially irreversible in vivo by conversion of the sucrose-P produced to sucrose by a specific phosphatase. Little is known about this enzyme, but it requires Mg^{2+} for activity, is specific for sucrose–P, and the maximum activity generally exceeds that of SPS (Hawker, 1985).

PPi is produced during formation of the uridine diphosphate glucose (UDP-Glc) required for sucrose formation. Until recently, it was thought that the PPi produced in the cytosol was rapidly hydrolysed by the inorganic pyrophosphatase that is present at high activities in leaves. However, it is now recognized that most, if not all, of the pyrophosphatase is confined to the plastid (Weiner et al., 1987) and that, in fact, the cytosol contains a significant pool of PPi. Turnover of PPi is required for continued sucrose synthesis (Quick et al., 1989a), but it is not known how this occurs. PPi might be utilized as substrate for the PPi–dependent proton pump of the tonoplast membrane (Rea and Sanders, 1987). The proton gradient generated could augment that of the proton-translocating ATPase of the membrane, and thereby contribute to the control of ion fluxes into, and out of, the vacuole (e.g. organic acid transport). Another possibility for PPi turnover involves the enzyme pyrophosphate:fructose-6-P:phosphotransferase (PFK(PPi)), which catalyses a near-equilibrium reaction in vivo (Edwards and ap Rees, 1986; Weiner et al., 1987). Thus, the PPi could be removed via a cycle between PFK(PPi) and the cytosolic Fru1,6Pase. The maximum activity of PFK(PPi) in leaves tends to decrease as leaves mature (Black et al., 1985), but the activity in a fully expanded leaf is still several-fold higher than that needed for turnover of the PPi generated during sucrose biosynthesis. These options for PPi turnover are shown schematically in relation to the sucrose formation pathway in Figure 1.1.

Having identified the major elements of the biosynthetic pathway, it becomes important to determine how the enzymes and metabolites interact in order to control flux through the pathway. There are several excellent recent reviews that deal with this subject in great detail (e.g. Copeland, 1990; Hawker, 1985; Stitt, 1990; Stitt *et al.* 1987a,b). Another important aspect, which is just beginning to receive attention, is that plant metabolism may respond differently when irradiance changes slowly (as in nature), or abruptly (as in most growth chambers). Differences in light conditions and rates of export appear to influence the regulatory mechanisms that control sucrose synthesis (Geiger *et al.*, 1991; Servaites *et al.*, 1989).

Sucrose synthesis must be regulated in order to ensure that sufficient Pi is returned to the chloroplast, but without being so rapid that the chloroplast is depleted of triose–P. In addition, sucrose synthesis is regulated so that partitioning of carbon between export (sucrose) and storage (starch) forms can be controlled without necessarily altering the rate of CO_2 fixation. The regulation of fluxes is often discussed in relation to 'feedback' and 'feedforward' conditions. The cytoplasmic Fru1,6Pase and SPS catalyse key regulatory steps and the activities of these two enzymes are controlled in a co-ordinated fashion in both situations.

1.1.2 *Feedforward control*

Stitt *et al.* (1987b) have described a model to explain how the cytoplasmic Fru1,6Pase only becomes active *in vivo* when the concentration of triose-P exceeds a 'threshold' concentration. Activity then increases sharply with a small increase in metabolite concentration. The apparent 'sigmoidal' kinetics arise because of the inverse relationship between $Fru1,6P_2$ (substrate) and $Fru2,6P_2$ (inhibitor). As the rate of photosynthesis increases, the concentration of $Fru2,6P_2$ decreases, while substrate increases, thereby allowing greater carbon flux into sucrose. As discussed by Stitt (1990), there are now at least two lines of evidence which suggest that the changes in $Fru2,6P_2$ concentration are the result of a rising 3-phosphoglycerate (3PGA):Pi ratio in the cytoplasm. It had previously been thought that the increase in concentration of both triose-P and 3PGA contributed to the decrease in $Fru2,6P_2$, but it now appears that 3PGA plays the major role. As the cytoplasmic Fru1,6Pase becomes more active *in vivo*, the concentration of fructose-6-phosphate (Fru6P, the product), and the hexose–phosphates in equilibrium with it, will also increase. In particular, a rising Glc6P:Pi ratio will result in allosteric activation of SPS (Doehlert and Huber, 1983). In addition, the sensitivity of SPS to effectors and substrates is modified by covalent modification in response to light/dark signals which reinforces the changes in metabolite levels. The importance of 'coarse control' (e.g. covalent modification) of enzyme activity is apparent when the rate of carbon flux into sucrose can increase dramatically without significant changes in the concentrations of the substrates (Fru6P and UDPGlc) of sucrose biosynthesis (Usuda *et al.*, 1987), or actually decreased concentrations of substrates (Servaites *et al.*, 1989).

1.1.3 *Feedback control*

When end-products of photosynthesis accumulate in a leaf, partitioning of carbon shifts to favour starch over sucrose while photosynthetic rate remains constant for some period of time. The notion is that the flux of carbon into sucrose is restricted, and thereby, relatively more carbon is allocated to starch within the chloroplast. All of the details of the mechanism(s) involved in 'feed-back control' of sucrose biosynthesis are not yet known. However, it appears that inactivation of SPS, in response to accumulation of sucrose (or some other metabolites) may be an initial response. The inactivation of SPS probably involves protein phosphorylation (covalent modification) and is discussed more fully in Section 1.2. Inactivation of SPS would result in a rise in cytosolic Fru6P concentration, which in turn would increase the level of Fru2,6P$_2$ (Stitt, 1990, and references therein). The increased Fru2,6P$_2$ would inhibit the cytosolic Fru1,6Pase and thereby restrict carbon flow through the pathway, with the result that more carbon would be retained within the chloroplast. Again, it is clear that mechanisms exist to co-ordinate the regulation of the cytosolic Fru1,6Pase and SPS. Metabolite interactions are complex within the pathway, as illustrated by the realization that at least three factors will act to modulate the increase of Fru2,6P$_2$ in response to a rise in Fru6P and are discussed in detail by Stitt (1990).

1.2 Properties of SPS and regulation by protein phosphorylation

1.2.1 *Properties of SPS*

Detailed information about the physical and biochemical properties of SPS is still extremely limited. The enzyme from spinach (Salvucci *et al.*, 1990; Walker and Huber, 1989a) and maize (Bruneau *et al.*, 1991) leaves (and presumably other sources) is composed of an approximately 130-kDa polypeptide, and the native enzyme is either a homotetramer (Walker and Huber, 1989a) or homodimer (Salvucci *et al.*, 1990). The difference in reported molecular size probably reflects the methods used, rather than purification state of the enzyme. Recent estimates of the apparent equilibrium constant of the enzyme partially purified from germinating pea seed ranged between 5 and 62, depending upon pH and Mg^{2+} concentration (Lunn and ap Rees, 1990). At pH 7.5 and 10 mM Mg^{2+}, the partially purified spinach leaf enzyme had an apparent equilibrium constant of 10 (J. L. Huber, unpublished data). The enzyme appears to be considerably displaced from equilibrium *in vivo* (Lunn and ap Rees, 1990; Stitt *et al.*, 1987a) indicating the potential for the control of carbon flux into sucrose. Kinetic properties of the enzyme have been reviewed in more detail by Stitt *et al.* (1987a) and Copeland (1990). Briefly, the enzyme displays hyperbolic saturation kinetics for both substrates, Fru6P and UDPGlc. The enzyme from spinach (Doehlert and Huber, 1985) and maize (Kalt-Torres *et al.*, 1987) leaves appears to be allosteric, and is activated by Glc6P and inhibited by Pi. The regulatory or allosteric site where Glc6P and Pi competitively interact appears to contain an essential and accessible sulphydryl group required for regulation by metabolites.

1.2.2 *Evidence for covalent modification of SPS*

SPS activity in leaves has been shown to fluctuate diurnally in many species, in particular with light/dark transitions. In general, extractable SPS activity increases within 10–20 min when leaves are illuminated, and decreases within 20–30 min when leaves are darkened. The occurrence, and the magnitude of changes in activity, depend upon the species and the conditions used to assay the enzyme (see Section 1.4.2). At least in spinach (Walker and Huber, 1989b) and maize (Bruneau *et al.*, 1991) it has been established unequivocally that activity changes as the result of covalent modification, because enzyme protein (measured immunochemically) remains constant. The mechanism has been identified as protein phosphorylation (J. L. A. Huber *et al.*, 1989). The enzyme, as extracted from spinach and maize leaves, is a mixture of 'forms' (Siegl and Stitt, 1990) that are interconverted by covalent modification of the protein. Substantial evidence suggests that these 'forms' correspond generally to different phosphorylation states of the enzyme.

Sucrose biosynthesis also occurs at high rates in certain non-photosynthetic tissues. For example, during germination of castor bean there is massive gluconeogenic conversion of stored fat reserves to sucrose in the endosperm. There is evidence that gluconeogenesis in the non-photosynthetic castor bean cotyledons can be controlled by $Fru2,6P_2$ (Kruger and Beevers, 1985) and by regulation of SPS activation state (Geigenberger and Stitt, 1991). When export from the cotyledons was blocked, sucrose synthesis was inhibited while breakdown was stimulated, thus producing a futile cycle of sucrose turnover. Importantly, the inhibition of sucrose synthesis was associated with an apparent inactivation of SPS analogous to that which occurs in leaves (Geigenberger and Stitt, 1991). Another example comes from ripening bananas. It is well known that the ethylene-induced ripening and sweetening of green bananas is associated with a conversion of stored starch to sucrose (followed by some hydrolysis to hexose sugars). In ripening bananas, sucrose accumulation rate was closely correlated with SPS activity assayed under conditions where covalent modification affects enzyme activity (i.e. limiting substrates plus the inhibitor Pi). Thus, the suggestion was made that kinetically different forms of SPS may exist at different stages of ripening (Hubbard *et al.*, 1990). In neither case (ripening bananas and germinating castor bean) was phosphorylation demonstrated to be responsible for the changes in apparent activation state of SPS, but the results are certainly suggestive of this mechanism. Thus, phosphorylation of SPS may occur in both photosynthetic and nonphotosynthetic tissues, although as discussed in Section

1.4, there is apparent genotypic variation in regulation of SPS even in photosynthetic tissues.

1.2.3 *Phosphorylation of SPS by SPS-kinase*

The net activity of SPS at any point in time is determined by the distribution between the phosphorylated (inactive) and dephosphorylated (active) forms of the enzyme. The greater the amount of phospho-SPS present, the lower the net activity. The distribution is established by the relative activities of the protein kinases and protein phosphatases present. This section will discuss some of the properties of the protein kinase that has been shown to phosphorylate and thus to inactivate SPS *in vitro*.

SPS-kinase can be assayed as the ATP-dependent inactivation of SPS, under conditions of limiting substrates plus Pi (often referred to as 'selective' conditions), or as the incorporation of label using $[\gamma-^{32}P]ATP$. At least some of the kinase co-purifies with spinach leaf SPS through multiple purification steps (S. C. Huber and J. L. Huber, 1991a), as measured by both assay systems. Additional purification has not been able to remove all kinase activity from spinach SPS. This result, along with dilution experiments with partially purified preparations, implies that the kinase has an affinity for SPS. However, SPS and SPS-kinase are distinct proteins because, with maize extracts, complete separation of SPS and SPS-kinase was achieved during Mono-Q ion exchange chromatography (S. C. Huber and J. L. Huber, 1991b).

Even though neither SPS nor SPS-kinase has been purified to homogeneity, it has been possible to utilize partially purified enzyme preparations to characterize SPS-kinase (S. C. Huber and J. L. Huber, 1991a). The kinase utilized MgATP (but not GTP) as substrate, the apparent $K_m(ATP)$ was approximately 5 μM, and the reaction was inhibited by various salts including NaCl. Inhibition of SPS-kinase by Glc6P has also been identified, which may be of physiological significance (Weiner *et al.*, 1992). Additionally, we have examined the substrate specificity of the kinase by introducing potential competing peptide substrates and inhibitors. None of these substrates, all of which are favoured by cAMP-dependent type protein kinases, had any effect on SPS phosphorylation under our assay conditions (Table 1.1). It is possible, however, to substitute adenosine 5'-O-(3-thiotriphosphate) (ATPγS) for ATP in the inactivation reaction (Table 1.1). Many, but not all, protein kinases will utilize ATPγS as substrate to phosphorylate substrate proteins. However, the thiophosphate linkage is not generally cleaved by protein phosphatases, making the phosphorylation reaction irreversible. It will be interesting to examine how phosphorylation of SPS with this analogue affects its stability toward phosphatases.

Although a significant amount of information on the kinase has been determined using partially purified SPS/SPS-kinase preparations, there are limitations to the usefulness of these fractions. We wanted to examine the factors that influence the activity of SPS-kinase *in vivo* as a basis for understanding the regulation of SPS activation state. Such investigations required a substrate for

Table 1.1. *The effect of various compounds on the ATP-dependent inactivation of SPS*

Additions	Relative kinase activity (%)[a]
50 μM ATP (control)	100
50 μM ATPγS	122
Peptide substrates	
50 μM ATP + 100 μM SP1[b]	87
50 μM ATP + 100 μM SP2[c]	108
50 μM ATP + 100 μM Kemptide[d]	103
Peptide inhibitor	
50 μM ATP + 1 μM inhibitor peptide[e]	89

SPS (and the associated kinase) was partially purified from spinach leaves by polyethyleneglycol fractionation and Mono-Q chromatography. The enzyme was pre-incubated for 30 min at 25°C prior to assay with limiting substrates plus Pi.

[a] 100% corresponds to a decrease in SPS activation state from 83 to 53%. Activation state is defined as SPS activity with limiting substrates plus Pi, as a percentage of the V_{max} activity (assayed with saturating substrates in the absence of Pi). SPS was assayed in reaction mixtures containing 50 mM MOPS–NaOH (pH 7.5), 15 mM $MgCl_2$, 10 mM UDPGlc, 3 mM Fru6P, 12 mM Glc6P (an activator), 10 mM Pi (an inhibitor), and 2.5 mM DTT ('limiting assay plus Pi'). For the V_{max} assay, Pi was omitted and the concentrations of Fru6P and Glc6P were 10 and 40 mM, respectively. Other details of the assay were as described by J. L. A. Huber *et al.* (1989).
[b] Synthetic peptide (SP) R-R-K-A-S-G-P (Sigma A-3651).
[c] G-R-G-L-S-L-S-R (Sigma G-1991).
[d] L-R-R-A-S-L-G (Sigma K-1127).
[e] Rabbit sequence (Sigma P-3294).

assaying SPS-kinase; i.e. highly activated SPS free of kinase. Since we have been unable to separate spinach leaf SPS and SPS-kinase, we attempted to inactivate the kinase with the ATP affinity analogue 5'-*p*-fluorosulphonylbenzoyladenosine (FSBA). This analogue binds covalently with certain protein kinases resulting in irreversible inactivation, and has been used to map the ATP-binding sites (Colman, 1983). Treatment of partially purified preparations of SPS/SPS-kinase with FSBA, followed by desalting to remove excess reagent, has provided a suitable 'kinase-free' SPS substrate. FSBA-treated SPS has been used to show that total extractable SPS-kinase activity in spinach leaves changes very little, if at all, throughout the day–night cycle, suggesting that coarse control of SPS-kinase is not a regulatory mechanism.

1.2.4 *Dephosphorylation of SPS by SPS-protein phosphatase (SPS-PP)*

SPS extracted from darkened leaves undergoes a time-dependent activation during a pre-incubation at 25°C which is believed to occur by dephosphorylation of SPS by SPS-PP (S. C. Huber and J. L. Huber, 1990b; Siegl *et al.*, 1990). While this 'spontaneous activation' of the enzyme *in vitro* is apparent in assays which contain limiting substrates plus P_i (selective conditions), SPS activity assayed with saturating substrates (V_{max}) remains constant. The assay conditions under which the activation is apparent, as well as the magnitude of the *in vitro* change in SPS activity, are similar to the *in vivo* light activation of SPS, suggesting a common biochemical mechanism for these changes in enzyme activity. One line of evidence which supports a role for dephosphorylation of SPS during *in vitro* activation is its sensitivity to the classical phosphatase inhibitors, Pi, molybdate, and vanadate as well as the very specific protein phosphatase (PP) inhibitor, okadaic acid (Bialojan and Takai, 1988). Moreover, *in vitro* release of ^{32}P from SPS labelled *in vivo* in the dark is inhibited by vanadate (S. C. Huber and J. L. Huber, 1990b). The spontaneous activation of SPS is reversed by subsequent addition of MgATP, implicating involvement of a protein kinase and providing additional evidence that the spontaneous activation of SPS *in vitro* involves dephosphorylation of the protein.

Okadaic acid has been shown to block the light activation of SPS *in vivo* (S. C. Huber and J. L. Huber, 1990b; Siegl *et al.*, 1990). Sensitivity of the partially purified endogenous SPS-PP to this inhibitor *in vitro* (complete inhibition at 5 nM, Weiner *et al.* 1992) indicates that the enzyme may be classified as a mammalian type 2A protein phosphatase (PP2A) (Cohen, 1989). Moreover, Siegl *et al.* (1990) have demonstrated that purified dark-SPS can be activated *in vitro* by exogenously supplied mammalian PP2A, but not PP1. This classification of the SPS-PP as type 2A is further supported by the demonstration that a mammalian proteinaceous inhibitor, specific for PP1 (termed inhibitor-2), does not inhibit the activation of phospho-SPS *in vitro*. Moreover, activation is Mg^{2+}-independent (which is essential for PP2A activity) (Cohen, 1989). Therefore, all of the endogenous protein phosphatase activity present in spinach leaves that acts on SPS appears to be of the type 2A.

In addition to the inhibitory phosphorylation sites, SPS appears to contain a residue(s) which, when phosphorylated, increases the activation state of the enzyme (J. L. Huber *et al.*, 1991). Highly activated SPS extracted from leaves fed mannose in the light was not stable when desalted crude extracts were pre-incubated at 25°C. The 'spontaneous inactivation' occurred *in vitro* in an ATP-independent manner and was mediated by an endogenous phosphatase(s). The loss in activation state was stimulated greatly by Mg^{2+} and was inhibited by Pi, fluoride, and molybdate (but not vanadate). The inactivating phosphatase(s) was not inhibited by okadaic acid and appears to be distinct from the PP responsible for the activation of phospho-SPS *in vivo* and *in vitro*. There is no evidence to suggest a role for this phosphatase activity in light/dark regulation of SPS, but the results point to a novel regulatory modification of SPS. The presence of two distinct phosphatases acting at different sites on the SPS protein is consistent with the postulate that SPS is phosphorylated at multiple sites *in vivo*.

1.2.5 *Evidence for multisite phosphorylation*

There are now several lines of evidence for phosphorylation of the SPS subunit protein at multiple seryl residues. However, it must be stressed that this will not be unequivocally demonstrated until the phosphorylation sites have been sequenced and shown to be different. When SPS is labelled *in vivo* by feeding excised leaves [^{32}P]Pi, tryptic peptide mapping resolves several ^{32}P-phospho-peptides, two of which appear to be responsible for the light/dark modulation of enzyme activity. These sites (designated phosphopeptides 5 and 7) were phos-phorylated *in vitro* by the endogenous protein kinase and resulted in inactivation of SPS (S. C. Huber and J. L. Huber, 1991a). Furthermore, the phosphorylation status of phosphopeptides 5 and 7 *in vivo* was higher in the dark (inactive) enzyme and lower in the light (active) enzyme. Okadaic acid inhibited the light- and mannose-activation of SPS *in vivo* and specifically prevented the dephos-phorylation of phosphopeptides 5 and 7. Taken together, the *in vitro* and *in vivo* data support an inhibitory role for these sites during light/dark regulation of SPS (J. L. Huber and S. C. Huber, 1992).

The phosphorylation status of the other SPS phosphopeptides which do not correlate with light/dark modulation of SPS activity may be regulated in response to other environmental signals (such as osmotic stress, see Section 1.3.1) or may simply be constitutive without apparent effect on SPS function. Also, some apparent phosphorylation sites may be artefacts of either the digestion or the thin layer electrophoresis/thin layer chromatography (TLE/TLC) system. Sequencing and localization of the phosphorylation regions will confirm the number of phosphorylation sites on SPS and provide clues to their regulatory function.

It is important to note two additional points. First, resolution of multiple ^{32}P-phosphopeptides from SPS labelled *in vivo* has been achieved using enzymatic (trypsin) and chemical (CNBr; R.W. McMichael, unpublished data) digestion of the SPS protein, and various chromatographic techniques to separate the peptides, including 2-D TLC/TLE, reverse phase chromatography (S. C. Huber and J. L. Huber, 1990a), and SDS–PAGE. A second point to note is that phos-phorylation sites not involved in light/dark regulation may well play a role under other conditions, and may influence the phosphorylation/dephosphorylation of a regulatory site(s). This is an important aspect for future study.

1.2.6 *Regulation of SPS phosphorylation* in vivo

It is not known with certainty what factors regulate the phosphorylation status, and hence the activity, of SPS *in situ*. In general, there may be at least two broad possibilities: either the activities of the interconverting enzymes (SPS-kinase and SPS-PP) are subject to fine control (e.g. by metabolites, etc.) or their maximum activities are regulated by some coarse control mechanism (covalent modifica-tion, steady-state level of protein, etc.). There may also be other possibilities. However, several lines of evidence suggest that metabolites play a role. First, the activation state of SPS correlates closely with photosynthetic rate (Battistelli *et al.*, 1991), which is usually accompanied by fluctuations in metabolic inter-mediates. Secondly, mannose-activation of SPS in darkness, first demonstrated by Stitt *et al.* (1988), hints at a critical role for either Pi (which decreases with mannose feeding) and/or mannose-6-P (which accumulates) as regulators of SPS-kinase and/or SPS-PP. In addition to fine control, there is also emerging evidence for coarse control of the SPS-PP in response to light/dark signals. Interestingly, the SPS-kinase does not appear to show this type of regulation. These specific mechanisms may help to explain how the phosphorylation status of SPS changes in response to light/dark transitions, and with time of day (i.e. decrease in activation state towards the end of photoperiod).

Our current working model of the signal transduction pathway which mediates the light activation/dephosphorylation of SPS is shown in Figure 1.2. Light appears to do two things. First, induction of photosynthesis must alter the cytoplasmic concentrations of Pi and P-esters. A decrease in cytoplasmic Pi concentration may be essential for the *in situ* activation of SPS, as Pi is an inhibitor of SPS-PP (S. C. Huber and J. L. Huber, 1990b; Weiner *et al.*, 1992). A slight decrease in Pi coupled with an increase in Glc6P (an inhibitor of SPS-kinase) would favour the dephosphorylation of SPS.

The second effect of light involves changes in the total extractable SPS-PP activity. These changes have an apparent dependence on protein synthesis (Weiner *et al.*, 1992; Figure 1.2). Thus, the light activation of SPS *in situ* can be reduced by pretreatment of leaves in the dark with cycloheximide (CHX). There is only a slight inhibition of maximum photosynthetic rate by CHX treatment, indicating that the effect of CHX is not mediated by metabolites. It is not known whether the dependence of light-activation of SPS-PP on protein synthesis is direct or indirect, i.e. whether some essential subunit (catalytic or regulatory) of the PP is being synthesized or whether some alteration in amino acid pools accompanying the inhibition of protein synthesis by CHX is the 'link' with protein synthesis. Until the molecular basis for the light activation of SPS-PP is known, the answers to these questions will remain open. However, it does appear that this coarse control of SPS-PP contributes to the rapid light activation of SPS *in situ*, and may also explain the progressive decline in SPS activation state

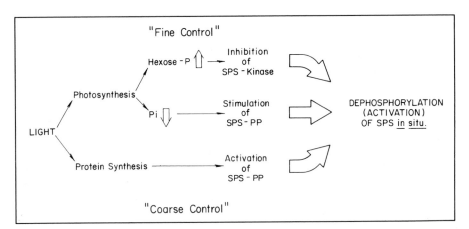

Figure 1.2. *Working model for the signal transduction pathway involved in the light activation of SPS by dephosphorylation. Elements of 'fine' and 'coarse' control are identified.*

in situ during the photoperiod (Weiner *et al.*, 1992). Clearly, much remains to be learned about the regulation of SPS-PP.

1.3 Impact of environmental factors on SPS

There has been considerable interest in studying the short- and long-term effects of various environmental stress conditions on photosynthesis and photosynthate partitioning both within the plant as a whole and among different compounds in the leaf. A short-term change in conditions (e.g. lowering of temperature) often results in dramatic shifts in metabolism that sometimes have long-term consequences. With specific regard to SPS, two environmental stress conditions have received recent attention and will be discussed here.

1.3.1 *Osmotic stress activation of SPS*

It is now recognized that the short-term effect of water stress is to inhibit photosynthesis as a result of stomatal closure (Daley *et al.*, 1989; Terashima *et al.*, 1988) rather than inhibition of the photosynthetic apparatus *per se*. It is also becoming clear that there are significant effects of osmotic stress on the partitioning of photosynthates. There are conflicting reports in the literature concerning the effects of water stress on the relative content of starch and sucrose in leaves. Undoubtedly, many factors and conditions may contribute to the different results reported. Numerous reports indicate that soluble sugars either remain constant or increase, while starch tends to decrease, in water-stressed plants (Quick *et al.*, 1992; Zrenner and Stitt, 1991, and references therein). In some species such as spinach, lupin and eucalyptus, there is a shift in partitioning to favour sucrose over starch, which may contribute to accumulation of sucrose for osmoregulation under stress conditions (Quick *et al.*, 1992).

Rapid effects of osmotic stress on photosynthate partitioning were first reported by Quick *et al.*, (1989b). When spinach leaf discs were incubated on hyperosmotic sorbitol solutions overnight, CO_2-saturated photosynthesis was not inhibited but partitioning was altered to favour sucrose. The shift in partitioning was correlated with an activation of SPS, and to an overall increase in the concentration of cytoplasmic metabolites as the cells plasmolysed (Quick *et al.*, 1989b). Subsequently, it was demonstrated that water-stressed leaves partitioned more carbon into sucrose and less into starch, not only when the stress was rapidly induced in leaf discs, but also when water stress developed slowly in whole plants (Zrenner and Stitt, 1992). What is most interesting, is that activation of SPS in response to osmotic stress occurs in darkness, and in leaves with a high concentration of sucrose that normally suppresses activation.

These results prompted us to examine further the osmotic stress activation of spinach leaf SPS in darkness, with specific emphasis on the mechanism involved. In our experimental system, osmotic stress was induced in excised leaves (detached from plants at the end of the dark period) by immersing the cut petioles in solutions containing increasing concentrations of sorbitol. Because the detached leaves were kept in darkness, light activation of SPS was prevented. We determined empirically that 0.2 M sorbitol was optimum for the osmotic stress activation of SPS in darkness; at this concentration, the leaves became slightly wilted. The osmotic stress activation of SPS had several interesting characteristics. First, it was time-dependent. After supplying 0.2 M sorbitol to excised leaves, about 90 min was required for maximum effect (generally an increase in SPS activation state from about 10% to about 25%).

The second interesting aspect concerns the mechanism involved. We already knew that light *per se* was not absolutely required for activation of SPS, and suspected that osmotic stress may trigger the partial dephosphorylation of SPS thereby activating the enzyme in darkness. In order to test this postulate, excised leaves were fed [^{32}P]Pi in the dark for 3 h, followed by exposure to osmotic stress for 90 min. As shown in Table 1.2, the osmotic stress treatment substantially increased the activation state of SPS in darkness (the V_{max} activity of SPS was unaffected) and, in contrast to the expectation, ^{32}P incorporation into SPS was increased relative to the unstressed control (Table 1.2). Thus, the osmotic stress activation of SPS cannot be explained simply as a result of dephosphorylation (i.e. the mechanism responsible for activation in response to light or mannose).

Unique sites may be phosphorylated on SPS during osmotic stress. As discussed above (see Section 1.2.5), SPS appears to be phosphorylated at multiple sites. Two-dimensional tryptic peptide mapping of SPS phosphorylated *in vivo* under different conditions has provided one approach to the identification of putative phosphorylation sites which may be involved in the activation/inactivation in response to light/dark transitions. Peptide mapping analysis with SPS labelled *in vivo* in the dark was either carried out with or without osmotic stress. When SPS is labelled *in vivo* by feeding leaves with [^{32}P]Pi, numerous tryptic peptides labelled with ^{32}P are resolved (J. L. Huber and S. C. Huber, 1992). Although complex, the pattern is very reproducible. Light/dark modulation of SPS activity is correlated with quantitative changes in the labelling of two tryptic phosphopeptides [previously designated as phosphopeptides 5 and 7 (J. L. Huber and S. C. Huber, 1992)] and are identified with arrowheads in Figure 1.3. The

Table 1.2. *Osmotic stress activates SPS in darkness and increases ^{32}P-labelling*

Leaf treatment	^{32}P-SPS (c.p.m./20 μg SPS)	SPS activation state (%)
Dark (control)	4900	9
Dark + osmotic stress	7400	23

Detached spinach leaves were fed [^{32}P]Pi via the transpiration stream in the dark for 3 h, followed by an additional 90 min in the dark in the absence (control) or presence (osmotic stress) of 0.2 M sorbitol. Leaf proteins were extracted, SPS activation state was measured as described in the legend of Table 1.1, and ^{32}P incorporated into SPS was measured after immunoprecipitation of the enzyme using monoclonal antibodies (J. L. A. Huber *et al.*, 1989).

phosphorylation of these putative regulatory sites was increased in the osmotically stressed leaf (Figure 1.3B) and hence, cannot explain the activation of SPS that occurs under these conditions. Rather, activation appears to be associated with the phosphorylation of several sites that are not (or are only weakly) phosphorylated in the unstressed control (phosphopeptides 8, 9, 10 and 13 in Figure 1.3B). Phosphorylation of these 'unique' sites accounted for about one-half of the total increase in ^{32}P-labelling of SPS after osmotic stress. Whether the four unique tryptic peptides are derived from the same peptide, or represent individual phosphorylation sites, can only be ascertained when the individual sites are sequenced. None the less, it appears that phosphorylation of some site(s) can partially antagonize the inhibitory effect of phosphorylation of the normal regulatory sites (phosphopeptides 5 and 7), and this may be responsible for osmotic stress activation of SPS.

There are at least two precedents for specific effects of stress on protein phosphorylation in plants. First, Srivastava *et al.* (1989) found that a hypertonic mannitol shock enhanced K$^+$ uptake by red beet storage tissue slices and that this was correlated with enhanced phosphorylation of a 39-kDa protein in the plasma membrane. It was suggested that the hypertonic stress induced activation of the inositol phospholipid cascade, which then affected the change in membrane protein phosphorylation (Srivastava *et al.*, 1989). Involvement of inositol phospholipids, and the role of enhanced phosphorylation of plasma membrane protein(s) remains to be established. A second example involves heat

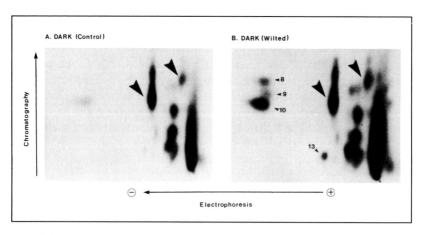

Figure 1.3. *Evidence for phosphorylation of unique sites on SPS in response to osmotic stress. The figure shows an autoradiogram of a two-dimensional separation of ^{32}P-labelled tryptic peptides derived from SPS immunopurified from leaves fed [^{32}P]Pi in the dark with petioles in (A) water (control) or (B) 0.2 M sorbitol (osmotic stress). The stress treatment was for 90 min before extraction and processing of the ^{32}P-SPS as described previously (Huber and Huber, 1992). The numbers in (B) designate peptide fragments that are more heavily labelled in the stressed leaf tissue. The arrowheads designate phosphopeptides 5 and 7 which are thought to be involved in light/dark regulation.*

shock, rather than osmotic stress. Krishnan and Pueppke (1987) reported that heat shock of soybean seedlings rapidly triggers phosphorylation of a set of proteins which did not appear to include the 'heat shock' proteins themselves. It was postulated that heat shock somehow modulates the activity of protein kinase(s), which in turn, alters the phosphorylation of *specific* proteins. Many stress factors can induce the classic 'heat shock' response, including water stress. It is quite possible that a rapid change in the phosphorylation status of specific proteins (e.g. SPS) is an early and general response to stress conditions.

Osmotic stress activation of SPS appears to involve gene expression, as do many stress responses in plants. For example, accumulation of abscisic acid (ABA) in dehydrated pea plants could be prevented by inhibitors of transcription (cordycepin or actinomycin D) and cytoplasmic protein synthesis (CHX) (Guerrero and Mullet, 1986). It has subsequently been shown that many genes are expressed in response to osmotic stress, and some can be induced in unstressed tissue by the application of ABA (Skriver and Mundy, 1990, and references therein). Thus, it was of interest to determine whether the osmotic stress activation of SPS could be blocked by inhibitors of gene expression. Preliminary experiments indicated that osmotic stress activation of SPS could be blocked almost completely by pretreating leaves with CHX (10 μM) for 3 h prior to imposition of the osmotic stress (data not shown), suggesting a requirement for cytoplasmic protein synthesis. As shown in Figure 1.4, transcription was also apparently required because pretreatment with cordycepin completely prevented the activation response. Thus, expression of specific genes appears to be required for the response, suggesting that it is a highly controlled process, and consequently, may be of physiological significance. We postulate that one or more of the genes specifically expressed under stress conditions encodes some component (either a catalytic or regulatory subunit) of a protein kinase and/or protein phosphatase that results in the phosphorylation of certain proteins at 'unique' sites, thereby modifying metabolism under the stress conditions. A highly speculative and simplified version of our working model is presented in Figure 1.5. As discussed above, under normal (nonstress) conditions, activation of SPS during a dark/light transition involves dephosphorylation, and inactivation upon darkening involves phosphorylation. One or two phosphorylation sites, in particular, appear to be of primary regulatory significance for light/dark modulation. In response to osmotic stress, we postulate that a unique protein kinase becomes activated, or the specificity is altered, such that new sites on SPS are phosphorylated. The effect is to activate SPS in darkness, possibly as a result of antagonism of the inhibitory effect caused by phosphorylation of other sites.

Does osmotic stress activation of SPS occur in intact plants when the stress is induced slowly over a period of days? As one approach to this problem, spinach plants were gradually water stressed. On the fourth day of withholding water, the treated plants were just beginning to show initial signs of wilting. On day 4, lights in the growth chamber were kept off, and SPS activation state and leaf relative water content (RWC) were monitored in the controls (stressed and nonstressed), and in stressed plants that were re-watered. As shown in Figure

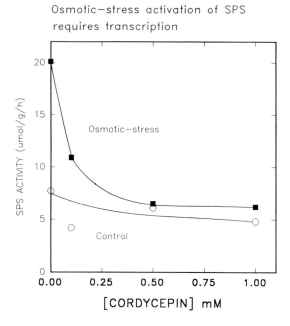

Osmotic–stress activation of SPS
requires transcription

Figure 1.4. *Cordycepin inhibits the osmotic stress activation of spinach leaf SPS in darkness. Leaves were excised from plants at the end of the dark period and fed the indicated concentration of cordycepin via the transpiration stream for 3 h in the dark. The osmotic stress treatment was then imposed by addition of 0.2 M sorbitol to the transpiration stream. Leaves were harvested after 90 min of the stress treatment and SPS was extracted and assayed.*

1.6, SPS activation state remained low (at about 10%) in well-watered plants that were maintained in the dark (Figure 1.6b). Water-stressed plants (Figure 1.6b, 'wilted control') had a considerably higher SPS activation state in the dark that was maintained at about 20% during the course of the experiment (about 8 h of extended darkness). Resupply of water to stressed plants at 0900 h resulted in a progressive decrease in SPS activation state in the dark (Figure 1.6b) that generally paralleled the increase in leaf RWC (Figure 1.6a). Several important conclusions emerged. First, water stress activation of SPS occurs *in situ* in whole plants subjected to slow drying of the soil. Thus, the phenomenon is not restricted to excised or artificial systems. Secondly, the process is reversible (within 4–5 h) and is clearly related to leaf RWC. It is quite possible (but not proven) that changes in cell turgor may be involved.

1.3.2 *Cold stress*

In addition to accumulating in water-stressed tissues, sucrose often accumulates in plants subjected to cold stress (for review, see Guy, 1990). While sugars such as the raffinose saccharides also accumulate, sucrose is the most widely found in freezing-tolerant plants. Sugar accumulation is absolutely critical to subsequent freezing tolerance (Guy, 1990). Despite the fact that numerous studies have documented the accumulation of soluble sugars in cold-stressed plants, few studies have focused on the enzymology of carbohydrate metabolism during cold acclimation. Sucrose accumulation at low temperatures is not restricted to leaves

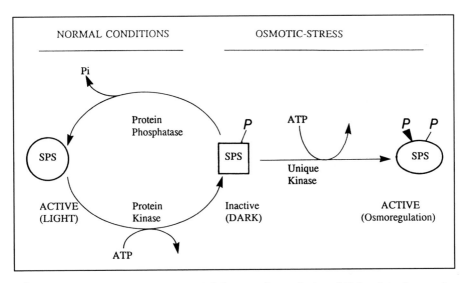

Figure 1.5. *Simplified working model showing the regulation of SPS activity by protein phosphorylation under normal (unstressed) and osmotic stress conditions. Light/dark regulation of SPS is thought to occur by dephosphorylation (activation) and phosphorylation (inactivation) as indicated. Although not proven, we postulate that a unique kinase becomes activated in response to osmotic stress and then phosphorylates phospho-SPS and partially activates the enzyme. See text for further discussion.*

of higher plants; for example, sweetening of potato tubers with cold storage is well known. Pollock and ap Rees (1975) studied the effect of cold storage on activities of various tuber enzymes and found no changes in the activity of sucrose-metabolizing enzymes, namely, SPS and sucrose synthase (SS), that might explain the phenomenon. In contrast, cold stress of *Chlorella* (Salerno and Pontis, 1989) and spinach (Holaday *et al.*, 1992) results in increased sucrose accumulation associated with increased SPS activities, while SS and invertase activities remained constant. Studies were conducted recently to determine the basis for increased SPS activity at low temperatures. It was found that increased SPS activity was paralleled by increased steady state levels of the approximately 130-kDa SPS subunit, as a result of an increased rate of synthesis (as opposed to increased turnover rate) (Guy *et al.*, 1992). It is not known whether primary control is at the level of transcription and/or translation. However, it appears that the elevated SPS activity may serve to increase synthesis of sucrose.

1.4 Species variation in partitioning strategies

1.4.1 *Sucrose storage*

Species vary dramatically in the partitioning of carbon between starch and sucrose (discussed above) and the accumulation of end-products in leaves during

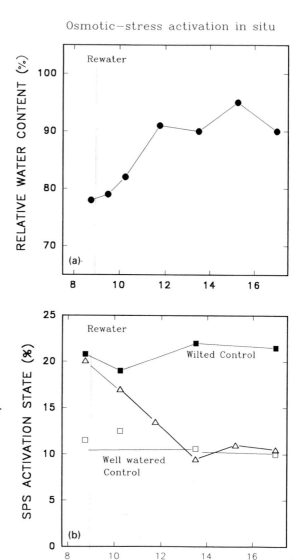

Figure 1.6. *Osmotic stress activation of spinach leaf SPS occurs in whole plants subjected to slow dehydration and can be reversed upon rewatering. Spinach plants were kept well watered or were allowed to dry over a period of 4 days. On the day of the experiment, leaves of the stressed plants were slightly wilted and all plants were kept in darkness. Some of the stressed plants were rewatered at 0900 h as indicated. At different times, leaf samples were harvested for measurement of (a) relative water content and (b) SPS activation state. Values are means of two determinations.*

the photoperiod. In particular, there is tremendous variation among species in the apparent capacity to accumulate sucrose as an end-product. Species such as soybean that tend to accumulate starch to high concentrations typically do not accumulate soluble sugars; there may be some increase in leaf sucrose concentration during the day, but quantitatively it is minor (Gordon, 1986; Huber, 1989). In contrast, species that do not form large amounts of starch (e.g. spinach, maize, and small grains), tend to accumulate substantial amounts of sucrose throughout the photoperiod. The accumulation, or lack thereof, of sucrose in leaves is not a

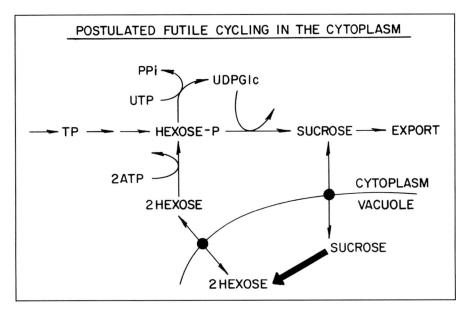

Figure 1.7. *Working model showing the synthesis of sucrose from triose-P (TP) exported from the chloroplast, and subsequent export of sucrose or movement into the vacuole followed by hydrolysis in high-invertase type plants. Adapted from S. C. Huber et al. (1990) with permission from Wiley-Liss.*

reflection of differences in the rate of biosynthesis; rather, some species are apparently able to accumulate sucrose [within the vacuole (Gerhardt *et al.*, 1987)] while other species are not. We have shown (S. C. Huber, 1989) that species which do not accumulate sucrose in leaves during photosynthesis retain high activities of soluble acid invertase, a vacuolar enzyme. We proposed that sucrose accumulation in vacuoles of high-invertase type plants was prevented by acid-invertase mediated hydrolysis. The acid invertase (from mature leaves of soybean and tobacco) had an apparent K_m (sucrose) of 2–3 mM, a pH optimum of 5.0–5.5, and thus seemed well suited for sucrose hydrolysis. A simplified scheme depicting our working model for sucrose turnover in high-invertase type plants is presented in Figure 1.7. As sucrose moves into the vacuole (probably via facilitated diffusion), it would be hydrolysed rapidly to glucose (Glc) and fructose (Fru). Because free hexose sugar concentrations are relatively low (under normal conditions), the released Glc and Fru must be rephosphorylated in the cytoplasm at the expense of ATP. The hexose-P could then re-enter metabolism and be utilized in the synthesis of sucrose. The turnover and resynthesis of sucrose as depicted in Figure 1.7 would constitute a futile cycle, because the net result is hydrolysis of ATP. Accurate measurements of the extent to which sucrose turnover occurs in leaves of high-invertase type plants are not available. However, it has been estimated (S. C. Huber *et al.*, 1990) that 75–85% of the sucrose synthesized in soybean and tobacco leaves is directly exported, while

15–25% may be 'recycled' through the vacuole. Thus, the energy cost of sucrose biosynthesis may be higher than generally thought.

Increased futile cycling of sucrose has been postulated to occur in leaves during feedback inhibition of photosynthesis. Foyer (1987, 1988) originally postulated that invertase in the cell wall may be involved, and this could certainly contribute to the feedback inhibition of photosynthesis in all species. Alternatively, there may also be increased sucrose turnover under feedback conditions involving vacuolar invertase. In fact, high-invertase type plants are apparently more sensitive to feedback inhibition (and loss of photosynthetic capacity) compared to low-invertase plants (Goldschmidt and Huber, 1992). The capacity for hexose phosphorylation in the cytoplasm via hexose kinases appears to be relatively low. Thus, when additional sucrose accumulates in the leaf cell (as a result of a restriction in phloem export), free hexose sugars begin to accumulate. This may be responsible for accumulation of hexose sugars in leaves of phloem-girdled plants (Goldschmidt and Huber, 1992) and in starchless mutants of the high-invertase type plants *Arabidopsis* (Caspar *et al.*, 1985) and *Nicotiana sylvestris* (S. C. Huber *et al.* 1990).

An analogous phenomenon has recently been described (Geigenberger and Stitt, 1991) in cotyledons of germinating *Ricinus communis* seedlings — a nonphotosynthetic tissue that is actively involved in the gluconeogenic conversion of fat to sucrose. Blockage of sucrose export from the cotyledons resulted in a futile cycle of sucrose synthesis/degradation, that apparently allowed for sensitive regulation of sucrose metabolism. The enzymes catalysing sucrose degradation in the cotyledon system could have involved sucrose synthase, and acid and alkaline invertase (Geigenberger and Stitt, 1991).

1.4.2 *Species differences in characteristics of SPS*

Differences in the mechanisms regulating SPS activity appear to exist, and altered properties of SPS may reflect different strategies for the regulation of sucrose synthesis and photosynthate partitioning, and the response of photosynthesis to feedback inhibition. There is still insufficient fundamental knowledge of SPS and its physical and biochemical properties from a variety of sources. However, given the limited knowledge to date, it is interesting to speculate about the significance of differences in properties of SPS. We originally proposed (S. C. Huber *et al.*, 1989) that at least three classes of SPS exist (designated I, II and III). Classifications were based on whether SPS activity underwent activation/inactivation *in vivo* in response to light/dark transitions and mannose feeding in darkness, and on some biochemical properties (Pi inhibition) of the SPS enzyme itself (Table 1.3). Group I contains largely monocots and is distinguished by the fact that light and mannose activation of SPS occurs, and involves a substantial increase (two to threefold) in V_{max} activity. Group II includes species where light and mannose activation occur, but does not involve increased SPS V_{max} activity. Rather, changes in regulatory and kinetic properties of the enzyme are altered, as originally reported for spinach by Stitt *et al.* (1988). Group III plants include

Table 1.3. *Characteristics of SPS in three groups of plants*

Parameter	Group I	Group II	Group III
Representative species	Maize	Spinach	Soybean
Ligh and mannose activation *in situ*	Yes (V_{max} and K_m)	Yes (K_m)	No
Protein phosphorylation	Yes	Yes	No(?)
Salt activation of phospho-SPS *in vitro*[a]	Yes	No	No
Allosteric regulation by Glc6P and Pi	Strong	Strong	Weak
Carbohydrate accumulated in leaves	Sucrose + starch	Sucrose + starch	Starch

[a] See S. C. Huber and J. L. Huber (1991c).
Adapted from S. C. Huber *et al.* (1990) with permission from Wiley-Liss.

species that apparently do not modulate SPS activity (at least under any of the assay conditions tested to date) in response to short-term light/dark transitions, or feeding of mannose to excised leaves. Rather, in soybean at least, diurnal changes in total SPS activity appear to occur that may reflect an inherent rhythm rather than rapid response via a covalent modification mechanism. Covalent modification of SPS has definitely been established only for maize (group I) and spinach (group II), and for both species the mechanism has been identified as protein phosphorylation (see Section 1.2). It is not known with certainty whether SPS from group III plants, such as soybean, is phosphorylated *in vivo* (but without kinetic effect) or whether some component for phosphorylation (e.g. the protein kinase) is missing.

Differences among the three groups in the SPS enzyme protein also appear to exist. Thus far, SPS has been partially purified (sufficient for kinetic analysis) from only one representative species of each group. However limited, it does appear that SPS from maize and spinach (groups I and II) are subject to allosteric regulation (Glc6P activation and P_i inhibition), whereas soybean SPS appears to be weakly regulated by metabolites (Nielsen and Huber, 1990). These differences in kinetic and regulatory properties suggest variation in the structure of the SPS subunit and/or holoenzyme. An additional line of evidence in support of this postulate is that antibodies raised against maize and spinach SPS tend to cross-react weakly, and cross-reactivity is generally restricted to other species of the same classification. Thus, immunological differences also seem to exist.

Species which light-activate SPS (groups I and II) tend to accumulate soluble sugars in leaves as an end-product of photosynthesis (and hence, are low-invertase type plants) whereas species of group III primarily accumulate starch as an end-product of photosynthesis. Group III plants tend to be high-invertase type species and, as discussed above, sucrose accumulation in leaves of these species appears to be prevented by the 'futile cycling' mechanism already discussed. In addition, group III plants tend to be more sensitive to feedback inhibition of photosynthesis compared to other species, although the exact basis for this difference remains unknown (Goldschmidt and Huber, 1992).

1.5 Summary and future perspectives

It is clear that the control of carbon flux into sucrose in the mesophyll cell cytoplasm during photosynthesis involves the co-ordinated regulation of Fru1,6Pase and SPS (and possibly other control points). Extensive flux analysis studies by Stitt and co-workers have established that changes in SPS activation state and fluctuations in Fru2,6P$_2$ concentration effectively control the flux rate of carbon into sucrose at low to moderate photosynthetic rates (i.e. rates attained by leaves under normal ambient conditions of CO$_2$ and light). Under conditions of maximum photosynthetic activity (as achieved experimentally with saturating CO$_2$ and saturating light), these restrictions can be overcome partially as a result of increased metabolite levels. In addition to these generalizations emerging from flux analysis studies (Neuhaus *et al.*, 1990), we have focused this contribution on four recent developments. The first concerns the recognition that protein phosphorylation is a major mechanism for the control of SPS activity *in situ*. However, the picture emerging is complex as the enzyme appears to be phosphorylated on multiple seryl residues and, therefore, may be substrate for more than one kinase and phosphatase. Phosphorylation sites that appear to be of regulatory significance have been identified by peptide mapping, but it is critical that these putative regulatory sites be sequenced. A major and novel aspect of the control of the phosphorylation status of SPS *in situ* apparently involves SPS-PP, which appears to be a specific form of PP type 2A. Regulation of the specificity of the SPS-PP, and the molecular mechanism responsible for the 'light activation' of the enzyme, will be important to determine. The second aspect concerns the osmotic stress activation of SPS, which may involve phosphorylation of unique sites on the protein (i.e. sites which are not phosphorylated under nonstress conditions). Gene expression is apparently required (either directly or indirectly) for synthesis/activation of the protein kinase involved. Thirdly, we are becoming aware that some sucrose turnover (synthesis/hydrolysis) may be a common feature of all plants cells — including those whose major function is sucrose synthesis and also those heterotrophic cells where sucrose breakdown is the net flux. The energetic significance of the futile cycling, and the regulatory significance in leaves remains to be determined but may be related to feedback inhibition of photosynthesis (Goldschmidt and Huber, 1992). Lastly, there appear to be significant differences among species in the regulatory properties of SPS. The molecular basis for these differences in kinetic and regulatory properties needs to be established, as does the impact (if any) on photosynthetic metabolism *in vivo*. As the SPS gene has now been cloned from maize (Worrell *et al.*, 1991), it is likely that our knowledge of this important enzyme at the molecular level will increase rapidly.

References

Battistelli, A., Adcock, M.D. and Leegood, R.C. (1991) The relationship between the activation state of sucrose phosphate synthase and the rate of CO$_2$ assimilation in spinach leaves. *Planta*, 183, 620–622.

Bialojan, C. and Takai, A. (1988) Inhibitory effect of a marine-sponge toxin, okadaic acid, on protein phosphatases. *Biochem. J.* 256, 283–290.

Black, C.C. Jr, Carnal, N.W. and Paz, N. (1985) Roles of pyrophosphate and fructose 2,6-bisphosphate in regulating plant sugar metabolism. In: *Regulation of Carbohydrate Partitioning in Photosynthetic Tissue* (eds R. L. Heath and J. Preiss). Waverly Press, Baltimore, pp. 76–92.

Bruneau, J.-M., Worrell, A.C., Cambou, B., Lando, D. and Voelker, T.A. (1991) Sucrose phosphate synthase, a key enzyme of sucrose biosynthesis in plants. Protein purification from corn leaves and immunological detection. *Plant Physiol.* 96, 473–478.

Caspar, T., Huber, S.C. and Somerville, C. (1985) Alterations in growth, photosynthesis and respiration in a starchless mutant of *Arabidopsis thaliana* (L.) deficient in chloroplast phosphoglucomutase activity. *Plant Physiol.* 79, 11–17.

Cohen, P. (1989) The structure and regulation of protein phosphatases. *Ann. Rev. Biochem.* 58, 453–508.

Colman, R. F. (1983) Affinity labeling of purine nucleotide sites in proteins. *Ann. Rev. Biochem.* 52, 67–91.

Copeland, L. 1990. Enzymes of sucrose metabolism. *Methods Plant Biochem.* 3, 73–85.

Daley, P.F., Raschke, K., Ball, J.T. and Berry, J.A. (1989) Topography of photosynthesis activity of leaves obtained from video images of chlorophyll fluorescence. *Plant Physiol.* 90, 1233–1238.

Doehlert, D.C. and Huber, S.C. (1983) Regulation of spinach leaf sucrose phosphate synthase by glucose-6-phosphate, inorganic phosphate, and pH. *Plant Physiol.* 73, 989–994.

Doehlert, D.C. and Huber, S.C. (1985) The role of sulfhydryl groups in the regulation of spinach leaf sucrose phosphate synthase. *Biochim. Biophys. Acta*, 830, 267–273.

Edwards, J. and ap Rees, T. (1986) Metabolism of UDP glucose by developing embryos of round and wrinkled varieties of *Pisum sativum*. *Phytochem.* 25, 2033–2039.

Flügge, U.-I. and Heldt, H.W. (1991) Metabolite translocators of the chloroplast envelope. *Ann. Rev. Plant Physiol. Plant Mol. Biol.* 42, 129–144.

Foyer, C.H. (1987) The basis of source-sink interactions in leaves. *Plant Physiol. Biochem.* 25, 649–657.

Foyer, C.H. (1988) Feedback inhibition of photosynthesis through source-sink regulation in leaves. *Plant Physiol. Biochem.* 26, 483–492.

Geigenberger, P. and Stitt, M. (1991) A 'futile' cycle of sucrose synthesis and degradation is involved in regulating partitioning between sucrose, starch and respiration in cotyledons of germinating *Ricinus communis* L. seedlings when phloem transport is inhibited. *Planta*, 185, 81–90.

Geiger, D.R., Shieh, W.-J., Lu, L.S. and Servaites, J.C. (1991) Carbon assimilation and leaf water status in sugar beet leaves during a simulated natural light regimen. *Plant Physiol.* 97, 1103–1108.

Gerhardt, R., Stitt, M. and Heldt, H.W. (1987) Subcellular metabolite levels in spinach leaves. Regulation of sucrose synthesis during diurnal alterations in photosynthesis. *Plant Physiol.* 83, 399–407.

Goldschmidt, E.E. and Huber, S.C. (1992) Regulation of photosynthesis by end-product accumulation in leaves of plants storing starch, sucrose and hexose sugars. *Plant Physiol.*, in press.

Gordon, A.J. (1986) Diurnal patterns of photosynthate allocation and partitioning among sinks. In: *Phloem Transport* (eds J. Cronshaw, W.J. Lucas and R.T. Giaquinta). Alan R.

Liss, New York, pp. 499–518.

Guerrero, F. and Mullet, J.E. (1986) Increased abscisic acid biosynthesis during plant dehydration requires transcription. *Plant Physiol.* 80, 588–591.

Guy, C.L. (1990) Cold acclimation and freezing stress tolerance: role of protein metabolism. *Ann. Rev. Plant Physiol. Plant Mol. Biol.* 41, 187–223.

Guy, C.L., Huber, J.L.A. and Huber, S.C. (1992) Sucrose phosphate synthase and sucrose accumulation at low temperature. *Plant Physiol.*, in press.

Hawker, J.S. 1985. Sucrose. In: *Biochemistry of storage carbohydrates in green plants* (eds P.M. Dey and R.A. Dixon). Academic Press, London, pp. 1–51.

Holaday, A.S., Martindale, W., Alred, R., Brooks, A. and Leegood, R.C. (1992) Changes in activities of enzymes of carbon metabolism in leaves during exposure of plants to low temperatures. *Plant Physiol.*, in press.

Hubbard, N.L., Pharr, D.M. and Huber, S.C. (1990) Role of sucrose phosphate synthase in sucrose biosynthesis in ripening bananas and its relationship to the respiratory climacteric. *Plant Physiol.* 94, 201–208.

Huber, J.L. and Huber, S.C. (1992) Seryl-specific multisite phosphorylation of spinach leaf sucrose-phosphate synthase. *Biochem. J.* 283, 877–882.

Huber, J.L.A., Huber, S.C. and Nielsen, T.H. (1989) Protein phosphorylation as a mechanism for regulation of spinach leaf sucrose-phosphate synthase activity. *Arch. Biochem. Biophys.* 270 681–690.

Huber, J.L., Hite, D.R.C., Outlaw, W.H. Jr and Huber, S.C. (1991) Inactivation of highly activated spinach leaf sucrose-phosphate synthase by dephosphorylation. *Plant Physiol.* 95, 291–297.

Huber, S.C. (1989) Biochemical mechanism for regulation of sucrose accumulation in leaves during photosynthesis. *Plant Physiol.* 91, 656–662.

Huber, S.C. and Huber, J.L.A. (1990a) Regulation of spinach leaf sucrose-phosphate synthase by multisite phosphorylation. *Curr. Top. Plant Biochem. Physiol.* 9, 329–343.

Huber, S.C. and Huber, J.L. (1990b) Activation of sucrose-phosphate synthase from darkened spinach leaves by an endogenous protein phosphatase. *Arch. Biochem. Biophys.* 282, 421–426.

Huber, S.C. and Huber, J.L. (1991a) In vitro phosphorylation and inactivation of spinach leaf sucrose-phosphate synthase by an endogenous protein kinase. *Biochim. Biophys. Acta* 1091, 393–400.

Huber, S.C. and Huber, J.L. (1991b) Regulation of maize leaf sucrose-phosphate synthase by protein phosphorylation. *Plant Cell Physiol.* 32, 319–326.

Huber, S.C. and Huber, J.L. (1991c) Salt activation of sucrose-phosphate synthase from darkened leaves of maize and other C-4 plants. *Plant Cell Physiol.* 32, 327–333.

Huber, S.C., Nielsen, T.H., Huber, J.L.A. and Pharr, D.M. (1989) Variation among species in light activation of sucrose-phosphate synthase. *Plant Cell Physiol.* 30, 277–285.

Huber, S.C., Huber, J.L. and Hanson, K.R. (1990) Regulation of the partitioning of the products of photosynthesis. In: *Perspectives in Biochemical and Genetic Regulation of Photosynthesis* (ed. I. Zelitch). Alan R. Liss, New York, pp. 85–101.

Kalt-Torres, W., Kerr, P.S. and Huber, S.C. (1987) Isolation and characterization of multiple forms of maize leaf sucrose phosphate synthase. *Physiol. Plant.* 70, 653–658.

Krishnan, H.B. and Pueppke, S.G. (1987) Heat shock triggers rapid protein phosphorylation in soybean seedlings. *Biochem. Biophys. Res. Commun.* 148, 762–767.

Kruger, N.J. and Beevers, H. (1985) Synthesis and degradation of fructose-2,6-bisphosphate in the endosperm of castor bean seedlings. *Plant Physiol.* 77, 358–364.

Lunn, J.E. and ap Rees, T. (1990) Apparent equilibrium constant and mass-action ratio for sucrose-phosphate synthase in seeds of *Pisum sativum*. *Biochem. J.* 267, 739–743.

Neuhaus, H.E., Quick, W.P., Siegl, G. and Stitt, M. (1990) Control of photosynthate partitioning in spinach leaves. Analysis of the interaction between feedforward and feedback regulation of sucrose synthesis. *Planta*, 181, 585–592.

Nielsen, T.H. and Huber, S.C. (1989) Unusual regulatory properties of sucrose-phosphate synthase purified from soybean (*Glycine max*) leaves. *Physiol. Plant.* 76, 309–314.

Pollock, C.J. and ap Rees, T. (1975) Activities of enzymes of sugar metabolism in cold-stored tubers of *Solanum tuberosum*. *Phytochemistry*, 14, 613–617.

Pollock, C.J. and Housley, T.H. (1985) Light induced increase in sucrose phosphate synthase activity in *Lolium temulentum*. *Ann. Bot.* 55, 593–596.

Quick, P., Neuhaus, E. and Stitt, M. (1989a) Increased pyrophosphate is responsible for a restriction of sucrose synthesis after supplying fluoride to spinach leaf discs. *Biochim. Biophys. Acta*, 973, 263–271.

Quick, W.P., Siegl, G., Feil, R. and Stitt, M. (1989b) Short-term water stress leads to stimulation of sucrose synthesis by activating sucrose-phosphate synthase. *Planta*, 177, 535–546.

Quick, W.P., Chaves, M.M., Wendler, R., David, M., Rodrigues, M.L., Passaharinho, J.A., Pereira, J.S., Adcock, M.D., Leegood, R.C. and Stitt, M. (1992) The effect of water stress on photosynthetic carbon metabolism in four species grown under field conditions. *Plant Cell Environ.* 15, 25–35.

Rea, P.A. and Sanders, D. (1987) Tonoplast energization: two H^+ pumps, one membrane. *Physiol. Plant.* 71, 131–141.

Salerno, G.L. and Pontis, H.G. (1989) Raffinose synthesis in *Chlorella vulgaris* cultures after a cold shock. *Plant Physiol.* 89, 648–651.

Salvucci, M.E., Drake, R.R. and Haley, B.E. (1990) Purification and photoaffinity labeling of sucrose phosphate synthase from spinach leaves. *Arch. Biochem. Biophys.* 281, 212–218.

Servaites, J.C., Fondy, B.R., Li, B. and Geiger, D.R. (1989) Sources of carbon for export from spinach leaves throughout the day. *Plant Physiol.* 90, 1168–1174.

Sicher, R.C. and Kremer, D.F. (1984) Changes in sucrose phosphate synthase in barley primary leaves during light/dark transitions. *Plant Physiol.* 76, 910–912.

Siegl, G. and Stitt, M. (1990) Partial purification of two forms of spinach leaf sucrose-phosphate synthase which differ in their kinetic properties. *Plant Sci.* 66, 205–210.

Siegl, G., Stitt, M. and Mackintosh, C. (1990) Sucrose-phosphate synthase is dephosphorylated by protein phosphatase 2A in spinach leaves. *FEBS Lett.* 270, 198–202.

Skriver, K. and Mundy, J. (1990) Gene expression in response to abscisic acid and osmotic stress. *Plant Cell*, 2, 503–512.

Srivastava, A., Pinnes, M. and Jacoby, B. (1989) Enhanced potassium uptake and phosphatidylinositol-phosphate turnover by hypertonic mannitol shock. *Physiol. Plant.* 77, 320–325.

Stitt, M. (1990) Fructose-2,6-bisphosphate as a regulatory molecule in plants. *Ann. Rev. Plant Physiol. Plant Mol. Biol.*, 41, 153–185.

Stitt, M. and Quick, W.P. (1989) Photosynthetic carbon partitioning: its regulation and possibilities for manipulation. *Physiol. Plant.* 77, 633–641.

Stitt, M., Huber, S.C. and Kerr, P. (1987a) Control of photosynthetic sucrose formation. In: *The Biochemistry of Plants*, Vol. 10, *Photosynthesis* (eds M.D. Hatch and N.K. Boardman). Academic Press, New York, pp. 327–408.

Stitt, M., Gerhardt, R., Wilke, I. and Heldt, H.W. (1987b) The contribution of fructose 2,6-bisphosphate to the regulation of sucrose synthesis during photosynthesis. *Physiol. Plant.* 69, 377–386.

Stitt, M., Wilke, I., Feil, R. and Heldt, H.W. (1988) Coarse control of sucrose phosphate synthase in leaves: alterations of kinetic properties in response to the rate of photosynthesis and the accumulation of sucrose. *Planta,* 174, 217–230.

Terashima, I., Wong, S.C., Osmond, C.B. and Farquhar, G.D. (1988) Characterisation of non-uniform photosynthesis induced by abscisic acid in leaves having different mesophyll anatomies. *Plant Cell Physiol.* 29, 143–155.

Usuda, H., Kalt-Torres, W., Kerr, P.S. and Huber, S.C. (1987) Diurnal changes in maize leaf photosynthesis. II. Levels of metabolic intermediates of sucrose synthesis and the regulatory metabolite fructose 2,6-bisphosphate. *Plant Physiol.* 83, 289–293.

Walker, J.L. and Huber, S.C. (1989a) Purification and preliminary characterization of sucrose phosphate synthase using monoclonal antibodies. *Plant Physiol.* 89, 518–524.

Walker, J.L. and Huber, S.C. (1989b) Regulation of sucrose-phosphate synthase in spinach leaves by protein level and covalent modification. *Planta,* 177, 116–120.

Weiner, H., Stitt, H., and Heldt, H.W. (1987) Subcellular compartmentation of pyrophosphate and alkaline pyrophosphatase in leaves. *Biochim. Biophys. Acta,* 893, 13–21.

Weiner, H., McMichael Jr, R.W. and Huber, S.C. (1992) Identification of factors regulating the phosphorylation status of sucrose-phosphate synthase *in vivo. Plant Physiol.,* in press.

Worrell, A.C., Bruneau, J.-M., Summerfelt, K., Boersig, M. and Voelker, T.A. (1991) Expression of a maize sucrose phosphate synthase in tomato alters leaf carbohydrate partitioning. *Plant Cell,* 3, 1121–1130.

Zrenner, R. and Stitt, M. (1991) Comparison of the effect of rapidly and gradually developing water-stress on carbohydrate metabolism in spinach leaves. *Plant Cell Environ.* 14, 939–946.

Chloroplasts in animals: photosynthesis and assimilate distribution in chloroplast 'symbioses'

M.L. Williams and A.H. Cobb

2.1 Introduction

'Chloroplast symbiosis' is the peculiar phenomenon by which certain marine molluscs, under natural conditions, continually acquire and retain functional algal chloroplasts within their digestive cells. Such sequestered chloroplasts remain structurally and metabolically intact for significant periods of time (days to weeks) yet release a major proportion of their photosynthetic products to be utilized by the animal host.

In reality, such relationships are not true symbioses in that they are short-lived and apparently wholly biased towards the mollusc (Hinde, 1980). The sequestered chloroplasts cannot divide within their new environment as this would require cytoplasmic participation from the algal cells. Similarly, no metabolite transport from the mollusc to the chloroplast has been reported other than an assumed diffusion of respiratory CO_2. It is therefore difficult to perceive any selective advantage offered by such an association to either the chloroplast or to the parent plant. As such, the label 'chloroplast symbiosis' would appear over ambitious; terms such as 'foreign organelle retention' (Blackbourn *et al.*, 1973), 'temporary retention of plastids' (Smith and Douglas, 1986) and the more succinct but ungainly 'chloroplast farming' (Hinde, 1980) have been proposed to describe this phenomenon more accurately.

Although the subject of several reviews (e.g. Hinde, 1980; Muscatine, 1973; Smith and Douglas, 1986; Taylor, 1976; Trench, 1975, 1979, 1980), our understanding of the physiology of such relationships is limited. Indeed, it would not be too cynical to summarize the collected information on assimilate partitioning within 'chloroplast symbioses' in a single sentence; triose and hexose phosphates are probably made available to the animal tissue. There is obviously the need for further research. Furthermore, it is disappointing to note that, at the time of writing, only one laboratory would appear to be working on this subject area (cf. Rumpho *et al.*, 1991). A fundamental aim of this Chapter is therefore to re-stimulate interest into what can only be described as an important biological phenomenon having the potential to provide a model for the establishment of true symbioses. The authors are aided in this endeavour by recent information concerning assimilate distribution within the participating algal species and the development of a means of isolating relatively pure chloroplasts from the animal tissue. Hence, this short review will be structured primarily as a comparison between the metabolism of the chloroplast within the alga and within the animal host.

2.2 Participating species

Although the presence of algal cells within certain marine molluscs has been known for over 100 years (e.g. Brandt, 1883), their identification as algal chloro-plasts was not realized until the mid 1960s (cf. Kawaguti and Yamasu, 1965; Taylor, 1968; Trench, 1969). Participating species were initially thought to be limited to elysioid sacoglossans and chloroplasts of siphonaceous alga (Greene, 1970a; Hinde and Smith, 1974; Table 2.1), although there is now evidence to suggest a more widespread occurrence (Clark and Bussaca, 1978). Indeed, subse-quent to the discovery of plastid retention in sacoglossan molluscs, a similar

Table 2.1. *Some participating species in chloroplast 'symbioses'*

Source of plastids	Sacoglossan host	References
Chlorophyceae		
Order Caulerpales		
Family Codiaceae		
Codium australicum	*Elysia moaria*	Brandley, 1981
Codium fragile	*Elysia hedgpethi*	Greene, 1970b
	Elysia viridis	Trench *et al.*, 1973a
	Placobranchus sp.	Muscatine and Greene, 1973
Codium tomentosum	*Elysia viridus*	Taylor, 1968
Codium spp.	*Elysia furvacauda*	Brandley, 1984
Family Caulerpaceae		
Caulerpa racemosa	*Volvatella bermudae*	Clark, 1982
Caulerpa setularoides	*Tridachia crispata*	Trench *et al.*, 1969
	Tridachiella diomedia	Trench *et al.*, 1969
Family Udoteaceae		
Avrainvillea nigricans	*Costasiella lilianae**	Clark *et al.*, 1981
Xanthophyceae		
Order Vaucheriales		
Family Vaucheriaceae		
Vaucheria compacta	*Elysia chlorotica*	West, 1977
Vaucheria sp.	*Limapontia depressa*	Hinde and Smith, 1974

* *Costasiella lilianae* is the only eolodiform mollusc reported in the literature to retain functional algal plastids for significant periods of time.

phenomenon is now thought to operate within a range of marine protists (e.g. Helioza and Foraminifera; Blackbourn *et al*, 1973; Lee and McEnery, 1983; Lee *et al.*, 1988; Patterson and Duerrschmidt, 1987). However, a total lack of physiological information concerning plastid/protist associations excludes them from this review.

The majority of plastids retained by sacoglossan molluscs originate from siphonaceous algae because these represent the common food source. From a review of algal species commonly grazed by sacoglossan molluscs, Greene (1970a) suggests that algae of the order Siphonales (now Caulerpales according to Bold and Wynne, 1985) constitute the major part. There are, however, notable exceptions; *Elysia viridis* (Montagu) (Opisthobranchia) occurs naturally on species of both *Codium* (Caulerpales) and *Chaetomorpha* (Cladophorales) and evidence is now available to suggest that physiological conditioning is involved in diet selection (Jensen, 1989). When populations of *E. viridis* associated with *Codium* are transferred to *Chaetomorpha* only 50% are able to feed, whilst 100% of the population of *E. viridis* associated with *Chaetomorpha* learn to accept *Codium* as their new food source. Furthermore, this latter population undergoes conditioning to its new food plant, in that re-transfer of the molluscs back to *Chaetomorpha* involves a re-adaptation to feed (Jensen, 1989). This may also be the case for *Elysia furvacauda* which retains at least three types of functional plastids; *Codium*-type plastids are retained in the spring, *Microdictyon*-type plastids within the autumn and a mixture of rhodoplasts and *Microdictyon*-type plastids in the winter. With each shift in plastid type a period is observed in which no functional plastids are retained (Brandley, 1984).

The majority of physiological data of algal plastid/sacoglossa associations are derived from that between chloroplasts of *Codium fragile* (Suringar) Hariot and the mollusc *E. viridis*, and from one population in particular; that occurring in the intertidal zone at Bembridge, Isle of Wight, UK. This colony has been the subject of extensive research for over 20 years and has yielded important data on both assimilate partitioning within siphonaceous alga as well as that occurring between the chloroplast and the mollusc. It is this relationship that will form the basis of this review.

2.3 *Codium fragile*: the food source

2.3.1 *Frond morphology and chloroplast stability*

The morphology of siphonaceous algae and the stability of their chloroplasts in general appear ideal in facilitating the establishment of the algal plastid/sacoglossan relationship. The order Caulerpales represents a coenocytic condition in which the cells are multinucleate and possess few or no cross walls. This has resulted in the development of particular morphological structures which confer a form of support otherwise provided by the septa. Most species of the Codiaceae contain a compact thallus composed of closely interwoven coenocytic filaments or siphons. The thallus of *C. fragile* is a dichotomous structure, being cylindrical in cross-section with the siphons arranged in a colourless, vertically aligned medullary region, the outer layer of which is composed of the dilated tips or utricles of the coenocytic siphons. It is here where the chloroplasts are located, surrounded by a large central vacuole (Figure 2.1; Hawes, 1979). Such a coenocytic condition is thought to favour the suctoral feeding habit of saco-glossan molluscs where puncturing the thick cell wall of the utricle enables the cell contents to be sucked rapidly into the digestive cells, thereby yielding large amounts of cell sap (Muscatine, 1973).

Chloroplasts of many siphonaceous alga are distinct from those of higher plants. Primarily, they are smaller and possess a less elaborate arrangement of photosynthetic membranes. In chloroplasts of *C. fragile*, the thylakoids cross the stroma in pairs or triplets revealing none of the granal stacking characteristic of

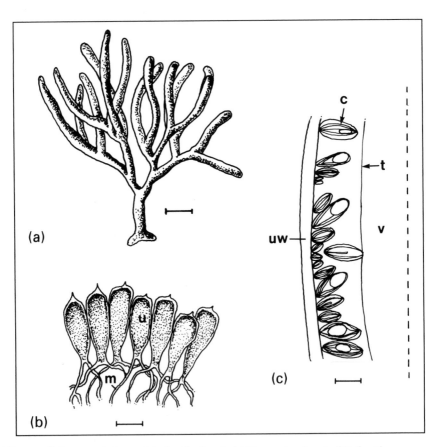

Figure 2.1. *Morphology of* C. fragile *frond. (a) Vegetative frond of* C. fragile *(bar = 1 cm). (b) Cross-section of thallus (u, chloroplast-containing utricles; m, colourless siphons within the medullary region; bar = 100 μm). (c) Diagram of the chloroplast arrange-ment within the utricle (uw, utricle cell wall; c, chloroplasts; t, tonoplast membrane; v, vacuole; bar = 3 μm).*

higher plant chloroplasts, yet show changes in fluorescence yield typical of the redistribution of excitation energy from photosystem II to photosystem I (Sealey *et al.*, 1990). Of primary importance to the plastid/sacoglossan relationship is the renowned stability of siphonaceous chloroplasts *in vitro*. Isolated chloroplasts of *Caulerpa sedoides* still maintain photosynthetic activity after 27 days from injection into hens' eggs (Giles and Sarafis, 1972). Isolated chloroplasts of *C. fragile* show no decline in $^{14}CO_2$ fixation for at least 6 h after isolation, even when stored in the light at 19°C, whilst those of spinach show a marked decline in $^{14}CO_2$ fixation over the same period (Gallop *et al.*, 1980). In addition, *Codium* chloroplasts reveal a high osmotic requirement for photosynthesis (Figure 2.2; Cobb, 1977; Cobb and Rott, 1978; Schonfeld, *et al.*, 1973), yet are remarkably resistant to osmotic shock. Isolated chloroplasts of *C. fragile* retain 50–70% of their original photosynthetic rate after resuspension in medium excluding an osmotic support (Cobb and Rott, 1978), whilst Grant and Borowitzka (1984a) reported that subjecting *C. fragile* chloroplasts to a reduction in osmolarity from 1.3 to 0.26 osmol had no effect on subsequent rates of photosynthesis at the

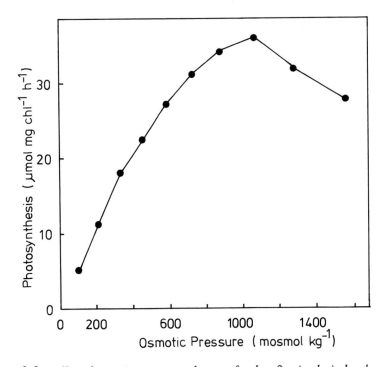

Figure 2.2. *Effect of osmotic pressure on the rate of carbon fixation by isolated chloroplasts of* C. fragile. *Chloroplasts were pre-incubated for 5 min in varying sucrose concentrations prior to addition of* $NaH^{14}CO_3$. *Data recalculated from Cobb and Rott (1978).*

original osmotic pressure. Similarly, chloroplasts of the siphonaceous algae *Caulerpa simpliciuscala*, *C.germinata* and *C.cactoides* even show resistance to mild detergent activity (Borowitzka, 1976; Grant and Borowitzka, 1984a,b; Wright and Grant, 1978). Such robust structural stability of the chloroplasts must be an important determinant in the establishment of plastid retention in sacoglossans. Reasons for this stability remain unknown and a major target for further study.

2.3.2 *Assimilate partitioning within* C. fragile

Although our knowledge of assimilate partitioning within the algal frond is limited, basic similarities exist between that occurring within *C. fragile* and leaves of higher plants. Labelling and enzymatic studies on *C. fragile* reveal the synthesis of cytoplasmic sucrose and chloroplast starch (Kremer, 1976; Trench *et al*, 1973a; Williams and Cobb, 1985; 1988a), the presence of a Pi:3PGA (3-phosphoglycerate) translocator on the chloroplast inner envelope membrane (Rutter and Cobb, 1983b) and the operation of a ferrodoxin/thioredoxin system stimulating fructose-1,6-bisphosphatase (Fru1,6P$_2$)activity within the isolated algal chloroplast (Yu and Pederson, 1989). There are, however, two notable differences in assimilate production and distribution within the frond when compared with leaves of higher plants; fronds of *C. fragile* may excrete significant amounts of glycollate into the aquatic medium (Benson, 1983; Brinkhuis and Churchill, 1972), whilst isolated chloroplasts of *C. fragile* show Pi:glucose-6-phosphate (Glc6P) translocatory activity (Rutter and Cobb, 1983b).

Glycollate excretion. Initial studies on isolated *C. fragile* chloroplasts were performed by Trench *et al.* (1973a). A comparison of the products of photosynthetic ^{14}C reduction between fronds and crudely isolated chloroplasts of *C. fragile* revealed sucrose to be the major labelled product within the frond, whilst in the case of the chloroplasts radioactivity principally appeared in glycollate. Using a more refined chloroplast extraction method, Hinde (1978) reported a 6% release of assimilate from *C. fragile* chloroplasts over a 24 h period. In this case, glucose monophosphate and glycollate were the only labelled products released. An unusual aspect of algal metabolism is the excretion of glycollate and other organic compounds into the aquatic medium, particularly under conditions facilitating high rates of photorespiration (Samuel *et al.*, 1971). Light-induced glycollate excretion by *C. fragile* fronds has been reported by Benson (1983), whilst a similar study by Brinkhuis and Churchill (1972) found *C. fragile* to excrete 40% of its assimilated carbon on exposure to supra-optimal irradiance. The importance of such excretion is uncertain; the photorespiratory pathway is present in many Chlorophyceae (Kremer, 1981) except that glycollate is oxidized to glyoxylate in the absence of oxygen by glycollate dehydrogenase (Frederick *et al.*, 1973). However, *C. fragile* is exposed at low tide to high irradiance for considerable periods of time, rendering the alga susceptible to photo-inhibition and photo-oxidative damage (see Section 2.3.3). It has been suggested that this excretion of organic compounds may represent a means of disposing of excess glycollate when photorespiratory rates are particularly high (Lorimer and Andrews, 1981).

Chloroplast starch. The principal storage form of carbohydrate within the *Codium* chloroplast is similar to higher plant starch. Starch extracts from *Cladophera rupestris*, *Enteromorpha compressa* and *C. fragile* were analysed by Love *et al.* (1963) and found to contain amylose and amylopectin in ratios similar to those of higher plants, with chemical properties and enzyme degradation patterns similar to those from potato starch, save for a marked susceptibility to enzymic degradation (Meeuse and Smith, 1962).

Maximum rates of starch synthesis within isolated chloroplasts of *C. fragile* are similar to those observed for spinach, occurring at irradiances facilitating saturating and super-saturating rates of photosynthesis (Williams and Cobb, 1985). However, chloroplast starch content of *C. fragile* is inversely proportional to the concentration of stromal Pi (Figure 2.3; Williams and Cobb, 1988a), an observation considered to reflect the gross regulation of ADPGlc pyrophosphorylase activity within the chloroplast. This enzyme is inhibited by high stromal Pi:3PGA ratios which may be produced *in vitro* by increasing the concentration of Pi external to the chloroplast (Ghosh and Preiss, 1966; Heldt *et al*, 1977; Kaiser and Bassham, 1979a,b). Although the presence of this enzyme within this alga has yet to be verified, net starch synthesis in isolated chloroplasts of *C. fragile* appears to be regulated in a similar manner to that of vascular plants, being markedly stimulated at optimal irradiance by exogenously applied 3PGA and inhibited by exogenous Pi (Williams and Cobb, 1988b). However, unlike vascular plant chloroplasts, exogenously applied Glc6P stimulates both photosynthesis at optimal irradiance and also net starch synthesis within isolated chloroplasts of *C. fragile* (Williams and Cobb, 1988b).

Phosphate translocators. The development of an improved method for isolating chloroplasts from *C. fragile* fronds in which final preparations were essentially free of cytoplasts (see Section 2.4.1) and mitochondria (Cobb, 1977), has enabled the transport properties of the chloroplast envelope membranes to be investigated. Surprisingly, such studies have indicated the presence of two separate and distinct phosphate translocators on the inner envelope membrane of *Codium* chloroplasts (Rutter and Cobb, 1983b). One is similar to the Pi:3PGA translocator of higher plant chloroplasts; uptake of ^{32}Pi being light-stimulated and strongly inhibited by *p*-chloromercuribenzene sulphonic acid (PCMBS, an inhibitor of the Pi:3PGA translator) with exogenous 3PGA facilitating the release of label from chloroplasts pre-labelled with ^{32}Pi. However, uptake of [^{14}C]Glc6P into isolated *Codium* chloroplasts also occurs which is not light-stimulated, is slower than the uptake of Pi and is less sensitive to PCMBS. Similarly, exogenous Pi stimulates the release of label from chloroplasts pre-loaded with [^{14}C]Glc6P and vice versa. From these observations Rutter and Cobb (1983b) have proposed an apparently unique Pi:Glc6P translocator to be present on the inner-envelope

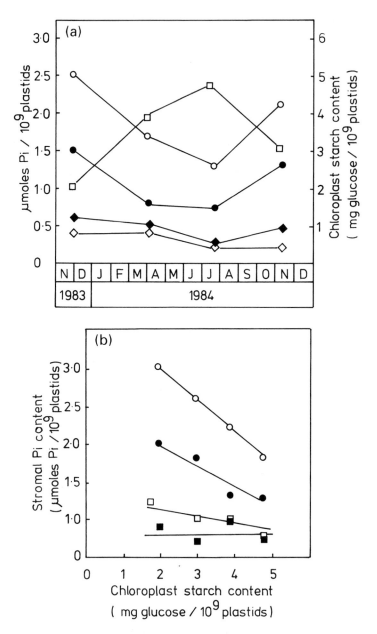

Figure 2.3. *Chloroplast starch and Pi content of C.* fragile *fronds. (a) Seasonal variation in chloroplast starch and Pi content (□, starch; ○, total Pi; ●, stromal Pi; ◆, alkali-soluble polyphosphate; ◇, acid-soluble polyphosphate). (b) Relationship between chloroplast starch and Pi content (○, total Pi; ●, stromal Pi; □, alkali-soluble polyphosphate; ■, acid-soluble polyphosphate). Data recalculated from Williams and Cobb (1988a).*

membrane of *Codium* chloroplasts which would appear similar in function to the pea root amyloplast Glc6P translocator (Bowsher *et al.*, 1989). Such import of Glc6P into the chloroplast stroma would therefore explain the stimulatory effect of exogenous Glc6P on both bicarbonate-dependent O_2 evolution and net starch synthesis observed for *C. fragile* chloroplasts illuminated at optimal and supra-optimal irradiance (Williams and Cobb, 1988b). The role of a specific Glc6P translocator on the chloroplast envelope membrane is unclear, although Glc6P may be an important intermediate in the synthesis of hexitols and hexitol phosphates for use in maintaining cell turgor and cell wall synthesis; the thick cell walls of *C. fragile* contain substantial amounts of mannan (Huizing *et al.*, 1979), whilst mannitol functions in osmotic adaption in many algal species (Helleburst, 1976). One explanation would be the presence of separate populations of chloroplasts within the algal frond (cf. Lüttke *et al.*, 1976), of which the presence of the Glc6P translocator would be indicative of starch production. However, although heteroplastidy (i.e. the presence of both chloroplasts and starch-filled leucoplasts) is used as a taxonomic feature to characterize both the Udoteaceae and Caulerpaceae (Caulerpales; Bold and Wynne, 1985), it is not thought to occur within the Codiaceae.

2.3.3 *Effect of irradiance on photosynthesis by* C. fragile

C. fragile exhibits shade adaptations enabling efficient photosynthesis at the low photosynthetic flux densities experienced within the intertidal zone at high tide. Photosynthesis in both fronds and isolated chloroplasts saturates at low irradiances (Cobb and Rott, 1978; Cobb *et al.*, 1990). Chloroplasts contain high levels of siphonoxanthin and siphonein (Anderson, 1983, 1985; Benson and Cobb, 1981, 1983), reveal both low chlorophyll a:b and high photosystem II:photosystem I ratios (Anderson, 1985; Benson and Cobb, 1983) and possess 75% of the total pigment content within the light harvesting complexes (Benson and Cobb, 1983). However, the intertidal habitat of *C. fragile* is regularly exposed to wide extremes of photosynthetic photon flux density (PPFD) on both a seasonal and daily basis. Hence chloroplasts within the frond and also within the digestive cells of the mollusc are especially at risk of photo-inhibition and photo-oxidation at the higher PPFDs realized at low tide. However, in a comparative study of the effect of irradiance on photosynthesis by Chlorophyta of differing morphology, full sunlight (1400–1960 μmol m^{-2} s^{-1}) inhibited photosynthesis in fronds of *Chaetomorpha linum*, *Enteromorpha intestinalis*, *Ulva lobata* and *U. rigida*, but not in fronds of *C. fragile* (Arnold and Murray, 1980). Such resistance to photo-inhibition in fronds of *C. fragile* has also been shown by Sealey *et al.* (1990; Figure 2.2). In contrast, isolated chloroplasts and thylakoids of *C. fragile* photoinhibit at relatively low irradiance (Cobb *et al.*, 1990; Williams and Cobb, 1985; Figure 2.4). Photoprotective mechanisms must therefore exist *in situ* that are absent *in organello* and presumably do not operate within the mollusc. Ramus (1978) proposed that the presence of a large central vacuole within the utricle serves to reflect incident light around a thin column of cytoplasm in which the

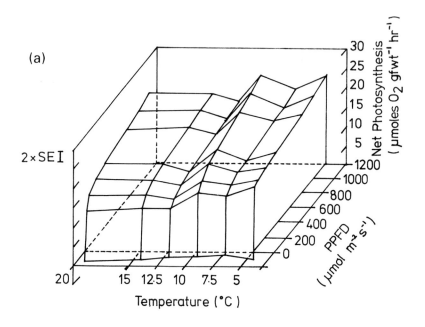

(a)

$2 \times SE$ I

Net Photosynthesis (μmoles O_2 gfwt^{-1} hr^{-1})

PPFD (μmol m^{-2}s^{-1})

Temperature (°C)

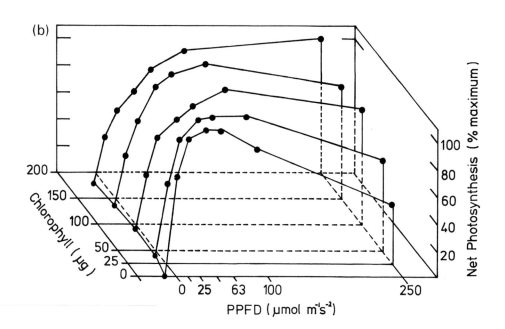

(b)

Chlorophyll (μg)

Net Photosynthesis (% maximum)

PPFD (μmol m^{-1}s^{-2})

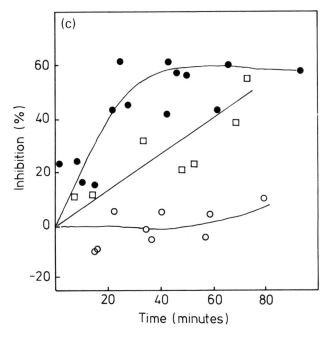

Figure 2.4. *Photo-inhibition within isolated chloroplasts of* C. fragile. *(a) Light saturation response for photosynthesis by* C. fragile *frond trips. No light-induced inhibition of O_2 evolution is observed at high PPFDs. (b) Light saturation response for isolated chloroplasts of* C. fragile. *Light-induced inhibition of O_2-evolution is observed for chloroplast populations of 25 and 50 μg chlorophyll cm^{-3} at low PPFDs. Increasing the population of chloroplasts to approximately that calculated for the utricle produces a saturation response similar to that observed within the intact frond. (c) Light-induced inhibition of photosynthetic electron transport by crude thylakoid preparations of* C. fragile *illuminated at 100 μmol $m^{-2} sec^{-1}$ PPFD (●, PS II activity; ○, PS I activity; □, PS II and PS I activity in series). Data redrawn from Sealey et al. (1990).*

chloroplasts are arranged in stacks, thus increasing the chance of chloroplast/ photon interaction. The probability of self-shading of the chloroplasts within the utricle thus occurs enabling an extended light saturation response for photo-synthesis (Ramus *et al.*, 1976). Indeed, increasing the chlorophyll concentration within an oxygen electrode vessel from 20 to 200 μg cm^{-3} effectively results in a light saturation response for photosynthesis by isolated chloroplasts of *C. fragile* similar to that observed for the algal frond (Cobb *et al.*, 1990).

2.3.4 *Seasonal variation in photosynthesis and assimilate distribution*

The production and distribution of assimilate within *C. fragile* and *E. viridis* is dependent upon seasonal variations in both the physical and biotic aspects of their intertidal zone environment, particularly light intensity and quality, and phosphate status. Highest rates of photosynthesis have been found in *Codium* chloroplasts isolated from fronds sampled early in the growth cycle during late autumn when irradiance values are less extreme (Cobb and Rott, 1978; Rutter, 1982). It is at this time that vegetative growth, pigment content and chloroplast volume are at a maximum (Figure 2.5; Benson *et al.*, 1983; Williams *et al.*, 1984). Moreover, the increased nutrient status of the intertidal zone environment, due to decay of the previous season's algal growth, enables *C. fragile* to accumulate reserves of both nitrogen and phosphate (Benson *et al.*, 1983; Hanisak 1979a,b; Hanisak and Harlin, 1978; Rutter and Cobb, 1983a; Williams and Cobb, 1988a).

Unlike higher plant chloroplasts, those of *C. fragile* possess inorganic poly-phosphates (poly Pi) which are considered to function as phosphate reserves and vary in both amount and type during the growth cycle of this alga (Benson *et al.*, 1983; Williams and Cobb, 1988a, Figure 2.3). Such variations in internal phosphate status are influential on photosynthetic rate and determine the response of isolated *Codium* chloroplasts to exogenously applied phosphate. In higher plant chloroplasts, high external concentrations of Pi inhibit photosynthesis. However, exogenous concentrations of 10 mM Pi only inhibit photosynthesis in *C. fragile* chloroplasts of a relatively high internal phosphate status (Rutter and Cobb, 1983a).

During the summer months the intertidal zone is more extreme with respect to prolonged tidal exposure, increased competition from neighbouring species and a consequential decrease in nutrient availability. It is at this stage in the growth cycle that *C. fragile* enters the reproductive phase, accompanied by a decrease in isolated chloroplast photosynthesis, chloroplast volume and pigment content, a mobilization of chloroplast Pi reserves and the development of frond hairs to aid nutrient uptake (Figure 2.5; Benson *et al.*, 1983). Surprisingly, it is during the late spring and summer months, when the majority of *C. fragile* colonies at Bembridge are transitional between vegetative and reproductive stages of growth, that the largest populations of *E. viridis* are found.

2.4 *Elysia viridis:* the host

The order Sacoglossa is a large group of opisthobranch molluscs comprising over 200 described species of which the majority appear green in colour. They all feed suctorally and with few exceptions are herbivorous. A restricted diet is apparent in that green algae of the order Caulerpales constitute the food of the majority of species (Greene, 1970a; Jenson, 1980, 1983). The distinctive coloration of the molluscs is due to the retention of chloroplasts within the epithelial cells of the digestive diverticulum (cf. Muscatine, 1973 for a histological account).

2.4.1 *Uptake and retention of plastids*

Many species of sacoglossan mollusc possess anatomical adaptations to preferred food plants (Jenson, 1980). *E. viridis* possesses a radula which is able to puncture the thick cell walls of *C. fragile*, and a buccal mass that enables the algal cell contents to be sucked into the digestive diverticula where they are believed to be

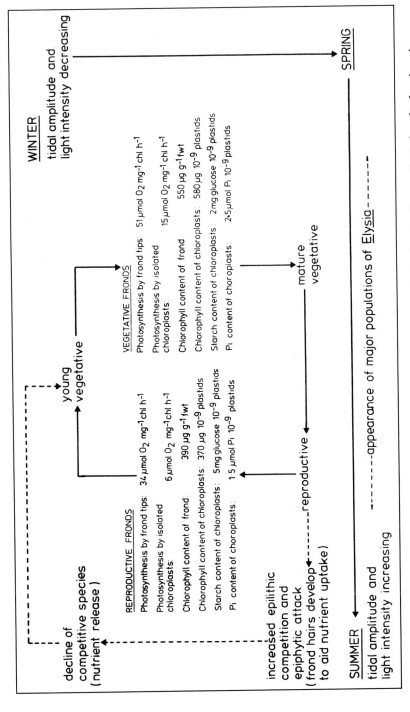

Figure 2.5. *Schematic diagram showing photosynthetic growth strategy of C. fragile fronds (photosynthetic data for fronds and chloroplasts illuminated at optimal irradiance recalculated from Seley et al. 1990).*

sequestered within the epithelial cells by phagocytosis (McLean, 1976; Trench, 1975).

When cells of coenocytic algae are wounded, either by slicing or penetration, the cytoplasmic organization rapidly changes to form cytoplasts; tonoplast-bound vesicles containing varying populations of chloroplasts and mitochondria (Gibor, 1965; Grant and Borowitzka, 1984b). The cytoplasts of siphonaceous alga ingested by herbivorous sacoglossa contain nuclei and mitochondria in addition to chloroplasts (Trench, 1975). However, the majority of ultrastructural studies on elysioid and eolodiform sacoglossans reveal only chloroplasts within the digestive cells with occasional contaminatory cytoplasm attached (Hawes, 1979; McLean, 1976; Trench, 1975; West, 1977). Hence, it would appear likely that an initial selection of chloroplasts within the digestive cell occurs.

C. fragile chloroplasts appear densely packed within the epithelial cells, occupying much of the cell cytoplasm and estimates of 10^7–10^8 chloroplasts per animal have been recorded (Cobb and Rott, unpublished data). Electron microscope studies reveal no structural damage or alteration of the chloroplasts within the digestive cells except for a general change in shape from ellipsoid in the alga to ovoid within the animal, presumably resulting from an altered osmotic environment (Hawes, 1979; Taylor, 1968; Trench *et al.*, 1973a; 1973b) and/or a physical change in the nature of the envelope membranes (Williams and Cobb, 1989).

A point of contention is the proposed existence of a surrounding host membrane forming a barrier between chloroplasts and host cytoplasm. Hawes (1979) reported that all chloroplasts observed within *E. viridis* were bound by a closely appressed host membrane, although in some cases this was ruptured. Similarly, Hawes and Cobb (1980) revealed the persistence of this host membrane throughout the existence of the chloroplasts within the digestive cells. However, an earlier study by Trench *et al.* (1973b) reported that an appreciable proportion of chloroplasts within *E. viridis* were free within the digestive cell cytoplasm. It has even been hypothesized that the host phagosome membrane is reabsorbed by the digestive cells (Muscatine *et al.*, 1975; Trench *et al.*, 1973b). However, such an absence of host membrane would be deleterious to the maintenance of the chloroplasts within the digestive cells, rendering them susceptible to lysosomal and/or other enzymic damage from the host.

Experimentally it is very difficult to ascertain the persistence of functional chloroplasts within the digestive cells of sacoglossans. However, the molluscs would appear capable of distinguishing between newly ingested, photosynthetically active plastids and those showing a decline in photosynthetic rate (Gallop *et al.*, 1980). It has been proposed that the size of the population of chloroplasts within *E. viridis* is controlled by the host, at a chlorophyll concentration of 0.01–0.02 μg chlorophyll a μg^{-1} host protein (Douglas, 1985). Moreover, the persistence of chloroplasts within the digestive diverticula is dependent upon the supply of food available to the host (Clark and Busacca, 1978; Clark *et al.*, 1981; Hinde and Smith, 1975). Under conditions of continual replenishment, *Codium* chloroplasts may be retained within the digestive cells of *E. viridis* for approximately 2 weeks, yet if the mollusc is starved of its food plant and maintained within the light, the sequestered chloroplasts remain photosynthetically active for approximately 3 months (Hinde and Smith, 1972). Under similar conditions chloroplasts within *Placobranchus ianthobapsus* are still viable after 27 days (Greene, 1970b), those within *Costasiella lilianae* for up to 65 days (Clark *et al.*, 1981), whilst those in *Elysia hedgpethi* become ineffective after only 10 days (Greene, 1970b).

2.4.2 *Photosynthesis and assimilate partitioning within* E. viridis

Finite rates of carbon fixation within intact *E. viridis* will ultimately depend upon the availability, age and phosphate status of the sequestered chloroplasts and therefore comparison with rates observed for the intact alga must proceed with caution. Trench *et al.*, (1973b) cite rates of photosynthesis for the mollusc ranging from approximately 40% to 70% of that observed for frond segments of *C. fragile* when measured on a chlorophyll basis.

An exhaustive survey of $^{14}CO_2$ assimilation in both intact *C. fragile* and *E. viridis* has revealed distinct differences in the labelling of alcohol-soluble and -insoluble compounds between the two (Kremer, 1976). Primarily, the percentage of total ^{14}C in phosphate esters, free sugars and citric acid cycle intermediates was higher in *E. viridis* and suggests a more rapid turnover of photosynthetically fixed carbon than that observed within the alga. Also, a comparatively higher amount of label was transferred into the insoluble fraction, mainly into poly-saccharides (sixfold increase) other than chloroplast starch, the percentage distribution of label into this latter fraction being decreased by a factor of 18. Such a decrease in starch content and synthesis has also been observed for chloroplasts isolated from *E. viridis* (Williams and Cobb, 1989) and is indicative of a high percentage release of photosynthate from the chloroplasts *in situ*.

Increasing the temperature from 5°C to 20°C alters assimilate production in *Elysia tuca* by doubling the photosynthetic rate observed (Stirts and Clark, 1980). Under such conditions the proportion of fixed carbon retrieved from the alcohol-insoluble fraction increases and is thought to reflect a gradual change from maintenance metabolism at sub-optimal temperatures to the synthesis of long chain molecules required for somatic growth (Stirts and Clark, 1980).

Comparative studies of $^{14}CO_2$ fixation by intact and homogenized *E. viridis* have revealed a substantial incorporation of label into galactose within whole animals, but not in the crude *Elysia* homogenate (Trench *et al.*, 1973b). It was deduced that precursors of galactose were released from the chloroplasts within the digestive cells but that the capacity for galactose synthesis was lost on homogenization. Unlike *Codium* chloroplasts isolated from the algal frond, where 2–6% of assimilate may be released to the surrounding medium (Hinde, 1978; Trench *et al.*, 1973b), those isolated from the mollusc release up to 55% of fixed CO_2 (Trench *et al.*, 1973b; Williams and Cobb, 1989). By calculating the incorporation of label into galactose it has been proposed that at least 36% of fixed carbon is released to the digestive cells in *E. viridis* and as much as 50% in both *Tridachia crispata* and *Tridachiella diomedia* (Trench *et al.*, 1969, 1973b). In the case of *P. ianthobapsus*, the release of ^{14}C-labelled photosynthate from chloroplast-retaining to chloroplast-free tissue has been monitored with time. A light saturation response was observed with a maximum of approximately 25% of assimilate being released to the mollusc after 24 h (Greene, 1970b).

The reasons for such a high release of photosynthate from the sequestered chloroplasts are unclear. Certainly, the nature of the chloroplast envelope membrane may be altered in some way within the digestive cells (Brandley, 1981; Williams and Cobb, 1989) which could account for a potential release of photosynthate of approximately 50% calculated for chloroplasts isolated from the mollusc (Williams and Cobb, 1989). An often-quoted explanation is the presence of a thermo-labile 'factor' within *E. viridis* that stimulates the release of photosynthate to the host tissue (Brandley, 1981; Gallop, 1974; Hinde, 1978). Unfortunately, experiments proposing to support this hypothesis merely involved the addition of an unqualified, crude homogenate of *E. viridis* to chloroplasts isolated from the algal frond or a subjective analysis of freeze-fractured sections of sacoglossan mollusc. That the high percentage release of photoassimilate may merely reflect the semi-digestion of an otherwise robust chloroplast envelope membrane is yet to be disproven.

The precise identity of the compounds released from the sequestered chloroplasts to the digestive cells is still uncertain. Trench *et al.*, (1973b) suggested that glucose was the major product released from crude homogenates of *Elysia*, whilst Hinde (1978) reported glucose monophosphate as the only major labelled compound released into the medium by chloroplasts isolated from *C. fragile* fronds. The problem with identification centres on the failure of past workers to isolate relatively pure chloroplasts from the mollusc, this process proving difficult due to the copious amounts of mucus generated by sacoglossan molluscs (i.e. 400 μg mucus h^{-1} g^{-1} fwt; Trench *et al.*, 1972). Although a technique now exists to isolate and purify functional chloroplasts from *E. viridis* (Williams and Cobb, 1989), the necessary labelling experiments have yet to be achieved. However, the identification of the two phosphate translocators on the inner envelope membrane of *Codium* chloroplasts (Rutter and Cobb, 1983b) suggests the likelihood that both triose (Hinde, 1978) and glucose phosphates may be the primary compounds released within the mollusc.

Photosynthesis would appear to play an important role in both the physiology and nutrition of the mollusc. The ion balance of *E. viridis* is regulated by photosynthesis; light stimulates ion fluxes both in and out of the mollusc but more specifically Na$^+$ efflux and K$^+$ influx (Deutch, 1978). With regard to nutrition, chloroplast reserves of polyphosphate are mobilized and made available to the mollusc (Figure 2.6; Cobb, 1978). This is of some importance in the intertidal zone habitat of *E. viridis* where ambient levels of P$_i$ are extremely low, especially during the late spring to summer months when competition for nutrients is increased.

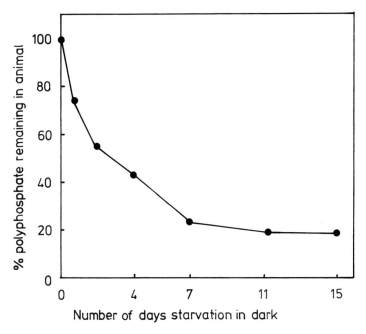

Figure 2.6. *Effect of starvation in the dark on the amount of chloroplast poly P_i within* E. viridis. *Data recalculated from Cobb (1978).*

Monitoring the body weight of *E. viridis* has also been used as a means of assessing the importance of photosynthesis in the diet of the mollusc. *E. viridis* maintained in the dark in the presence or absence of *C. fragile* lose weight more rapidly than those maintained in the light (Hinde and Smith, 1975). However, photosynthesis may not be of vital importance to all chloroplast-retaining saco-glossans. The rate of weight loss from *Placobranchus ocellatus* when starved of its food plant in the dark is little different from that observed when starved in the light (Switzer-Dunlap, 1975); photosynthesis within the *Vaucheria/Limapontia depressa* association is much lower when compared with that for *E. viridis* (Hind and Smith, 1975); *Calliopaea bellula, Hermea bifida, Limapontia capitata, Lobiger serradifalca, Oxynoe olivaceae, Placida dendricata* and burrowing species of *Volvatella* are not thought to maintain chloroplasts in a functional condition (Clarke, 1982; Hinde, 1980; Greene, 1970b; Marin and Ross, 1989), whilst mature *Elysia furvacauda* may exist for periods of its life cycle without any chloroplasts being retained (Brandley, 1984).

Autoradiographic studies on tissue sections of ^{14}C-labelled sacoglossans reveal a major proportion of the photosynthetically reduced carbon accumulating within the pedal mucus glands of *E. viridis*, *P. ianthobapus* and *Tridachia crispata* (Greene, 1970b; Taylor, 1968; Trench *et al.*, 1969). Considering the copious amounts of mucus produced by sacoglossans; then such secretory activity must represent a major sink for photosynthate.

2.4.3 *Chloroplast starch and phosphate reserves*

A continued release of a high proportion of assimilate from the sequestered chloroplasts to the mollusc would impose severe restrictions on the duration of photosynthesis. Photosynthetic carbon reduction (PCR) cycle intermediates would be depleted, and if the released metabolites were indeed triose and hexose phosphates then Pi would become limiting within the chloroplast. One way of driving photosynthesis under such conditions would be to mobilize chloroplast reserves of starch and polyphosphate. Mobilization of chloroplast polyphosphate within intact *E. viridis* has been shown (Figure 2.6; Cobb, 1978). However, the chloroplast starch content of *E. viridis* when sampled in late spring and summer is only 10% of that observed for chloroplasts isolated from the algal frond (Williams, 1986; Williams and Cobb, 1989). It is unlikely that mobilization of the remainder would provide sufficient PCR cycle intermediates to support pro-longed rates of photosynthesis for more than a few hours. This is of minor con-sequence under conditions of constant chloroplast renewal, but does not explain the observation that chloroplasts within *E. viridis* may continue to photo-synthesize, albeit with a declining rate, for up to 25 days when starved in the light (Hinde and Smith, 1972, 1975). It would therefore seem likely that some form of carbon and Pi import into the chloroplasts must occur and that the maximum release of assimilate may merely be a function of host sink activity, that is, the proportion of assimilate made available to the host at any one time would be governed by the metabolic activity of the mollusc.

Whilst fronds of *C. fragile* appear resistant to photo-inhibition and photo-oxidation, isolated chloroplasts of this alga readily photo-inhibit at relatively low irradiance values (Figure 2.4), and in the case of isolated thylakoids, reveal symptoms of photo-oxidative damage (Sealey *et al.*, 1990). Hence, although chloroplast-retaining sacoglossans show photoreactive behaviour (principally the avoidance of supra-optimal irradiance, Rahat and Monselise, 1979; Weaver and Clark, 1981), if the density of their population of *Codium* chloroplasts is less than that within the utricle of the algal frond then photo-oxidation may occur. A metabolic means of preventing photo-oxidation in chloroplasts is to increase the input of carbon into the PCR cycle. Evidence for the involvement of carbon turnover in the prevention of photo-oxidation in *E. viridis* is only available from one ultrastructural study (Hawes and Cobb, 1980). *E. viridis* were starved under conditions of both light and dark for up to 28 days and the appearance of the chloroplasts monitored by electron microscopy. Chloroplasts within the diges-tive cells of the dark-starved animals only showed symptoms of a redundant photosynthetic apparatus and were still intact. In contrast, animals starved in the light revealed a progressive degeneration of the chloroplasts observed as both a significant swelling of the plastids and disintegration of the thylakoids. Such symptoms of photo-oxidative damage, however, only occurred following the depletion of the chloroplast starch reserves. Similarly, in *Euglena gracilis*, the reserve polysaccharide paramylon is decreased during glycollate production, with one-third of such mobilized carbon being metabolized to glycollate. Under these conditions, symptoms of photo-oxidation are only evident after total depletion of the polysaccharide reserve (Yokata and Kitoaka, 1982). Hence it would seem likely that the extent of mobilization of chloroplast starch and stromal reserves of polyphosphate within the mollusc may dictate the longevity of chloroplast retention.

2.5 Summary

Temperate populations of *E. viridis* are to be found during the late spring and summer months where the majority of *C. fragile* appear transitional between vegetative and reproductive growth forms. This stage of growth is characterized by high levels of chloroplast starch within the algal frond and a reduction in the stromal phosphate status (Figure 2.5). At optimal irradiance approximately 2–6% of assimilate is released from the chloroplasts within the utricle (assuming low levels of cytoplasmic Pi) (Figure 2.7), but this proportion may increase at low tide where supra-optimal irradiance would facilitate the release of glycollate from the chloroplast and the excretion of other organic compounds from the frond. In contrast, the release of assimilate from chloroplasts retained within the digestive cells of *E. viridis* is increased and may vary from 20 to 50% due to an apparent change in the nature of the chloroplast envelope membrane. Presumably, triose and hexose phosphate constitute the bulk of assimilate released to the mollusc, whilst the release of glycollate from the chloroplasts may be significant at supra-

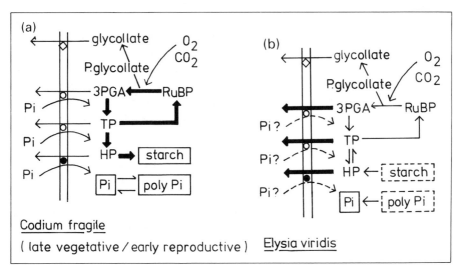

Figure 2.7. *Schematic diagram showing differences between PCR, photosynthate release, starch and polyphosphate metabolism for chloroplasts within C.* fragile *(a) and* E. viridis *(b) (see Section 2.5 for details).*

optimal irradiance (Figure 2.7). The major sink for photosynthetically reduced carbon within the mollusc would appear to be mucus production.

Under conditions of continual replenishment, chloroplasts are retained within the epithelial cells of the digestive diverticula for approximately 2 weeks, the mollusc maintaining the population of its chloroplasts at a chlorophyll concentration of between 0.01 and 0.02 μg chlorophyll a μg^{-1} host protein. Under conditions of food scarcity, chloroplasts may be retained within the mollusc for significantly longer periods of time.

Photosynthesis, starch synthesis, starch and polyphosphate content of the sequestered chloroplasts are less than that observed for chloroplasts within the algal frond. Complete mobilization of these reserves may predispose the chloroplasts to photo-oxidation at relatively low irradiance values and it is plausible that the ultimate fate of the sequestered chloroplasts is not one of digestion but of photo-oxidation.

Acknowledgement

Dedicated to David Smith for his enthusiasm and encouragement.

References

Anderson, J.M. (1983) Chlorophyll-protein complexes of a *Codium* species, including a light-harvesting siphonoxanthin-chlorophyll a/b protein complex, an evolutionary relic of some chlorophyta. *Biochim. Biophys. Acta*, 724, 370–380.

Anderson, J.M. (1985) Chlorophyll-protein complexes of a marine, green alga, *Codium* species (Siphonales). *Biochim. Biophys. Acta*, 806, 145–153.

Arnold, K.E. and Murray, S.N. (1980) Relationships between irradiance and photosynthesis for marine benthic green alga (Chlorophyta) of differing morphologies. *J. Exp. Mar. Biol. Ecol.* 43, 183–192.

Benson, E.E. (1983) *Studies on the structure and function of Codium fragile chloroplasts.* Ph.D. Thesis, Nottingham Polytechnic.

Benson, E.E. and Cobb, A.H. (1981) The separation, identification and quantitative determination of photopigments from the siphonaceous marine alga *Codium fragile. New Phytol.* 88, 627–632.

Benson, E.E. and Cobb, A.H. (1983) Pigment/protein complexes of the intertidal alga *Codium fragile* (Suringar) Hariot. *New Phytol.* 95, 581–594.

Benson, E.E., Rutter, J.C. and Cobb, A.H. (1983) Seasonal variation in frond morphology and chloroplast physiology of the intertidal alga *Codium fragile* (Suringar) Hariot. *New Phytol.* 95, 569–580.

Blackbourn, D.J., Taylor, F.J.R. and Blackbourn, J. (1973) Foreign organelle retention by cilliates, *J. Protozool.* 20, 451–460.

Bold, H.C. and Wynne, M.J. (1985) *Introduction to the Algae*, 2nd Edn. Prentice Hall Inc., New Jersey.

Borowitzka, M.A. (1976) Some unusual features of the ultrastructure of the chloroplasts of the green algal order Caulerpales and their development. *Protoplasma*, 89, 129–147.

Bowsher, C.G., Hucklesby, D.B. and Emes, M.J. (1989) Specific transport of phosphate, glucose-6-phosphate, dihydroxyacetone phosphate and 3-phosphoglycerate into amyloplasts from pea roots. *FEBS Lett.* 253, 183–186.

Brandley, B.K. (1981) Ultrastructure of the envelope of *Codium australicum* (Silva) chloroplasts in the alga and after acquisition by *Elysia moaria* (Powell). *New Phytol.* 89, 679–686.

Brandley, B.K. (1984) Aspects of the ecology and physiology of *Elysia* cf. *furvacuada* (Mollusca: Sacoglossa). *Bull. Mar. Sci.* 34, 207–219.

Brandt, K. (1883) Uber die morphologische und physiologische Bedertung des Chlorophylls bie Thieren. *Mitteil. Zool. Stat. Neapel*, 4, 191–302.

Brinkhuis, B.H. and Churchill, A.C. (1972) Primary productivity of *Codium fragile. J. Phycol.* 8 (Suppl.), 15.

Clark, K.B. (1982) *Volvatella bermudae*, new species (Mollusca: Ascoglossa) from Bermuda, with comments on the genus. *Bull. Mar. Sci.* 32, 112–120.

Clark, K.B. and Bussaca, M. (1978) Feeding specificity and chloroplast retention in four tropical ascoglossa, with a discussion of the extent of chloroplast symbiosis and the evolution of the order. *J. Moll. Studies*, 44, 272–282.

Clark, K.B., Jenson, K.R., Stirts, H.M. and Fermin, C. (1981) Chloroplast symbiosis in a non-elysiid mollusc, *Costasiella lilianae* Marcus (Hermaeidae: Ascoglossa (=Sacoglossa): effects of temperature, light intensity and starvation on carbon fixation rate. *Biol. Bull.* 160, 43–54.

Cobb, A.H. (1977) The relationship of purity to photosynthetic activity in preparations of *Codium fragile* chloroplasts. *Protoplasma*, 92, 137–146.

Cobb, A.H. (1978) Inorganic polyphosphate involved in the symbiosis between chloroplasts of the alga *Codium fragile* and mollusc *Elysia viridis. Nature*, 272, 554–555.

Cobb, A.H. and Rott, J. (1978) The carbon fixation characteristics of isolated *Codium fragile* chloroplasts. Chloroplast intactness, the effect of photosynthetic carbon reduction cycle intermediates and the regulation of RuBP carboxylase *in vitro. New Phytol.* 81, 527–541.

Cobb, A.H., Hopkins, R.M., Williams, M.L. and Sealey, R.V. (1990) Photoinhibition in thylakoids and intact chloroplasts of *Codium fragile* (Suringar) Hariot. In: *Current Research in Photosynthesis*, Volume 11 (ed. M. Baltscheffsky). Kluwer, The Hague, pp. 451–454.

Deutch, B. (1978) Light regulated body ion balance in marine slug *Elysia viridis. Nature*, 274, 159–160.

Douglas, A.E. (1985) Relationship between chlorophyll a content and protein content of invertebrate symbioses with algae or chloroplasts. *Experientia*, 41, 280–282.

Frederick, S.E., Gruber, P.J. and Tolbert, N.E. (1973) The occurrence of glycollate dehydrogenase and glycollate oxidase in green plants. An evolutionary survey. *Plant Physiol.* 52, 318–323.

Gallop, A. (1974) Evidence for the presence of a 'factor' in *Elysia viridis* which stimulates photosynthate release from its symbiotic chloroplasts. *New Phytol.* 73, 1111–1117.

Gallop, A., Bartrop, J. and Smith, D.C. (1980) The biology of chloroplast acquisition by *Elysia viridis. Proc. R. Soc. London B, Biol. Sci.* 207, 335–349.

Ghosh, H.P. and Preiss, J. (1966) Adenosine diphosphate glucose pyrophosphorylase — a regulatory enzyme in the biosynthesis of starch in spinach leaf chloroplasts. *J. Biol. Chem.* 241, 4491–4504.

Gibor, A. (1965) Surviving chloroplasts *in vitro. Proc. Natl Acad. USA*, 54, 1527–1531.

Giles, K. and Sarafis, V. (1972). Chloroplast survival and division *in vitro. Nature*, 236, 56–57.

Grant, B.R. and Borowitzka, M.A. (1984a) Changes in structure of isolated chloroplasts of *Codium fragile* and *Caulerpa filiformis* in response to osmotic shock and detergent treatment. *Protoplasma*, 120, 155–164.

Grant, B.R. and Borowitzka, M.A. (1984b) The chloroplasts of giant-celled coenocytic algae: biochemistry and structure. *Bot. Rev.* 50, 267–305.

Greene, R.W. (1970a) Symbiosis in sacoglossan opisthobranchs: symbiosis with algal chloroplasts. *Malocologia*, 10, 357–368.

Greene, R.W. (1970b) Symbiosis in sacoglossan opisthobranchs: translocation of photosynthetic products from chloroplast to host tissue. *Malacologia*, 10, 369–380.

Hanisak, M.D. (1979a) Growth patterns of *Codium fragile* ssp. *tomentosoides* in response to temperature, irradiance, salinity and nitrogen source. *Mar. Biol.* 50, 319–332.

Hanisak, M.D. (1979b) Nitrogen limitation of *Codium fragile* ssp. *tomentosoides* as determined by tissue analysis. *Mar. Biol.* 50, 333–337.

Hanisak, M.D. and Harlin, M.M. (1978) Uptake of inorganic nitrogen by *Codium fragile* ssp. *tomentosoides* (Chlorophyta). *J. Phycol.* 14, 450–454.

Hawes, C.R. (1979) Ultrastructural aspects of the symbiosis between algal chloroplasts and *Elysia viridis*. *New Phytol.* 84, 445–450.

Hawes, C.R. and Cobb, A.H. (1980) The effects of starvation on the symbiotic chloroplasts in *Elysia viridis*: a fine structural study. *New Phytol.* 84, 375–379.

Heldt, H.W., Chon, C.J., Maronde, D., Herold, A., Stankovic, Z.S., Walker, D.A., Kraminer, A., Kirk, M.R. and Heber, U. (1977) Role of orthophosphate and other factors in the regulation of starch formation in leaves and isolated chloroplasts. *Plant Physiol.* 59, 1146–1155.

Helleburst, J.A. (1976) Osmoregulation. *Ann. Rev. Plant Physiol.* 27, 485–507.

Hinde, R. (1978) The metabolism of photosynthetically fixed carbon by isolated chloroplasts from *Codium fragile* (Chlorophyta: Siphonales) and by *Elysia viridis* (Mollusca: Sacoglossa). *Biol. J. Linn. Soc.* 10, 329–342.

Hinde, R. (1980) Chloroplast 'symbiosis' in sacoglossan molluscs. In: *Endocytobiology* Volume 1 (eds W. Schwemmler and H.A.A. Schenk). Walter de Gruyter and Co., New York, pp. 729–736.

Hinde, R. and Smith, D.C. (1972) Persistance of functional chloroplasts in *Elysia viridis* (Opisthobranchia, Sacoglossa). *Nature*, 239, 30–31.

Hinde, R. and Smith, D.C. (1974) 'Chloroplast symbiosis' and the extent to which it occurs in Sacoglossa (Gastropoda: Mollusca). *Biol. J. Linn. Soc.* 6, 349–356.

Hinde, R. and Smith, D.C. (1975) The role of photosynthesis in the nutrition of the mollusc *Elysia viridis*. *Biol. J. Linn. Soc.* 7, 161–171.

Huizing, H.J., Rietema, H. and Sietsma, J.H. (1979) Cell wall constituents of several siphonous green algae in relation to morphology and taxonomy. *Br. Phycol. J.* 14, 25–32.

Jensen, K.R. (1980) A review of Sacoglossan diets, with comparative notes on radular and buccal anatomy. *Malacol. Rev.* 13, 55–77.

Jensen, K.R. (1983) Factors affecting the feeding selectivity in herbivorous Ascoglossa. (Mollusca: Opisthobranchia). *J. Exp. Mar. Biol. Ecol.* 66, 135–148.

Jensen, K.R. (1989) Learning as a factor in diet selection by *Elysia viridis* (Montague: Opisthobranchia). *J. Moll. Studies*, 55, 79–88.

Kawaguti, S. and Yamasu, T. (1965) Electron microscopy on the symbiosis between an elysioid gastropod and chloroplasts of a green algae. *Biol. J. Okayama Univ.* 12, 81–92.

Kaiser, W.M. and Bassham, J.A. (1979a) Light–dark regulation of starch metabolism in chloroplasts. 1. Levels of metabolites in chloroplasts and medium during light–dark transition. *Plant Physiol.* **63**, 105–108.

Kaiser, W.M. and Bassham, J.A. (1979b) Light–dark regulation of starch metabolism in chloroplasts. II. Effect of chloroplastic metabolite levels on the formation of ADP–glucose by chloroplast extracts. *Plant Physiol.* **63**, 109–113.

Kremer, B.P. (1976) Photosynthetic carbon metabolism, of chloroplasts symbiotic with a marine opisthobranch. *Z. Pflanzenphysiol.* **77**, 139–145.

Kremer, B.P. (1981) Carbon metabolism. In: *The Biology of Seaweeds. Botanical Monographs*, Volume 17 (eds C.S. Lobban and M.J. Wynne). Blackwell Scientific Publications, London, pp. 493–533.

Lee, J.L. and McEnery, M.E. (1983) Symbiosis in foraminifera. In: *Algal Symbiosis: A Continuum of Interactive Strategies* (ed. L.J. Goff). Cambridge University Press, Cambridge, pp. 37–68.

Lee, J.L., Lanners, E. and Kuile, B.T. (1988) The retention of chloroplasts by the foraminifer *Elphidium crispum. Symbiosis*, **5**, 45–59.

Lorimer, G.H. and Andrews, J.T. (1981) The C_2 chemo- and photorespiratory carbon oxidation cycle. In: *The Biochemistry of Plants: A Comprehensive Treatise*, Volume 8 (eds M.D. Hatch and N.K. Boardman). Academic Press, London, pp. 330–375.

Love, J., Mackie, W., McKinnell, J.W. and Percival, E. (1963) Starch–type polysaccharide isolated from the green seaweeds *Enteromorpha compressa, Ulva lactuca, Cladophora rupestris, Codium fragile* and *Chaetomorpha capillaris. J. Chem. Soc.* 4177–4182.

Lüttke, K., Rahmsdorf, U. and Schmid, R. (1976) Heterogenicity in chloroplasts of siphonaceous algae as compared with higher plant chloroplasts. *Z. Naturforsch.* **31**, 108–110.

Marin, A. and Ross, J. (1989) The Sacoglossa Mollusca Opisthobranchia of southeast Iberian peninsular. A catalog of species and presence of algal chloroplasts in them. *Iberus*, **8**, 25–50.

McLean, N. (1976) Phagocytosis of chloroplasts in *Placida dendricata* (Gastropoda: Sacoglossa). *J. Exp. Zool.* **197**, 321–330.

Meeuse, B.J.D. and Smith, B.N. (1962) A note on the amolytic breakdown of some raw algal starches. *Planta*, **57**, 624–635.

Muscatine, L. (1973) Chloroplasts and algae as symbionts in molluscs. *Int. Rev. Cytol.* **36**, 137–169.

Muscatine, L., Pool, R.R. and Trench, R.K. (1975) Symbiosis of algae and invertebrates: aspects of the symbiont surface and the host–symbiont interface. *Trans. Am. Micro. Soc.* **94**, 450–469.

Patterson, D.J. and Deurrschmidt, M. (1987) Selective retention of chloroplasts by algivorous Heliozoa: fortuitous chloroplast symbiosis? *Eur. J. Protistol.* **23**, 51–55.

Rahat, M. and Monselise, E.B. (1979) Photobiology of the chloroplast hosting mollusc *Elysia timida* (Oposthobranchia). *J. Exp. Biol.* **79**, 225–233.

Ramus, J. (1978) Seaweed anatomy and photosynthetic performance: the ecological significance of light guides, heterogenous absorption and multiple scatter. *J. Phycol.* **14**, 352–362.

Ramus, J., Beale, S.I. and Mauzerall, D. (1976) Correlation of changes in pigment content with photosynthetic capacity of seaweeds as a function of water depth. *Mar. Biol.* **37**, 231–238.

Rumpho, M.E., Pierce, S.K. and Bowersox, J.A. (1991) Chloroplast biochemistry of a marine slug-algal symbiosis. *Plant Physiol.* **96** (Suppl. 1), 50.

Rutter, J.C. (1982) *A study of the translocation properties of the chloroplasts of the alga Codium fragile.* Ph.D. Thesis, Nottingham Polytechnic.

Rutter, J.C. and Cobb, A.H. (1983a) Photosynthesis by isolated *Codium fragile* chloroplasts of varying internal phosphate status. *New Phytol.* **95**, 549–557.

Rutter, J.C. and Cobb, A.H. (1983b) Translocation of orthophosphate and glucose-6-phosphate in *Codium fragile* chloroplasts. *New Phytol.* **95**, 559–568.

Samuel, S., Shah, N.M. and Fogg, G.E. (1971) Liberation of extracellular products of photosynthesis by tropical phytoplankton. *J. Mar. Biol. Assoc. (UK)*, **51**, 793–798.

Schonfeld, M., Rahat, M. and Neumann, J. (1973) Photosynthetic reactions in the marine alga *Codium vermilara*, 1. CO_2 fixation and the Hill reaction in isolated chloroplasts. *Plant Physiol.* **52**, 283–287.

Sealey, R.V., Williams, M.L. and Cobb, A.H. (1990) Adaptations of *Codium fragile* (Suringer) Hariot fronds to photosynthesis at varying flux density. In: *Current Research in Photosynthesis*, Volume 2 (ed. M. Baltscheffsky). Kluwer, The Hague, pp. 455–458.

Smith, D.C. and Douglas, A.E. (1986) *The Biology of Symbiosis*. Edward Arnold, London.

Stirts, H.M. and Clark, K.B. (1980) The effects of temperature on products of symbiotic chloroplasts in *Elysia tuca* Marcus (Opisthobranchia: Ascoglossa). *J. Exp. Mar. Biol. Ecol.* **47**, 39–47.

Switzer-Dunlap, M. (1975) *Development and chloroplast symbiosis in Placobranchus ocellatus*. Ph.D. Thesis, University of Hawaii.

Taylor, D.L. (1968) Chloroplasts as symbiotic organelles in the digestive gland of *Elysia viridis* (Gastropod: Opisthobranchia). *J. Mar. Biol. Assoc. (UK)*, **48**, 1–15.

Taylor, D.L. (1976) Chloroplast endosymbiosis. *Port. Acta Biol.* **14**, 385–396.

Trench, R.K. (1969) Chloroplasts as functional endosymbionts in the mollusc *Tridachia crispata* (Bergh), (Opisthobranchia, Sacoglossa). *Nature*, **222**, 1071–1072.

Trench, R.K. (1975) Of 'leaves that crawl'. Functional chloroplasts in animal cells. In: *Symbiosis: Symposia of the Society of Experimental Biology* Volume 29, (eds D. H. Jennings and D. L. Lee). Cambridge University Press, Cambridge, pp. 229–265.

Trench, R.K. (1979) The cell biology of plant–animal symbiosis. *Ann. Rev. Plant Physiol.* **30**, 485–531.

Trench, R.K. (1980) Uptake, retention and function of chloroplasts in animal cells. In: *Endocytobiology*, Volume 1 (eds W. Schwemmler and H.E.A. Schenk). Walter de Gruyter and Co., New York, pp. 703–727.

Trench, R.K., Boyle, E. and Smith, D.C. (1973a) The association between chloroplasts of *Codium fragile* and the mollusc *Elysia viridis*. 1. Characteristics of isolated *Codium* chloroplasts. *Proc. R. Soc. London B, Biol. Sci.* **184**, 51–61.

Trench, R.K., Greene, R.W. and Bystrom, B.G. (1969) Chloroplasts as functional organelles in animal tissues. *J. Cell. Biol.* **42**, 404–417.

Trench, R.K., Trench, M.E. and Muscatine, L. (1972). Symbiotic chloroplasts: their photosynthetic products and contribution to mucus synthesis in two marine slugs. *Biol. Bull.* **142**, 335–349.

Trench, R.K., Boyle, E. and Smith, D.C. (1973b) The association between chloroplasts of *Codium fragile* and the mollusc *Elysia viridis*. 1. Chloroplast ultrastructure and photosynthetic carbon fixation in *E. viridis*. *Proc. R. Soc. London B, Biol. Sci.* **184**, 63–81.

Weaver, S. and Clark, K.B. (1981) Light intensity and colour preferences in five ascoglossan (= sacoglossan) molluscs (Gastropoda: Opisthobranchia): A comparison of chloroplast-symbiotic and asymbiotic species. *Mar. Behav. Physiol.* **7**, 297–306.

West, H. (1977) Chloroplast symbiosis and development in *Elysia chlorotica*. *Am. Zool.* **17**, 968.

Williams, M.L. (1986) *Metabolic studies of chloroplast symbiosis.* Ph.D. Thesis. Nottingham Polytechnic.

Williams, M.L. and Cobb, A.H. (1985) Effect of irradiance and light quality on starch synthesis by isolated chloroplasts of *Codium fragile. New Phytol.* 101, 79–89.

Williams, M.L. and Cobb, A.H. (1988a) Observations on chloroplast starch content and synthesis by isolated chloroplasts of *Codium fragile. New Phytol.* 108, 285–290.

Williams, M.L. and Cobb, A.H. (1988b) Effect of inorganic phosphate, 3-phosphoglycerate and glucose-6-phosphate on net starch synthesis by isolated chloroplasts of *Codium fragile* (Suringar) Hariot. *New Phytol.* 108, 291–296.

Williams, M.L. and Cobb, A.H. (1989) The isolation of functional chloroplasts from the sacoglossan mollusc *Elysia viridis. New Phytol.* 113, 153–160.

Williams, M.L., Rutter, J.C., Benson, E.E. and Cobb, A.H. (1984) Photosynthesis by *Codium fragile* in an intertidal zone environment: a growth strategy. In: *Advances in Photosynthesis Research*, Volume 4 (ed. C. Sybesma). Martinus Nijhoff/W. Junk Publishers, Belgium, pp. 287–290.

Wright, S.W. and Grant, B.R. (1978) Properties of chloroplasts isolated from siphonous algae. Effect of osmotic shock and detergent treatment on intactness. *Plant Physiol.* 61, 768–771.

Yokata, A. and Kitoaka, S. (1982) Synthesis, excretion and metabolism of glycollate under highly photorespiratory conditions in *Euglena gracilis. Plant Physiol.* 70, 760–764.

Yu, S. and Pederson, M. (1989) Light–DTT activated fructose bisphosphatase in isolated chloroplasts from the marine, green macroalga, *Codium fragile* (Sur.) Hariot. *Bot. Mar.* 32, 79–84.

Pathways and mechanisms of phloem loading

A.J.E. van Bel

3.1 Conceptual evolution of the phloem loading mechanism

In the 1960s, research on photoassimilate transport from the mesophyll to the sieve tubes — a process later named phloem loading (Eschrich, 1970) — was boosted by the efforts of Kursanov and his collaborators (Kursanov, 1984; Kursanov and Brovchenko, 1970). Until then, there had been tacit agreement that the route of transport from the mesophyll to the veins was intercellular (Figure 3.1). Under the influence of Kursanov's group, the view gained ground that photoassimilates were released from the mesophyll prior to accumulation in the sieve tubes. The capacity for release was considered to be a general property of mesophyll cells, with the restriction that the release was concentrated in the vicinity of the veins (see review, Kursanov, 1984). In a rebuttal, advocates of symplastic transfer of photosynthate asserted that 'it was unlikely that a vital process such as phloem loading was fully dependent on movement through a compartment in which the transfer is directed by water flow, since fluctuations in transpiration would interfere with an efficient and well-directed loading' (Ziegler, 1971).

These criticisms diminished when the release was postulated to occur within the vein close to the sieve tube (Geiger *et al.*, 1973, 1974). Short-distance diffusion would greatly reduce the chance of transpirational interference (Figure 3.1). In addition, the plasmodesmatal frequency in the transfer zone appeared to be low which gave structural support to an apoplastic mechanism of phloem loading (Geiger *et al.*, 1973; Gunning *et al.*, 1974). Moreover, electron microscopy (EM) studies on leaf cell plasmolysis showed that the osmotic potential in the sieve element/companion cell complex (SE–CC complex) was much lower than that in the mesophyll cells (Evert *et al.*, 1978; Fisher and Evert, 1982; Geiger *et al.*, 1973). This was a very strong argument in favour of apoplastic phloem loading, as transfer against a concentration gradient (Figure 3.1) and symplastic transport were considered to be incompatible. Furthermore, when offered to stripped leaf

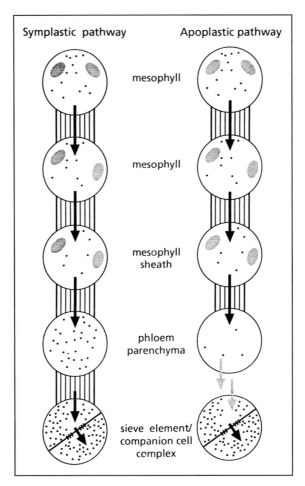

Figure 3.1. *Diagrams of the two potential pathways of phloem loading (from van Bel, 1992). In both cases, the sugar concentration in the SE–CC complexes is presumed to be higher than in the mesophyll (Evert et al., 1978; Fisher, 1986; Fisher and Evert, 1982; Geiger et al., 1973). The relative photoassimilate concentration is indicated by the density of the dotting. The cross-hatching between the cells represents symplastic continuity, the short stripes between SE and CC indicate the specialized branched plasmodesmata. Dark arrows indicate the symplastic transfer of photosynthate. Light arrows indicate apoplastic transfer of photosynthate which comprises consecutive release from the mesophyll symplast, diffusion via the apoplast and retrieval by the SE–CC complex symplast. The term companion cell includes intermediary cells. In several minor vein constructions, some cell types (e.g. mesophyll, phloem parenchyma) are absent. The site of apoplastic transfer is situated close to the SE–CC complex on grounds of the symplastic structure of the vein. Plasmodesmograms (van Bel, 1988; Bourquin et al., 1990) indicate a potential 'extra-veinal' apoplastic transfer in some species. The difference between the two pathways is the apoplastic or symplastic character of the SE loading (see Chapter 12). Reproduced from van Bel (1992a) with permission from Blackwell Scientific Publications.*

discs, sucrose was taken up preferentially by the veins (Delrot and Bonnemain, 1978; Geiger *et al.*, 1974). This fitted well the concept of sucrose uptake by the SE–CC complexes from the apoplast. Sucrose/proton co-transport, discovered at the end of the 1970s, was also fully consistent with apoplastic phloem loading in the minor veins (Giaquinta, 1977). An immense ΔpH as a driving force was expected (Giaquinta, 1983) on grounds of the high pH of phloem sap (about 8) and the low pH of the leaf apoplast (about 5).

These arguments led to overwhelming support for the concept of apoplastic phloem loading. Apoplastic phloem loading was believed to be a universal mechanism in all Angiosperms. Only some cucurbits were thought to have a different — but controversial and thermodynamically obscure — mechanism of phloem loading (Hendrix, 1968, 1977; Madore and Webb, 1981). In the 1980s, several new findings questioned the universality of apoplastic phloem loading (van Bel, 1987; Turgeon and Beebe, 1991). The controversy between the proponents of symplastic and apoplastic phloem loading has been reconciled by the recent concept of multiprogrammed phloem loading (van Bel and Gamalei, 1991). This concept proposes several modes of phloem loading, some of which may be quite versatile and able to react to changing environmental conditions.

3.2 The pathways of phloem loading

The multiprogrammed concept of phloem loading was based originally on ultra-structural studies of the phloem loading zone. These studies (Gamalei, 1985, 1989) showed an immense diversity of symplastic connectivity between the SE–CC complex and the mesophyll in the minor veins. Gamalei (1985, 1989, 1991) distinguished several gradations of plasmodesmatal connectivity, termed type 1, 1–2a, 2a, and 2b with, respectively, abundant (between 60 and 10 plasmodesmata per μm^2 interface), moderate, sporadic, and virtually no (less than 0.1 plasmodesmata per μm^2) plasmodesmatal contacts. At first sight, classification on the basis of plasmodesmatal density alone seems arbitrary. The discrimination between the vein types, however, is also based on differences in the vein architecture and ultrastructure of the companion cells (Gamalei, 1989, 1990). Universal features of the companion cells are the strongly branched mitochondrial network with many cristae (Figure 3.2) and the dense cytoplasmic matrix. Both indicate a high metabolic activity. The companion cells (intermediary cells) in minor veins with abundant symplastic connectivity (type 1) contain extensive endoplasmic reticulum (ER) labyrinths (Figure 3.2). Chloroplasts are absent and other plastids are small and scarce (Gamalei, 1989; Gamalei and Pakhomova, 1982). In minor veins with sporadic or virtually no plasmodesmata (types 2a and 2b), the companion cells are smaller. These cells contain several small vacuoles and chloroplasts embedded in an exceptionally dense cytoplasmic matrix (Figure 3.2). In minor veins with hardly any plasmodesmatal contacts (type 2b), the companion cells often possess conspicuous cell wall invaginations (transfer cells, Figure 3.2). The degree of symplastic connectivity of the SE–CC complex,

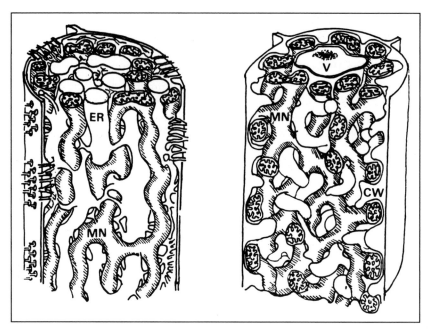

Figure 3.2. *Ultrastructure of the companion cells in minor veins with symplastic (type 1) and apoplastic (type 2) configuration (from Gamalei, 1990). The intermediary cells (companion cells) in type 1 veins (left) possess many plasmodesmatal connections with the adjacent vascular parenchyma cells. The companion cells in type 2 veins (right) have virtually no symplastic contacts with the parenchyma, directly connected with the mesophyll. In type 2b veins, the walls of the companion cells are strongly invaginated (transfer cells). ER, endoplasmic reticulum; MN, mitochondrial network; CW, cell wall; V, vacuole.*

together with the ultrastructure of the companion cell, has been used to define the minor vein configuration (type 1, 1–2a, 2a, 2b; van Bel and Gamalei, 1992).

The minor vein configuration is characteristic of taxonomic families. Only 12 (the polytypical families) out of 116 dicotyledonous families studied possessed more than one type of minor vein configuration (van Bel and Gamalei, 1991, 1992; Gamalei, 1989). Projection of the families onto the Takhtajan system for higher plant evolution showed that the vein configuration with abundant symplastic connectivity predominantly occurs in ancient families (Gamalei, 1989). The vein configuration with low plasmodesmatal frequency prevails in more recent families (Gamalei, 1989).

An obvious supposition is that the minor vein configuration may be correlated with diverse modes of phloem loading. Minor veins with abundant symplastic continuity may perform symplastic phloem loading, whilst veins with symplastic discontinuity may execute apoplastic phloem loading. A close structure/function relationship between vein configuration and the mode of phloem loading would lend credibility to the concept of multiprogrammed phloem loading.

3.3 Correspondence between minor vein configuration and the mode of phloem loading

3.3.1 Evidence obtained with leaf discs

Before the launching of the concept of multiprogrammed phloem loading, investigations had indicated that symplastic mesophyll–vein transfer was a feasible option for phloem loading in some species (Madore and Webb, 1981; Madore and Lucas, 1987; Schmitz *et al.*, 1987; Turgeon and Wimmers, 1988). The drawback of these elegant (but single-species) experiments is that symplastic phloem loading can easily be discounted as an extravagance which may occur in a limited number of specialized species. The only comparative paper (Turgeon and Wimmers, 1988) showed a delay in vein loading of sucrose in the apparent symplastic loader, *Coleus*, when compared to the apoplastic *Pisum*.

Now that the family-associated character of the minor vein configuration is established, different modes of phloem loading can be traced in a more straight-forward manner. Species from families with a distinct minor vein configuration (type 1 or type 2b) were selected and the plasmodesmatal connectivity of the SE–CC complex was checked by EM (van Bel *et al.*, 1992). The plasmodesmatal frequency (Table 3.1) was in accordance with the family classification of Gamalei

Table 3.1. *The plasmodesmatal connectivity between the SE–CC complex and adjacent cell in the minor veins of some dicotyledonous species*

	Vein type	Number of plasmodesmata per μm^2 interface
Epilobium montanum	1	34.1
Fuchsia hybrida	1	37.2
Hydrangea petiolaris	1	41.2
Oenothera biennis	1	37.0
Origanum majorana	1	31.8
Stachys sylvatica	1	43.1
Pelargonium zonale	2a	0.16
Centaurea montana	2b	0.03
Impatiens glandulifera	2b	0.12
Ligularia dentata	2b	0.08
Pisum sativum	2b	0.08
Symphytum officinale	2b	0.04
Acanthus mollis	1	38.6
	2b	0.05

The symplastic connectivity is expressed as the number of plasmodesmata per μm^2 interface and is counted according to the method of Gamalei and Pakhomova (1982). In *Acanthus*, two SE types were present in the minor veins, one with an abundant symplastic connectivity via an intermediary cell, the other with an apoplastic configuration. Table reproduced from van Bel *et al.* (1992) with permission from Springer-Verlag.

(1985). In a series of experiments, a correspondence between minor vein configuration and the physiological behaviour was investigated (van Bel *et al.*, 1992).

The phloem loading process was challenged by *p*-chloromercuribenzene sulphonic acid (PCMBS), an inhibitor of membrane transport of sugars. Application of PCMBS would be expected to block only apoplastic phloem loading (Figure 3.3). Symplastic transfer of photosynthate would be unaffected by PCMBS. Control and PCMBS-pretreated stripped leaf discs were fed with $^{14}CO_2$ and then floated on an incubation medium. After incubation, the discs were freeze-dried and autoradiographed (van Bel *et al.*, 1992).

The experimental set-up resembles that of Madore and Lucas (1987), with the difference that here the plants were predarkened for 5–7 days (van Bel *et al.*, 1992). Discs from plants grown under a normal day/night regime hardly showed any accumulation by the minor veins. In discs of predarkened plants, ^{14}C accumulated in the veins, possibly as result of photoassimilate depletion in the sieve tubes (van Bel *et al.*, 1992). Essential features of this approach are as follows:

(a) The unconventional orientation of the discs, with the stripped side uppermost, enabling maximal entry of $^{14}CO_2$ and a correspondingly high rate of photosynthesis. Moreover, this positioning protects against any indirect effects of PCMBS inducing stomatal closure.

(b) Feeding with $^{14}CO_2$ — and not with ^{14}C-sugars — guarantees the production of the natural sugar mixture (see Section 3.4.1).

The quantified contrast between mesophyll and minor vein was taken as a measure of the SE loading (Figures 3.4, 3.5; van Bel *et al.*, 1992). The autoradiographs indicated no effect of PCMBS on phloem loading (Figure 3.4) in *Epilobium*, *Fuchsia*, and *Oenothera* (members of the Onagraceae), *Origanum* and *Stachys* (Lamiaceae) and *Hydrangea* (Hydrangeaceae). These families posses a symplastic minor vein configuration (Gamalei, 1989). In contrast, a significant PCMBS effect (Figure 3.5) showed up in representatives of families with an apoplastic minor vein configuration (*Pisum* — Fabaceae; *Pelargonium* —

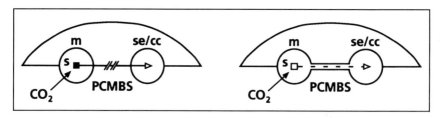

Figure 3.3. *Putative actioin of PCMBS in leaves with symplastic or apoplastic phloem loading. In 'apoplastic loaders' (left), phloem loading is blocked by PCMBS either at the level of the mesophyll release or the accumulation by the SE–CC complexes. Phloem loading by the highest-order veins is not prevented by PCMBS in 'symplastic loaders' (right). m, mesophyll; s, sugar; se/cc, SE–CC complex. Reproduced from van Bel (1992b) with permission from Elsevier.*

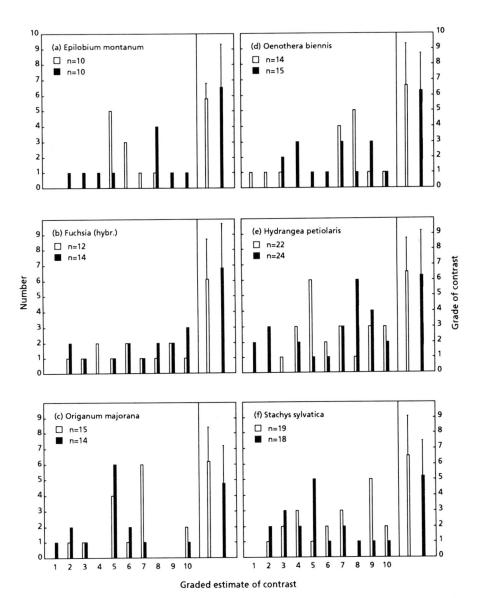

Figure 3.4. *Effect of PCMBS on phloem loading in stripped discs from predarkened leaves of* Epilobium *(a)*, Fuchsia *(b)*, Origanum *(c)*, Oenothera *(d)*, Hydrangea *(e)*, and Stachys *(f)*. *After feeding with* $^{14}CO_2$, *discs were floated on a standard medium and chased for 2 h. The bars represent the graded estimate of contrast between mesophyll and minor veins in control (□) and PCMBS-treated (■) discs. The left–hand part of the graphs gives the number of individuals (n = total number of discs) with a certain grade of contrast. The right–hand part of the graphs represents the average contrast (± SE) between mesophyll and vein in control (□) and treated (■) discs. There was no statistically significant difference in contrast between control and treatment in any of the species. Reproduced from van Bel et al. (1992) with permission from Springer-Verlag.*

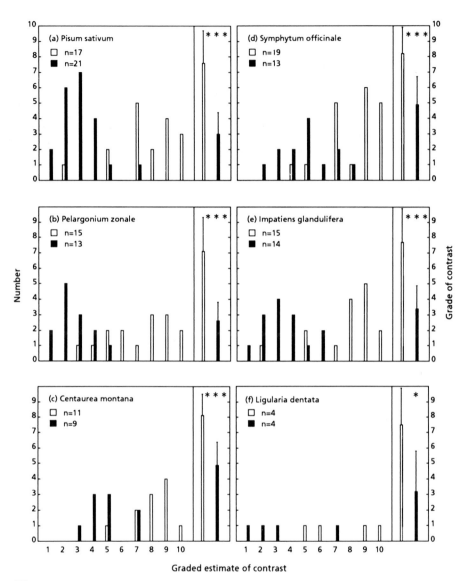

Figure 3.5. *Effect of PCMBS on phloem loading in epidermally stripped discs from predarkened leaves of Pisum (a), Pelargonium (b),* Centaurea *(c),* Symphytum *(d),* Impatiens *(e), and* Ligularia *(f). After feeding with* $^{14}CO_2$, *discs were floated on standard medium and chased for 2 h. The bars represent the graded estimate of contrast between mesophyll and minor veins in control (□) and PCMBS-treated (■) discs. The left-hand part of the graphs gives the number of individuals (n = total number of discs) with a certain grade of contrast. The right-hand part of the graphs represents the average contrast (± SE) between mesophyll and vein in control (□) and treated (■) discs.* $* = 0.05 \geqslant P \geqslant 0.01,$ $*** = P \leqslant 0.001,$ *values that are statistically different at the 95% and 99.9% confidence levels, respectively. Reproduced from van Bel* et al. *(1992) with permission of Springer-Verlag.*

Geraniaceae; *Centaurea, Ligularia* — Asteraceae; *Impatiens* — Balsaminaceae; *Symphytum* — Boraginaceae).

3.3.2 *Evidence obtained with whole leaves*

Use of whole leaves instead of discs has some advantages.

(a) The number of species that can be used for work with leaf discs is restricted to those with a strippable leaf epidermis. Species (including nearly all trees) with a strong sclerenchymatous attachment to the epidermis are unsuitable.

(b) Intact leaves do not require predarkening for sufficient ^{14}C accumulation by the veins (van Bel *et al.*, 1992) which makes the system less artificial.

(c) Artefacts as result of cutting and stripping are absent.

(d) Last but not least, arrival of ^{14}C-labelled photosynthate in phloem exudate, collected at the cut end of the petiole, unequivocally demonstrates loading of photosynthate into the sieve tubes. Vein labelling in discs may provide misleading evidence for phloem loading, because the radioactivity may be resident in the mestome sheath, not in the SE–CC complex.

PCMBS applied via the transpiration stream was therefore employed to discriminate between symplastic and apoplastic phloem loading in intact leaves. Of course, the uneven distribution of PCMBS over the leaf, the potential PCMBS effect on stomatal opening and the deregulation of retrieval mechanisms along the phloem pathway are potential sources of error with this procedure. In spite of these apparent drawbacks, the method has appeared to be successful in whole leaves (Turgeon and Gowan, 1990). A modification to this approach was used to distinguish between different modes of phloem loading (van Bel *et al.*, unpublished results). After feeding with $^{14}CO_2$, petioles were placed in EDTA in order to collect phloem exudate (King and Zeevaart, 1974). EDTA prevents the formation of callose and enables the release of phloem sap, the composition of which is similar to that acquired with other methods (Girousse *et al.*, 1991; Weibull *et al.*, 1990). The difference in ^{14}C content of the exuded sap from control and PCMBS-treated leaves indicates to what extent phloem loading is affected by PCMBS. If this ratio is approximately 1, PCMBS presumably has no effect. However, even in symplastic loaders, some PCMBS effect is expected due to interference with sugar retrieval along the phloem pathway.

In the representatives of type 1 families (Lamiaceae, Hydrangeaceae, Lythraceae, Onagraceae), the ^{14}C ratio between the phloem exudate of control and treated leaves varied between 0.5 and 3.0 (Table 3.2). Representatives of type 2b families (Asteraceae, Balsaminaceae, Dipsacaceae, Linaceae, Tropaeolaceae, Valerianaceae) or type 2b members of polytypical families (Fabaceae, Scrophulariaceae) generally showed much higher ratios of up to 75 (Table 3.2).

Due to the presence of the epidermis, the thickness of the higher order veins and the wrinkling of whole leaves during freeze-drying, there was less contrast between minor veins and mesophyll in autoradiographs of whole leaves than in

CH. 3. PATHWAYS AND MECHANISMS OF PHLOEM LOADING

Table 3.2. *Effect of PCMBS on phloem loading in intact leaves of dicotyledonous species with a symplastic (type 1) or apoplastic (type 2) minor vein configuration*

Type	Family	Species	I	II	III
1	Ginkyaceae	*Ginkyo biloba*		0	
1	Lamiaceae	*Coleus blumei*	+	0	0
		Galeobdolon maculatum	0	0	0
		Lamium album	0		
		Origanum majus	0	+	0
1	Hydrangeaceae	*Philadelphus coronarius*	0		
1	Lythraceae	*Cuphea ignea*	0	0	0
		Lythrum salicaria	0		
1	Onagraceae	*Epilobium montanum*	0	0	0
		Fuchsia hybrida	+	0	0
		Oenothera biennis	+		0
2a	Gentianaceae	*Exacum affine*	+	+	+
2b	Asteraceae	*Bellis perennis*	+ +		
		Chrysanthemum sp.	+ + +	+ +	+
		Senecio vulgaris	+ + +	+ +	+ +
		Taraxacum officinale	0	+ +	+ +
2b	Balsaminaceae	*Impatiens sultani*	+ + +	+ +	+ +
2b	Dipsacaceae	*Succisa pratensis*	0		
2b	Fabaceae	*Pisum sativum*		0	+ +
2b	Linaceae	*Linum flavum*	+ +	+	
2b	Scrophulariaceae	*Antirrhinum majus*	+ +	+	+
2b	Tropaeolaceae	*Tropaeolum majus*	+		
2b	Valerianaceae	*Centranthus ruber*	+ +	+	+ +
		Valerianella locusta	+ +	+ +	+

One hour before pulse-labelling with $^{14}CO_2$, the petioles of excised leaves were placed in EDTA-containing media with or without PCMBS. The media were transported to the minor vein endings by the transpiration stream for 1 h. After feeding with $^{14}CO_2$ for 30 min with the petioles in EDTA solution, the petioles were kept in the EDTA solutions for 2 h. Radioactivity in the phloem exudate was counted by scintillation spectrophotometry. The leaves were freeze-dried and autoradiographed. The contrast between mesophyll and major veins (main vein and second-order vein) in the autoradiographs was measured by densitometry. Radioactivity of the mesophyll and major vein zones was counted by scintillation spectrophotometry after separate digestion of the tissues (see van Bel *et al.*, 1992). The PCMBS effect is expressed as the ratio between the control and PCMBS-treated plants. The ratios between the phloem exudates (column I) are represented by 0 (ranging from 0.5 to 2.0), + (2.0–5.0), + + (5.0–24.9), and + + + ($\geqslant 25.0$); the ratios between the vein/mesophyll contrasts on the autoradiographs (column II) by 0 (0.80–1.25), + (1.25–1.99), and + + ($\geqslant 2.00$); the ratios between the vein/mesophyll ^{14}C ratios (column III) by 0 (0.70–1.25), + (1.25–1.99), and + + ($\geqslant 2.00$). Blank spaces indicate that determinations were not carried out (e.g. insufficient exudation, lack of contrast for densitometry).

those of discs. Only the contrast between mesophyll and major veins was detectable. Densitometric measurements demonstrated that the contrast between the mesophyll area and the higher order veins was equal in PCMBS-treated and control leaves of type 1 species (Table 3.2). PCMBS reduced the contrast in type 2b species (Table 3.2). The ^{14}C contents in equivalent parts of dissected leaves confirm this observation (Table 3.2). In conclusion, the three-step procedure using whole leaves revealed a consistent reduction of phloem loading by PCMBS in type 2b species and not in type 1 species (Table 3.2).

3.3.3 *Minor vein ultrastructure and the mode of phloem loading*

The results with discs and whole leaves complement each other, indicating a coincidence between the mode of phloem loading and the minor vein configuration. Whether or not PCMBS affects phloem loading corresponds exactly with the demarcation between the symplastic and apoplastic minor vein configuration (Figures 3.4 and 3.5, and Table 3.2). However, the results may not be as logical as they appear for the following reasons.

(a) The pictograms of Gamalei (1985, 1989) only pertain to the SE–CC complex and the mesophyll. The plasmodesmatal contacts with cells other than those in the direct route between mesophyll and SE have not been taken into consideration. The role of the vascular parenchyma cells in the loading process is still completely obscure (cf. Fisher, 1991; Giaquinta, 1983).

(b) Minor veins often contain more than one SE (*Coleus* — Fisher, 1986; *Cucumis* — Schmitz *et al.*, 1987; *Commelina* — van Bel *et al.*, 1988; *Acanthus* — van Bel *et al.*, 1992). Plasmodesmograms — diagrammatical presentations of plasmodesmatal frequencies (van Bel *et al.*, 1988) — illustrate distinctly dissimilar connectivity of the SEs within the same vein. It is not entirely clear whether these vein cross-sections actually constitute the finest minor veins. If they do, the simultaneous occurrence of apoplastic and symplastic phloem loading is structurally conceivable (for review see van Bel, 1992a). A two-way loading path, however, contrasts with Gamalei's classification — based on the plasmodesmatal connectivity of one single SE (Gamalei, 1989) — which was the cornerstone of the experiments presented here (Figures 3.4 and 3.5, and Table 3.2).

(c) Plasmodesmograms (van Bel and Gamalei, 1991) also show a potential for apoplastic passage of photoassimilates outside the vein (e.g. Bourquin *et al.*, 1990). This implies that species with a symplastic vein configuration — presumed to carry out symplastic SE loading — may also include an apoplastic step in the phloem loading pathway. As a result, phloem loading in such species would be strongly suppressed by PCMBS.

Given all these caveats, the observed tight relationship between the vein typology (Gamalei, 1989) and the mode of phloem loading is surprising. The lack of effect of PCMBS in leaves with a symplastic minor vein configuration (van Bel *et al.*, 1992) seemingly rules out any significant loading via a 'concealed' apoplastic pathway. It appears that the metabolism and associated ultrastructure of the companion cell and the symplastic connectivity of the SE–CC complex are the key factors in determining the mode of phloem loading.

3.4 The mechanisms of phloem loading

3.4.1 *Transport sugars in the phloem*

The presumptive pathways of photosynthate transfer (Figure 3.1) imply that the corresponding mechanisms of SE loading must be quite different. It is noteworthy that sucrose is the only transport sugar in type 2 families (Gamalei, 1985; Zimmermann and Ziegler, 1975). In type 1 families, mixtures of sucrose and oligosaccharides occur in variable proportions in the phloem exudate (Gamalei, 1985; Zimmermann and Ziegler, 1975). It is plausible that sucrose and oligosaccharides and sucrose alone are loaded in minor veins with symplastic and apoplastic configuration, respectively.

3.4.2 *The mechanism of apoplastic phloem loading*

Apoplastic phloem loading seems to have a sound thermodynamic basis (Delrot, 1987; Giaquinta, 1983). According to the concept of apoplastic phloem loading, sucrose is released from the mesophyll symplast prior to active loading into the SE–CC complexes (Figure 3.1). The selective carrier-mediated uptake of sucrose is driven by an appreciable proton-motive force, which allows, in theory, an accumulation by a factor of approximately 10^5 (Giaquinta, 1983). Although this model seems to be correct in broad outline, some comments are appropriate.

(a) The location and mechanism of photosynthate release from the mesophyll symplast is a matter of debate. The low plasmodesmatal frequency between SE–CC complexes and other vascular elements in numerous species (Gamalei, 1985) is supportive of photoassimilate release close to the SE–CC complex. Moreover, experiments with dyes have identified 'stagnant-water zones' in the apoplast around the phloem in the minor veins (J.S. Pate, personal communication). Transpiration will have little effect on solute movement in this unstirred space. As a result, two important conditions for apoplastic phloem loading (few plasmodesmata and insensitivity to transpiration) are fulfilled. On the other hand, the frequency distribution in a few plasmodesmograms (e.g. Bourquin *et al.*, 1990) shows potential extra-veinal photosynthate release which would be exposed to transpirational flow.

(b) Further, there is no certainty about the mechanism of release. Diffusional loss as part of a pump-leak system (Turgeon, 1984) and active release, especially in legume leaves (Anderson, 1983; M'Batchi and Delrot, 1988), have been postulated, both based on results from treatments with PCMBS.

(c) Assertions with regard to the proton-motive force — as likely and attractive as they may be — are founded on the behaviour of transport phloem and the composition of transport phloem exudate. Local measurements of membrane potential and ΔpH of the phloem in the minor vein have never been performed. This lack of information is important as the sugar uptake characteristics of stripped leaf discs may be attributed mainly to the mesophyll (van Bel, 1992a).

(d) The contribution of substances other than sugars (e.g. amino acids and potassium) to the osmotic potential of the sieve tube exudate is high (e.g. Girousse *et al.*, 1991; Hayashi and Chino, 1986; Weibull *et al.*, 1990). Nevertheless, the differences in osmotic potential between the mesophyll and the SE–CC complexes are commonly ascribed to differences in sugar concentration.

3.4.3 *The mechanism of symplastic phloem loading*

The principal obstacle to accepting a symplastic mode of phloem loading (Delrot, 1987; Giaquinta, 1983) is the presumptive incompatibility of symplastic transport with the apparent sugar accumulation by the SE–CC complex (Evert *et al.*, 1978; Fisher and Evert, 1982; Geiger *et al.*, 1973). This paradox may be explained by the 'polymerization trap mechanism' (Figure 3.6) introduced by Turgeon (1991). It proposes that sucrose and galactinol produced by the mesophyll diffuse along their concentration gradient to the intermediary cells (Figure 3.6). There, they are synthesized to raffinose, and eventually stachyose or verbascose. The diameter of raffinose exceeds the molecular exclusion limit of the plasmodesmata between intermediary cells and the neighbouring cells, preventing back-flow to the mesophyll. The diameter of the branched plasmodesmata between intermediary cells

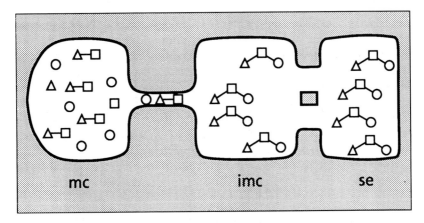

Figure 3.6. *Model of the polymerization trap mechanism (constructed after Turgeon, 1991) to explain symplastic phloem loading against the photoassimilate gradient. Galactinol (○) and sucrose (glucose △ and fructose □) diffuse from mesophyll to the intermediary cells where they are synthesized to raffinose. The molecular size of the raffinose prevents symplastic back-flow to the mesophyll, but allows transfer to the SE via broader plasmodesmata. mc, mesophyll cell; imc, intermediary cell; se, sieve element. Reproduced from van Bel (1992a) with permission of Blackwell Scientific Publications.*

and SEs may be large enough to allow unhindered passage of the galactosyl-oligo-saccharides towards the SEs (Figure 3.6).

The presence of oligosaccharides exclusively in the veins seems therefore an absolute prerequisite for the acceptance of the 'polymerization trap mechanism'. The model (Figure 3.6) is compatible with the activity of raffinose-6-galactosyl-transferase (= stachyose synthase) in *Cucurbita* leaves (Gaudreault and Webb, 1981) and with the presence of oligosaccharides in their intermediary cells (Pristupa, 1983). Furthermore, considerable amounts of oligosaccharides were present in the vein tissue of *Cucumis*, whereas the mesophyll was virtually devoid of sucrosylgalactosides (Schmitz and Holthaus, 1986). This finding is corroborated by immunolocalization of sucrosylgalactosidases (stachyose synthase) in the intermediary cells of *Cucumis* (Holthaus and Schmitz, 1991a,b). That the uptake parameters for sucrose and stachyose uptake were nearly identical in *Coleus* leaf discs (Madore, 1990) does not conflict with the model of Turgeon. That mesophyll and veins, both contributing to the uptake from the medium (Fondy and Geiger, 1977; Madore and Webb, 1981), are equipped with uptake systems for sucrose and stachyose does not rule out the 'polymerization mechanism'.

Not consistent with the model is stachyose synthesis in isolated mesophyll cells of *Cucurbita pepo* (Madore and Webb, 1982) and the relatively high content of raffinose, stachyose and verbascose in the mesophyll of the succulent cucurbit *Xerosicyos danguyi* (Madore *et al.*, 1988).

The production of oligosaccharides is an integral part of the 'polymerization trap model' for symplastic phloem loading (Figure 3.6). Therefore, high quantities of sucrose in the phloem sap of presumptive symplastic loaders (Gamalei, 1985; Zimmermann and Ziegler, 1975) seemingly conflict with this proposal (Figure 3.6). The discrepancy may be explained by the fact that phloem sap collected at some distance from the point of phloem loading does not account for the sieve tube content at the actual loading site. To bring Turgeon's model into conformity with the presence of appreciable amounts of sucrose in the phloem exudate, we must assume that the oligosaccharides are converted to sucrose along the transport path or that sucrose is loaded in parallel with the galactosyl-oligosaccharides into the SEs (Turgeon, 1991).

3.5 The ecophysiological concept of phloem loading

Analysis of the global distribution of plant families (with the associated vein con-figuration) indicates that disparity in the mode of loading may have an eco-physiological significance (van Bel and Gamalei, 1992; Gamalei, 1989). Certain minor vein configurations show a preference for specific climatic zones (Gamalei, 1991). Provided that the coincidence between minor vein configuration and the mode of phloem loading (van Bel *et al.*, 1992) is universal, apoplastic phloem loading prevails in the temperate and arid zones (van Bel and Gamalei, 1992; Gamalei, 1991).

Coincidentally, this concept explains why apoplastic phloem loading was

Table 3.3. *Principal families of outdoor crops and fruits of the temperate zone and the minor vein configuration*

Family	Minor vein type	Representatives
Asteraceae	2b	Lettuce, artichoke, chicory, endive
Brassicaceae	2a–2b	Radish, Brussels sprouts, cauliflower, red cabbage, kale
Chenopodiaceae	2a	Spinach, beet, sugar beet
Faboideae	2b	Bushbean, broadbean, pea
Gramineae	2a	Maize, wheat, oats, barley, rye
Liliaceae	2b	Onion, asparagus, leek, salsify
Rosaceae	1–2a	Apple, pear, cherry, plum, rose-hip
Solanaceae	2a/2b	Potato

The minor vein typology is defined in the text (see Section 3.2). Table reproduced from van Bel (1992b) with permission from Elsevier.

dominant in the species investigated until recently. Most investigations used crops of the temperate zone. The prominent dicotyledonous crops of the temperate zone are all members of families with a 2a- or 2b-minor vein configuration (Table 3.3). These are plants which produce and distribute massive amounts of photosynthesis products in a relatively short growing season. The corresponding need for efficient translocation is probably best met by an apoplastic mode of phloem loading (van Bel, 1989; van Bel and Gamalei, 1992). The mode of phloem loading in the Gramineae has not been investigated sufficiently. On account of the high variability of their minor vein structure, the graminaceous crops may have different and very flexible modes of phloem loading (e.g. Chonan *et al.*, 1985).

The present geographical location of the families invite speculation on the selective pressures on the evolutionary changes in minor vein configuration (van Bel and Gamalei, 1992; Gamalei, 1991). The symplastic-to-apoplastic transition may have been provoked by temperature and water stresses (van Bel, 1992a; van Bel and Gamalei, 1992; Gamalei, 1989, 1990, 1991). Likely shortcomings of the symplastic mode of phloem loading under these conditions are: (a) the closure of plasmodesmata, possibly induced by Ca^{2+}, at low temperature (Erwee and Goodwin, 1983; Gamalei, 1989; Minorsky, 1985), (b) the restricted transport capacity of the intermediary cells, particularly at temperatures below 10°C (Gamalei, 1990), (c) the inability to set up an appreciable pressure gradient (Richardson *et al.*, 1984), although this is not recognized as a general feature (Fisher, 1986; Turgeon and Hepler, 1989), and (d) the relatively high viscosity of oligosaccharide solutions (van Bel, 1992a; Lang, 1978).

References

Anderson, J.M. (1983) Sucrose release from soybean leaf slices. *Physiol. Plant.* **66**, 319–327.

van Bel, A.J.E. (1987) The apoplast concept of phloem loading has no universal validity. *Plant Physiol. Biochem.* **25**, 677–686.

van Bel, A.J.E. (1989) The challenge of symplastic phloem loading. Bot. Acta, 102, 183–185.

van Bel, A.J.E. (1992a) Different phloem loading machineries correlated with the climate. Acta Bot. Neerl. 41, 121–141.

van Bel, A.J.E. (1992b) Mechanisms of sugar translocation. In: Crop Photosynthesis: Spatial and Temporal Determinants (eds N.R. Baker and H. Thomas). Elsevier, Amsterdam, pp. 177–210.

van Bel, A.J.E. and Gamalei, Y.V. (1991) Multiprogrammed phloem loading. In: Recent Advances in Phloem Transport and Assimilate Compartmentation (eds J.-L. Bonnemain, S. Delrot, W.J. Lucas and J. Dainty). Ouest Editions, Nantes, pp. 128–139.

van Bel, A.J.E. and Gamalei, Y.V. (1992) Ecophysiology of phloem loading in source leaves. Plant Cell Environ. 15, 265–270.

van Bel, A.J.E., Van Kesteren, W.J.P. and Papenhuijzen, C. (1988) Ultrastructural indications for coexistence of symplastic and apoplastic phloem loading in Commelina benghalensis leaves. Differences in ontogenic development, spatial arrangement and symplastic connections of the two sieve tubes in the minor vein. Planta, 176, 159–172.

van Bel, A.J.E., Gamalei, Y.V., Ammerlaan, A. and Bik, L.P.M. (1992) Dissimilar phloem loading in leaves with symplasmic or apoplasmic minor vein configurations. Planta, 186, 518–525.

Bourquin, S., Bonnemain, J.-L. and Delrot, S. (1990) Inhibition of loading of ^{14}C assimilates by p–chloromercuribenzenesulfonic acid. Localization of the apoplastic pathway in Vicia faba. Plant Physiol. 92, 97–102.

Chonan, N., Kawahara, H. and Matsuda, T. (1985) Ultrastructure of transverse veins in relation to phloem loading in the rice leaf. Jap. J. Crop Sci. 54, 160–169.

Delrot, S. (1987) Phloem loading: apoplastic or symplastic? Plant Physiol. Biochem. 25, 667–676.

Delrot, S. and Bonnemain, J.-L. (1978) Etude du mécanisme de l'accumulation des produits de la photosynthèse dans les nervures. Compt. Rend. Acad. Sci. Paris, 287 D, 125–130.

Erwee, M.G. and Goodwin, P.B. (1983) Characterisation of the Egreria densa Planch. leaf symplast I. Inhibition of intercellular movement of fluorescent probes by group II ions. Planta 158, 320–328.

Eschrich, W. (1970) Biochemistry and fine structure of phloem in relation to transport. Ann. Rev. Plant Physiol. 21, 193–214.

Evert, R.F., Eschrich, W. and Heyser, W. (1978) Leaf structure in relation to solute transport and phloem loading in Zea mays L. Planta, 138, 279–294.

Fisher, D.G. (1986). Ultrastructure, plasmodesmatal frequency, and solute concentration in green areas of variegated Coleus blumei Benth. leaves. Planta, 169, 141–152.

Fisher, D.G. (1991) Plasmodesmatal frequency and other structural aspects of assimilate collection and phloem loading in leaves of Sonchus oleraceus (Asteraceae), a species with minor vein transfer cells. Am. J. Bot. 78, 1549–1559.

Fisher, D.G. and Evert, R.F. (1982) Studies on the leaf of Amaranthus retroflexus (Amaranthaceae): ultrastructure, plasmodesmatal frequency, and solute concentration in relation to phloem loading. Planta, 155, 377–387.

Fondy, B.R. and Geiger, D.R. (1977) Sugar selectivity and other characteristics of phloem loading in Beta vulgaris L. Plant Physiol. 59, 953–960.

Gamalei, Y.V. (1985) Characteristics of phloem loading in woody and herbaceous plants. Sov. Plant Physiol. 32, 656–665.

Gamalei, Y.V. (1989) Structure and function of leaf minor veins in trees and herbs. *Trees* 3, 96–110.

Gamalei, Y.V. (1990) *Leaf Phloem* (in Russian). Nauka, Leningrad.

Gamalei, Y.V. (1991) Phloem loading and its development related to plant evolution from trees to herbs. *Trees* 5, 50–63.

Gamalei, Y.V. and Pakhomova, M.V. (1982) Distribution of plasmodesmata and parenchyma transport of assimilates in the leaves of several dicots. *Sov. Plant Physiol.* 28, 649–661.

Gaudreault, P.-R. and Webb, J.A. (1981) Stachyose synthesis in leaves of *Cucurbita pepo*. *Phytochemistry* 20, 2629–2633.

Geiger, D.R., Giaquinta, R.T., Sovonick, S.A. and Fellows, R.J. (1973) Solute distribution in sugar beet leaves in relation to phloem loading and translocation. *Plant Physiol.* 52, 585–589.

Geiger, D.R., Sovonick, S.A., Shock, T.L. and Fellows, R.J. (1974) Role of free space in translocation of sugar beet. *Plant Physiol.* 54, 892–898.

Giaquinta, R.T. (1977) Phloem loading of sucrose. pH dependence and selectivity. *Plant Physiol.* 59, 750–755.

Giaquinta, R.T. (1983) Phloem loading of sucrose. *Ann. Rev. Plant Physiol.* 34, 347–387.

Girousse, C., Bonnemain, J.-L., Delrot, S. and Bournoville, R. (1991) Sugar and amino acid composition of phloem sap of *Medicago sativa*: a comparative study of two collecting methods. *Plant Physiol. Biochem.* 29, 41–48.

Gunning, B.E.S., Pate, J.S., Minchin, F.R. and Marks, I. (1974) Quantitative aspects of transfer cell structure in relation to vein loading in leaves and solute transport in legume nodules. *Symp. Soc. Exp. Biol.* 18, 87–126.

Hayashi, H. and Chino, M. (1986) Collection of pure phloem sap from wheat and its chemical composition. *Plant Cell Physiol.* 27, 1387–1393.

Hendrix, J.E. (1968) Labeling pattern of translocated stachyose in squash. *Plant Physiol.* 43, 1631–1636.

Hendrix, J.E. (1977) Phloem loading in squash. *Plant Physiol.* 60, 567–569.

Holthaus, U. and Schmitz, K. (1991a) Stachyose synthesis in mature leaves of *Cucumis melo*. Purification and characterization of stachyose synthase (EC 2.4.1.67). *Planta*, 184, 525–531.

Holthaus, U. and Schmitz, K. (1991b) Distribution and immunolocalization of stachyose synthase in *Cucumis melo* L. *Planta*, 185, 479–486.

King, R.W. and Zeevaart, J.A.D. (1974) Enhancement of phloem exudation from cut petioles by chelating agents. *Plant Physiol.* 53, 96–103.

Kursanov, A.L. (1984) *Assimilate Transport in Plants*. Elsevier, Amsterdam.

Kursanov, A.L. and Brovchenko, M.I. (1970) Sugars in the free space of leaf plates: their origin and possible involvement in transport. *Can. J. Bot.* 48, 1243–1250.

Lang, A. (1978) A model of mass flow in the phloem. *Aust. J. Plant Physiol.* 5, 535–546.

Madore, M.A. (1990) Carbohydrate metabolism in photosynthetic and nonphotosynthetic tissues of variegated leaves of *Coleus blumei* Benth. *Plant Physiol.* 93, 617–622.

Madore, M.A. and Webb, J.A. (1981) Leaf free space analysis and vein loading in *Cucurbita pepo*. *Can. J. Bot.* 59, 2550–2557.

Madore, M.A. and Webb, J.A. (1982) Stachyose synthesis in isolated mesophyll cells of *Cucurbita pepo*. *Can. J. Bot.* 60, 126–130.

Madore, M.A. and Lucas, W.J. (1987) Control of photoassimilate movement in source-leaf tissues of *Ipomoea tricolor* Cav. *Planta*, 171, 197–204.

Madore, M.A., Mitchell, D.E. and Boyd, C.M. (1988) Stachyose synthesis in source leaf tissues of the CAM plant *Xerosicyos danguyi* H. Humb. *Plant Physiol.* 87, 588–591.

M'Batchi, B. and Delrot, S. (1988) Stimulation of sugar exit from leaf tissues of *Vicia faba* L. *Planta*, 174, 340–348.

Minorsky, P.V. (1985) An heuristic hypothesis of chilling injury in plants: a role for calcium as the primary physiological transducer of injury. *Plant Cell Environ.* 8, 75–94.

Pristupa, N.A. (1983) Distribution of ketosugars among cells of conducting bundles of the *Cucurbita pepo* leaf. *Sov. Plant Physiol.* 30, 372–378.

Richardson, P.T., Baker, D.A. and Ho, L.C. (1984) Assimilate transport in cucurbits. *J. Exp. Bot.* 35, 1575–1581.

Schmitz, K. and Holthaus, U. (1986) Are sucrosyl–oligosaccharides synthesised in mesophyll protoplasts of mature leaves of *Cucumis melo*? *Planta*, 169, 529–535.

Schmitz, K., Cuypers, B. and Moll, M. (1987) Pathway and assimilate transport between mesophyll cells and minor veins of *Cucumis melo* L. *Planta*, 171, 19–29.

Turgeon, R. (1984) Efflux of sucrose from minor veins of tobacco leaves. *Planta*, 161, 120–128.

Turgeon, R. (1991) Symplastic phloem loading and the sink–source transition in leaves: a model. In: *Recent Advances in Phloem Transport and Assimilate Compartmentation* (eds J.-L. Bonnemain, S. Delrot, W.J. Lucas and J. Dainty). Ouest Editions, Nantes, pp. 18–22.

Turgeon, R. and Wimmers, L.E. (1988) Different patterns of vein loading of exogenous [^{14}C]sucrose in leaves of *Cucurbita pepo*. *Planta*, 179, 24–31.

Turgeon, R. and Hepler, P.K. (1989) Symplastic continuity between mesophyll and companion cells in minor veins of mature *Cucurbita pepo* leaves. *Planta*, 179, 24–31.

Turgeon, R. and Gowan, E. (1990) Phloem loading in *Coleus blumei* in the absence of carrier-mediated uptake of export sugar from the apoplast. *Plant Physiol.* 94, 1244–1249.

Turgeon, R. and Beebe, D.U. (1991) The evidence for symplastic phloem loading. *Plant Physiol.* 96, 349–354.

Weibull, J., Ronquist, F. and Brishammer, S. (1990) Free amino acid composition of leaf exudates and phloem sap. A comparative study in oats and barley. *Plant Physiol.* 92, 222–226.

Ziegler, H. (1971) Wasserumsatz und Stoffbewegungen. *Fortschr. Botanik*, 33, 63–84.

Zimmermann, M.H. and Ziegler, H. (1975) List of sugars and sugar alcohols in sieve–tube exudates. In: *Encyclopedia of Plant Physiology. Transport in Plants*, Volume I. *Phloem Transport* (eds M.H. Zimmermann and J.A. Milburn). Springer-Verlag, Berlin, pp. 480–503.

Sucrose and cell water relations

A.D. Tomos, R.A. Leigh, J.A. Palta and
J.H.H. Williams

4.1 Introduction

4.1.1 *Water relations, an old problem*

Higher plants are sophisticated hydraulic machines powered by linking biochemical energy to the transport of solutes across osmotically active membranes. This process generates considerable hydraulic pressure within cells. Turgor pressures of well-watered plant cells are typically of the range 0.3–1.0 MPa (Tomos, 1988). These pressures are of the same order of magnitude as those of the high pressure steam turbines of a nuclear power station: our local Atomfa Trawsfynydd operates at 1.8–5.24 MPa (Trydan Niwclear, 1991). The interaction between biochemistry and water relations is a very old one, going back to the first primeval cell. It began, however, not as a useful process but as a problem that had to be overcome in order to allow cellular life to evolve. Due to the essential selective-permeability of cell membranes (true semi-permeability is not compatible with life as it would not allow the access of metabolites to the cells) and to the ubiquitous presence of charged impermeant macromolecules within them, cells have always had to deal with the consequences of their water relations. The first cell, bathed in a solution of permeant ions, was faced with the problem of the Donnan equilibrium (Hempling, 1982). Counterions to the impermeant macromolecular electrolytes would have leaked in and filled the cell to a concentration greater than that outside. The resulting osmotic gradient would have driven a water flow into the cell, which would either have continued to expand indefinitely or, more likely, have ruptured. It really was a case of the phrase 'win or bust'! Two solutions to this dilemma appear to have been developed by plants. In their modern guises these are membrane transport mechanisms (initially to remove ions, principally Na^+, from the cells) and cell walls (to allow the development of a hydrostatic pressure to balance the osmotic pressure gradient). Cells were faced, therefore, with the dilemma of too much pressure before they had to cope with too little. This is also the dilemma facing cells that synthesize osmotica.

4.1.2 *Photosynthesis makes solutes*

Changes in solute concentration within a cell will have potentially dramatic effects on its hydrostatic pressure. Any net solute synthesis due to photo-synthesis calls for some measure to deal with the water relations consequences. The van't Hoff relationship ($\pi V = nRT$) dictates that for each 40 mM increase in sucrose concentration the osmotic pressure of a solution rises by approximately 0.1 MPa (1 bar) (Nobel, 1991). Homeostatic responses of the cells to this can be both passive and active (reviewed in Tomos, 1988).

4.1.3 *Passive responses to changes in water potential*

The walls of plant cells would appear to provide them with more flexibility than that enjoyed by their wall-less counterparts. In principle, osmotic imbalances can simply lead to water flow and turgor pressure change without the requirement for further action. With suitably robust walls this would allow for considerable changes in solute concentration, both inside and outside the cell, to be accommodated. For a cell with a totally rigid wall, an intracellular change of 100 mMolal concentration would be accompanied by a turgor change of some 0.25 MPa. Such a change in osmotic pressure would not be unusual during photo-synthesis. Plant cell walls, despite their thinness, are certainly capable of with-standing enormous changes in pressure. For example, turgor pressure in the cells of the intertidal alga *Cladophora* have been shown to oscillate between close to zero and more than 2.5 MPa under conditions that mimic the switch from full seawater to the freshwater of an estuary at ebb tide (Wiencke *et al.*, 1992). On the other hand, excess cell turgor certainly can be lethal. Leaf cells of the halophyte *Suaeda maritima* are generally protected from rainfall by the waxy cuticle of the leaves. Splitting the leaves and infiltrating them with dilute buffer (0.05 MPa osmotic pressure) results in the cells rupturing as their turgor pressures rise towards the 1.5 MPa value at water potential equilibrium of plants grown even under non-saline conditions (Clipson *et al.*, 1985; Tomos, 1988).

Cells, however, are not rigid and inelastic and will expand elastically with pressure. If no solutes cross the cell boundary, this results in internal dilution and a lowered equilibrium turgor pressure at higher water potential. In extreme cases, such as that of the marine giant-celled alga *Halicystis*, swelling is such that it is difficult to detect the turgor pressure change following alteration of the osmotic pressure of the medium (Zimmermann and Hüsken, 1980). The extent of cell expansion is a function of the volumetric elastic modulus (ε) of the cell (Philip, 1958). This relates changes in volume (ΔV) and pressure (ΔP) according to equation 1 and determines the trade off between volume changes and pressure changes:

$$\varepsilon = \frac{\Delta P}{\Delta V} V \qquad (1)$$

There are a number of reports in the literature of this being an important feature in the response to drought stress, where a lowered value of ε results in less turgor loss than would otherwise be obtained (Morgan, 1984). Less work has been published on the problems of excess pressure, although Steudle *et al.* (1980) performed a quantitative analysis of this situation during the malate accumulation of the CAM cycle of *Kalanchöe daigremontiana* where they emphasize the importance of water capacitance ($V/(\varepsilon + \pi_i)$) in defining the homeostatic ability of the cells.

In the case of the stomatal guard cell, however, an architecturally complex change in cell volume not only takes place, but is central to higher plant life. This is an example of a cell that exploits volume change to achieve a mechanical role which, considering the enormous pressures involved, can be very dramatic (Hill and Findlay, 1981).

If all cells were to respond to changes in solutes in such a spectacular way plants would be much more macroscopically mobile than they usually are. In addition, changes in cell volume in order to maintain turgor pressure have their drawbacks, especially in an organ with a fragile architecture. Splitting of fruit and leaves as a result of uncontrolled volume increase is generally seriously damaging, although presumably suitable architecture or material design could be found to accommodate it.

Do plants, therefore, take full advantage of the potential provided by their cell walls? With so few plants characterized at this level, an unequivocal answer is not possible. However, the norm would appear to be that cell turgor is actively maintained in the face of changes in the external environment (water and salt stress). In this paper we shall report work indicating that active turgor maintenance also occurs in the face of internal changes in water and solute relations parameters. The cell wall in these cases is not relied upon as a passive mechanical buffer, at least not in the longer term. Cells appear to use the biochemistry of membrane transport and polymerization to provide long-term solutions to their water relations dilemma rather than relying on wall mechanics. Moreover, they appear to do this in a way analogous to that used by multicellular animals — by regulating extracellular osmotic concentrations. In this respect the apoplast has much in common with the extracellular matrix of animals.

We have studied two such systems in our laboratory. In the context of this book these may be considered as representatives of source and sink organs of assimilate partitioning.

4.2 A source organ: wheat leaf cells

4.2.1 *Turgor regulation in mesophyll cells*

As part of our studies of the osmotic implications of sucrose accumulation we have studied the effect of sucrose synthesis on the water relations of wheat leaf cells. In order to maximize the accumulation of sucrose, its export was inhibited by excising the leaf of a wheat seedling while maintaining continued illumination. The leaves were kept with their cut ends bathed in distilled water. Turgor pressure was measured with the pressure probe (Hüsken *et al.*, 1978) in both epidermis and mesophyll cells. Osmotic pressure, inorganic ions and sugar concentrations were measured either in bulk tissue extract or in microdroplets obtained from single vacuoles *in situ* using the pressure probe as a sampling device (Malone *et al.*, 1989). Osmotic pressure was measured by picolitre osmometry (Malone *et al.*, 1989), inorganic ions by X-ray fluorescence (Malone *et al.*, 1991), and sugar concentration by enzyme-linked microscope fluorimetry [by a technique similar to that described in Zhen *et al.*, 1991, but using enzyme assays linked to glucose-phosphate dehydrogenase (Boehringer, 1989)]. Two summaries of these techniques are available (Tomos *et al.*, 1992 a,b).

Illumination of excised leaves, previously kept in darkness for 16 h to deplete the internal sucrose concentration, resulted in an increase of sucrose content of the mesophyll cells (Table 4.1). The concentration of sugars (principally sucrose) in individual cells increased dramatically from the approximately 10 mM in the leaves upon excision. To date, experimental variation between leaf preparations makes it difficult to generalize the magnitude of the increase, but values well in excess of 100 mM are obtained. [This is more than the increase in sucrose measured on a whole leaf basis (Figure 4.1a) and illustrates the importance of analysing individual cells in a heterogeneous system.]

During this period osmotic pressure was measured both in individual cells (Table 4.1) or on bulk tissue (Figure 4.1b). Despite the large increase in osmotic pressure observed both at the single cell and the whole leaf level, turgor pressure of the mesophyll cells remained constant (Table 4.1). What is the basis of this turgor homeostasis?

As discussed in the introduction, several options are available to the plant. These relate to wall elasticity and to the components of water potential within

Table 4.1. *Turgor pressure and solute parameters of single cells of lower epidermis and mesophyll of wheat leaves*

	0 h	24 h
Mesophyll cells		
Turgor pressure	0.35 MPa	0.38 MPa
Osmotic pressure	0.91 MPa	$\geqslant 1.0$ MPa
Water potential	-0.56 MPa	< -0.62 MPa
Sugars	10 mM	> 100 mM
Epidermal cells		
Turgor pressure	0.75 MPa	0.76 MPa
Osmotic pressure	0.86 MPa	1.26 MPa
Water potential	-0.1 MPa	-0.5 MPa
Sugars	< 2 mM	4 mM
K^+	140 mM	220 mM
Cl^-	75 mM	200 mM

After depletion of sucrose by 16 h in the dark, the leaves were excised, illuminated and measured at 0 and 24 h. (Data from a typical experiment rounded to significant figures.)

and outside the protoplast. Since, in this case, large changes in vacuolar osmotic pressure occur (Table 4.1), turgor maintenance due to high wall elasticity (low elastic modulus) is ruled out. If this had been the case the increase in volume would have resulted in dilution of the vacuolar contents to a value indistinguishable from the initial. Similarly, sucrose accumulation is not accompanied by a corresponding decrease in other cell osmotica.

The increase in sucrose concentration rules out polymerization as the basis of total turgor pressure maintenance. However, we have found that polymerization does contribute to the long-term maintenance of turgor pressure.

Sucrose/fructan interrelationship. Measured on a whole-leaf basis, after some 4 h sucrose concentration (Figure 4.1a) and osmotic pressure (Figure 4.1b) ceased to rise. This was not due to cessation of sucrose synthesis but was due to the initiation of sugar polymerization to fructan (Figure 4.1a). Fructan synthesis soon

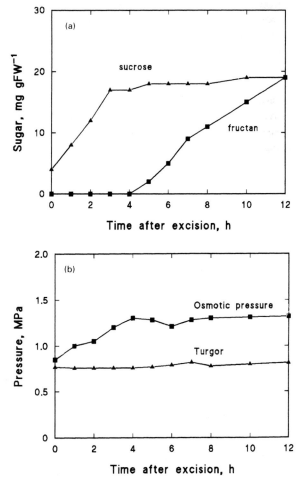

Figure 4.1. *Time course of (a) the bulk leaf content of sucrose and fructan and (b) epidermal cell tugor pressure and bulk leaf osmotic pressure following excision and illumination of wheat leaves. For sucrose 100 mM is equivalent to approximately 34 mg g^{-1} FW.*

reached a rate similar to the previous rate of synthesis of sucrose. This would appear to be an example of turgor adjustment, or at least maintenance, by polymerization. However, polymerization is not a prerequisite of turgor maintenance. Not only was the initial phase (0–4 h) of sucrose accumulation not accompanied by an increase in turgor pressure but also inhibition of the initiation of fructan synthesis by application of 10 μM cycloheximide had no effect on epidermal turgor despite continued sucrose synthesis and increasing osmotic pressure (Figure 4.2).

Cell wall osmotic adjustment. To explain the turgor homeostasis we must, there fore, turn to the cell wall and apoplast surrounding the cell. The hydraulic conductivity of cell membranes is so high (Tomos, 1988) that the water potential of the wall will be very close to that of the protoplast [for solutes of reflection coefficient approaching unity, i.e. most biologically important solutes (Tomos, 1988)]. The constancy of turgor pressure of the mesophyll cells shown in Table 4.1 indicates that wall water potential is regulated in step with sucrose accumulation and concurrent rise in osmotic pressure of the protoplasts. In principle, this regulation could be due to a change in either wall hydraulic tension or wall osmotic pressure.

Although a decrease in wall water potential by hydrostatic (transpiration-driven) tension was not ruled out in this experiment, it would seem unlikely that turgor pressure is kept low by increasing stomatal conductance. This would suggest a wall tension of minus several bars. Such a large negative value would require a significant hydraulic resistance in the apoplast between the cut end of the leaf (water supply) and the cells measured. In similar experiments on leaves of intact plants, immersion in water to stop transpiration had only a small effect on epidermal turgor pressure (Tomos *et al.*, 1989; Arif *et al.*, unpublished data), suggesting a wall tension near zero. The hydraulic resistance from the cut end of the leaf to the measured cells would be even lower than that of the intact plant

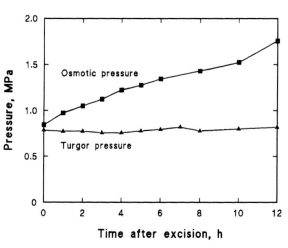

Figure 4.2. *Time course of epidermal cell turgor pressure and bulk leaf sucrose content following excision, illumination and treatment of a wheat leaf with 10 μM cycloheximide (cf. Figure 4.1b).*

and significant gradients of hydraulic pressure through the apoplast unlikely. Another explanation, a 30% drop in the value of the reflection coefficient of membranes for the internal solutes, whilst possible, has no precedent.

We are left with the conclusion that the turgor regulation observed in the mesophyll is the result of accumulation of osmotically active solutes in the free space of the cell wall. The precise magnitude of the increase in cell wall osmotic pressure remains unquantified due to an unexplained variation in single cell sucrose and osmotic pressure measurements from plant to plant. The data in Table 4.1, moreover, indicate that the water potential of the mesophyll cell walls is significantly less than zero even before the proposed wall osmotic adjustment following accumulation of sucrose. Upon excision of the low sucrose plants, the turgor pressure of the mesophyll is some 0.56 MPa below both single cell and bulk leaf osmotic pressure (Table 4.1). (Note that this is in contrast to the assumption of zero apoplast osmotic pressure widely used during interpretation of pressure chamber results.)

Low apoplast water potentials have been described in other systems (Cosgrove, 1986; Tomos, 1988). In our hands, this has been found for the leaves of the halophyte *S. maritima* (Tomos, 1988), the storage taproot of *Beta vulgaris* (Leigh and Tomos, 1983) and in the expanding zone of cereal leaves (Tomos *et al.*, 1989; Arif *et al.*, in preparation). In each of these examples a role of maintenance of a low turgor pressure in plant cells is suggested. This might appear paradoxical in the light of the large quantity of work that has been done on various strategies of plants to increase turgor in the face of water stress. Clearly water stress would be expected to decrease turgor further to a lower value. In the case of these cells, however, the lowered turgor pressure would be sustainable against a varying water potential by the movement of a small amount of solute across the plasma membrane in either direction (Tomos, 1988). Distilled water in the apoplast would only allow 'down regulation' of turgor. Turgor pressure can be maintained more economically by osmotically adjusting the wall free space rather than the protoplast. Movement of a unit of solute from protoplast to cell wall will change the respective concentrations according to the inverse of their volumes. If the wall comprises 1% of the cell volume, its concentration will change 99 times faster than that of the protoplast. In the expanding zone of wheat the evidence is consistent with turgor regulation breaking down at the point at which water stress exceeds the apparent level of wall solutes (Tomos *et al.*, 1989; Arif *et al.*, unpublished data). The nature of the wall solutes is discussed in Section 4.4.

Is sucrose synthesis regulated by turgor pressure? The observations following cycloheximide treatment (Figure 4.2) did not allow us to distinguish whether sucrose synthesis was limited by turgor pressure or not since turgor pressure is regulated independently of sucrose concentration. In order to assess the effect of changing turgor pressure on sucrose metabolism it is necessary to be able to alter the turgor pressure. This proved difficult. Mesophyll turgor pressure can indeed be altered by dipping the cut end of the leaf in a solution of osmoticum (such as

200 mM mannitol). However, on illumination, transpiration flow was such that mannitol levels within the tissue rapidly rose uncontrollably and the cells plasmolysed. It was, however, possible to observe the changes in amounts of sucrose and fructan following changes of turgor pressure in the mesophyll in non-illuminated excised tissue (that had previously been loaded with sucrose by illumination) when the turgor dropped transiently by some 0.1 MPa during the first 30 min of treatment and recovered during the following 30 min.

The initial results indicate that there was a reduction in the content of longer-chain fructan and an increase in the content of sucrose and the shortest fructan under these conditions (Table 4.2). This behaviour is consistent with sucrose synthesis from fructan being promoted by a decrease in turgor pressure. Depolymerization, therefore, may indeed contribute to turgor regulation. Clearly this analysis will need to be performed under more physiological conditions. It is worth noting that excised red-beet taproot tissue also depolymerizes sugars (from sucrose to monosaccharides in this case — see Section 4.3.2), but while the rate of hydrolysis is a function of the internal turgor, hydrolysis itself occurs in both hyper- and hypo-osmotic conditions (Perry *et al.*, 1987).

4.2.2 *Turgor-regulation in the epidermal cells*

The influence of photosynthetic sucrose synthesis on water relations extends beyond the mesophyll cells. Despite being non-photosynthetic, epidermal cells also undergo changes as a result of photosynthesis. These cells contain insignificant amounts of sucrose before and after illumination (Table 4.1). [Which, incidentally, provides evidence that any plasmodesmatal connections between wheat epidermis and mesophyll that might exist are not functional, in contrast to the situation in *Commelina* (Erwee *et al.*, 1985).] There is, therefore, no *a priori* reason why osmotic pressure should rise. Direct measurements show, however, that epidermal osmotic pressure did in fact rise by an amount similar to that of the bulk tissue and mesophyll (Table 4.1, Figure 4.1b). Why might this be? The answer would appear to be the precise converse of the situation in the mesophyll. The epidermal cells are being water stressed by the activity of the mesophyll (see

Table 4.2. *Concentrations of sucrose and fructans measured 0 and 1 h after reduction of turgor pressure of excised wheat leaves that had previously been illuminated*

Carbohydrate	0 h		1 h
Sucrose	65 mM		120 mM
Fructan DP3	40 mM		85 mM
Fructan DP4–6		No change	
Fructan DP7–8	Approx. 10 mM		Approx. 1 mM

Fructan classes were distinguished by thin layer chromatography. DP — degree of polymerization. Depolymerization may contribute to turgor maintenance.

below). Again turgor pressure was regulated (Table 4.1) indicating that the wall water potential decreased. It is difficult to see that this decrease could have come from anywhere else other than by propagation of the lowered mesophyll water potential to the epidermal walls.

The proposed sequence, therefore, is that sucrose synthesis in the mesophyll results in an increase in mesophyll protoplast osmotic pressure. Turgor in the mesophyll is maintained by osmotic adjustment of the cell wall. Apoplasmic continuity between mesophyll and epidermis allows solutes to move from the mesophyll walls to those of the epidermis. This results in a 'water stress' on the epidermis, the cells of which respond by osmotic adjustment of the protoplast to regulate their turgor. The basis of the epidermal osmotic adjustment is an increase of protoplast potassium and chloride concentration (Table 4.1).

Several questions arise from this sequence. From where do the K^+ and Cl^- ions come to the epidermis? Since the leaves are excised, they must come from other parts of the leaf. We have not yet measured ion concentrations in mesophyll cells in this system, but it is possible that they are the source of these ions. If this were the case, the lack of functioning symplastic pathway for sucrose between mesophyll and epidermis (see above) suggests that K^+ must enter the epidermal cells from the apoplast. It is, therefore, possible that the osmotic adjustment of the mesophyll wall is also achieved by regulating its KCl content. What is important to note here is that there is no necessity to propose a role for sucrose in turgor regulation. Indeed if sucrose were important in wall osmotic adjustment, why should the epidermis not take advantage of it in its own turgor regulation?

Water potentials of both epidermal and mesophyll apoplast drop by a similar amount during illumination suggesting a linkage between the two. However, the absolute values of water potential do not appear to be the same (Table 4.1). It is premature to speculate as to the reason for this and the relationship between epidermal and mesophyll apoplast remains unclear.

A final question concerns the source of metabolic energy for the epidermis. Epidermal vacuoles contain some 5 mM glucose together with traces of sucrose. Where do these come from, and how are their supplies regulated?

4.3 A sink organ: taproot of *Beta vulgaris*

4.3.1 *Constancy of turgor pressure during sucrose accumulation*

At the other end of the assimilate pathway (albeit in a different species) is the storage taproot of *Beta vulgaris*. During the growing season of the taproot of sugarbeet, sucrose accumulation leads to an increase in osmotic pressure in the storage cells, measured on a tissue basis (Figure 4.3). As in the case of the excised wheat leaves discussed above, no corresponding change in turgor pressure is observed. As with the wheat leaf cells, we have proposed that accumulation of solutes in the apoplast of the storage parenchyma is the basis of this behaviour. In the youngest measurable tissue (16 days after sowing) osmotic pressure is

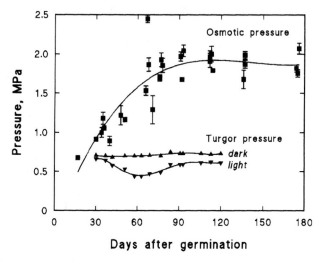

Figure 4.3. *Analysis of sugarbeet tissue during the growing season. Osmotic pressure (■), single-cell turgor pressure under transpiring (▼) and non-transpiring conditions (▲).*

equal to the stable long-term subsequent turgor pressure, indicating that the wall solutes are negligible at the beginning of the growing season and increase subsequently. In mature taproot of red-beet (also *Beta vulgaris*), efflux analysis of the solutes found in the cell wall indicated that free K^+ is present at concentrations sufficient to account for the wall osmotic pressure (Leigh and Tomos, 1983). Apoplasmic sucrose concentrations appear to be low. Sucrose itself would appear not to be involved in turgor regulation to any significant extent.

4.3.2 *Responses to imposed turgor changes*

We have studied the turgor regulation of beet taproot in a little more detail. Excised beet tissue bathed in mannitol solutions of varying osmotic pressure initially behave as ideal osmometers. Turgor pressure rises and falls in high and low concentrations, respectively, the value of turgor pressure corresponding to the osmotic pressure step across the plasma membrane in each case (Perry *et al.*, 1987; Tomos *et al.*, 1984). In this excised tissue a turgor adjustment subsequently occurs by a series of processes involving leakage and sucrose hydrolysis (Figure 4.4) (Perry *et al.*, 1987). After 2 days a stable turgor is achieved. This turgor pressure is approximately that expected of cells of the intact beet *in vivo* (Perry *et al.*, 1987). When expressed simply as a turgor pressure, this turgor adjustment appears incomplete (Figure 4.4a). When, however, the amount of adjustment over 3 days is plotted as a function of applied osmotic 'stress' (water potential) a clear linear correlation is observed (Figure 4.4b). The curves for red-beet and sugarbeet (Figure 4.4b) are displaced in such a way that the null point of osmotic adjustment (the mannitol concentration at which no change in turgor pressure is

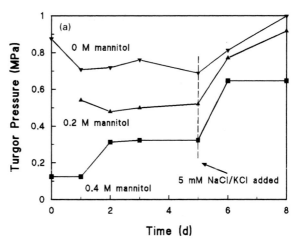

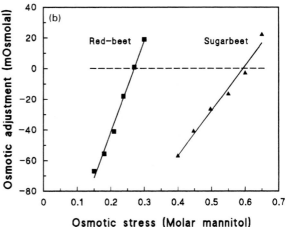

Figure 4.4. Turgor and osmotic adjustment in Beta vulgaris. *(a) Turgor adjustment of excised red beet discs bathed in 0, 200 mM and 400 mM mannitol solution. At 5 days, NaCl and KCl were added to the bathing medium (Perry* et al., *1987). (b) Osmotic adjustment of red-beet and sugarbeet over 3 days plotted as a function of the osmotic stress applied (Perry, unpublished data).*

noted after 3 days) corresponds to the turgor pressures expected in the intact red- and sugarbeet, respectively.

This example of turgor recovery, however, does not appear to be a good model for that observed in the long-term experiments in intact beet. The main difference here is that sucrose is rapidly hydrolysed during the first 48 h of the experiment. Clearly this is not an option in the intact beet, whose function is the storage of sucrose! After 48 h, however, sucrose concentrations are similar in all three treatments (being close to zero) and the final turgor pressure is determined not by the degree of hydrolysis but by respiration and the relative efflux from the cell of monosaccharides, of which glucose appears to be the one lost most readily (Perry *et al.*, 1987). The possibility of polymerization to starch was not investigated (Milling, 1990). This result calls for caution in interpreting the depolymerization of fructan noted above for excised leaves. After 5 days, 5 mM NaCl and KCl were added to the bathing medium and turgor rose in an apparently uncontrolled manner with no sign of turgor regulation. In their uptake of KCl, these cells resemble the epidermis of wheat (Section 4.2.5). Clearly, however, the turgor regulation mechanism is incomplete.

In a series of related experiments a situation closer to that proposed for the intact beet was observed. Freshly excised red-beet discs were bathed in silicone oil. Due to the thin layer of broken-cell aqueous sap (of iso-osmotic pressure) surrounding them, turgor pressure in these discs was low. Significant turgor recovery occurred at a rate much faster than that observed for the discs bathed in bulk osmoticum (Tomos, 1988). During this period no significant increase in bulk osmotic pressure was measured, indicating that sucrose hydrolysis had not begun (Perry *et al.*, 1987). An interpretation is that the cells rapidly take up solutes from the extracellular compartment (which in this case has rather a low volume relative to the intracellular volume). Such a system could be envisaged as a model for *in vivo* turgor recovery due to a decrease in wall osmotic pressure, rather than an increase in that of the protoplast.

4.3.3 Turgor-sensitive acid efflux

The basis of this turgor-dependent process in the response of *B. vulgaris* has been investigated. Many solutes, including sucrose, are believed to be imported into plant cells via a symport transport system linked to the movement of

Table 4.3. *The effect of various inhibitors on the acidification of the bathing medium by sugarbeet discs bathed in 800 mM manninol*

Inhibitor (μM)	Percentage of control acidification
CCCP	
0	100
0.1	89
1.0	30
10.0	-30^*
Monensin	
0	100
1	114
10	69
100	-1^*
1 mM	-26^*
PCMBS	
0	100
500	39
Anaerobiosis	
CO_2-free air	100
Nitrogen	3

*A net alkalinization. Data from Tomos (1989). These data suggest that the tugor-dependent acidification is due to an active proton pump.

protons down their electrochemical gradient (Poole, 1988). Excised beet discs acidify their bathing medium at a rate that is a function of the turgor pressure of the cells (Wyse *et al.*, 1986). This can be shown both by net change in pH of the bathing medium (Wyse *et al.*, 1986) or by back titration to the initial pH with KOH. The turgor pressure of maximum acid efflux is in the range of that found in the intact taproot. In comparing roots of red-, fodder- and sugarbeets, the turgor pressures of maximum acidification correlate not only with the turgor pressure of the intact root but also with the null points of zero osmotic adjustment observed in excised discs (Figure 4.4b). (This relationship remains to be characterized fully.)

Finally, we have attempted to identify an H^+/ATPase as the basis of this acid efflux. Interpretation of the results of experiments with inhibitors is complicated by questions of penetration into the cells. However, CCCP, monensin and PCMBS inhibit acidification of the medium (Table 4.3) (Tomos, 1989). The most unequivocal inhibitor is anoxia. Replacing aeration by bubbling with nitrogen rapidly and reversibly brings medium acidification to a halt (Figure 4.5) within 30 min. Acid efflux is certainly not an uncontrolled, passive process. In the intact system, if the same acidification process were to occur then wall pH would be altered very rapidly indeed since the intra-/extracellular volume ratio would be so much larger. Attempts to measure the apoplast pH of intact beets have not as yet been successful.

4.3.4 *Comparison of* in vitro *with* in vivo *behaviour*

All the components of a rapid turgor-regulating system in storage taproots are, therefore, demonstrable *in vitro* with excised tissue. It is therefore frustrating that, although in long-term experiments turgor is maintained constant (Figure 4.3), it is not over the short term. Sugar beet taproots undergo a diel shrinkage and swelling in response to transpiration demand (Figure 4.6). This can be measured as a change in taproot diameter, or as the change in turgor pressure of

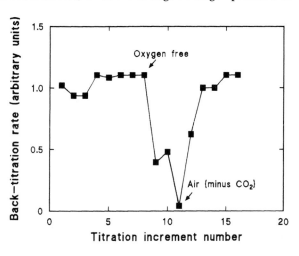

Figure 4.5. The influence of anoxia on medium acidification by excised sugarbeet discs. Anoxia was induced by bubbling the suspension with moist nitrogen. Acidification was measured by regular back-titration with KOH.

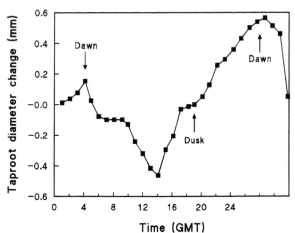

Figure 4.6. *Diel shrinking and swelling of sugarbeet taproot in response to cycles of transpiration. Dusk and dawn were reinforced by artificial lighting.*

the outer taproot cells (Figure 4.7a) (Palta *et al.*, 1987). Despite the proton-gradient based mechanism discussed above, on illumination of the leaves turgor drops in a manner that can be linearly related to transpiration rate (Figure 4.7b). Under continuous illumination turgor pressure remains low and constant (Figure 4.7a). This is not the behaviour expected if turgor pressure was regulated by shunting solutes into and out of the apoplast.

An explanation for these observations remains to be found. It is unlikely that the taproot is relatively anaerobic. It is possible that the turgor-regulated acid efflux observed *in vitro* is not related to turgor regulation. Apoplasmic sucrose concentrations may be so low that the increased proton motive force has osmotically insignificant sucrose to pump out of the apoplast. The putative H^+/ATPase could have a role in the leak salvage (Maynard and Lucas, 1982) or sucrose import systems, but no role in turgor regulation. An alternative role would be as the receptor in a leaf to root hydraulic signalling system along the lines recently described by Malone and colleagues (Malone and Stankovic, 1991). The turgor pressure of taproot cells is a linear function of transpiration rate (Figure 4.7b), which in turn is a complex signal relating to stomatal opening and potential CO_2 fixation. The apparent speed of propagation of the water potential gradient from leaf to taproot has been determined by simultaneously measuring transpiration rate, petiole wilting (related to turgor) and taproot cell turgor pressure following artificial illumination of intact sugarbeet plants (Figure 4.8). It can be shown in sugarbeet that such gradients move relatively slowly from the stomata (where they are generated) to the leaf petioles, then to the outer cells of the taproot. Some 30 min is required for the entire transmission along a pathway of about 50 cm (Figure 4.8).

The long-term control of turgor must be under a much more complex control since the taproot seems capable of 'knowing' that night-time turgor is to be maintained constant throughout the growing season, while day-time turgor can

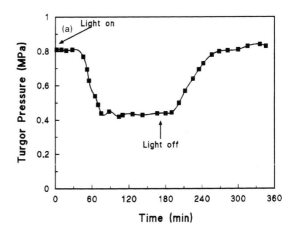

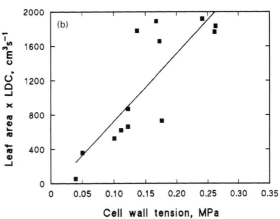

Figure 4.7. *The decrease in cell turgor pressure (the cell wall tension) of single cells of taproot of sugarbeet. (a) The response to increased transpiration. (b) Plotted as a function of the plant leaf area × leaf diffusive conductance. Data derived from the difference between turgor pressure under transpiring and non-transpiring conditions (Figure 4.3) and measured leaf areas.*

drop by up to 30% (Figure 4.3). This remarkable behaviour certainly demands an explanation.

4.4 Conclusions

Sucrose accumulation has a dramatic effect on cellular water relations. That its reflection and osmotic coefficients are near unity mean that it contributes fully to an increase in cell osmotic pressure. In taproots, and during the early stages of accumulation of fixed carbon in wheat leaves, sucrose accumulation does not appear to be accompanied by a fully compensating change in another vacuolar solute or by polymerization. This is in contrast to the observations in this laboratory of other systems in which loss of one inorganic solute can be compensated for by gaining another. Calcium appears partially to compensate for loss of K^+ as barley leaves age (Hinde *et al.*, 1992; Tomos *et al.*, 1992b) and Cl^- partially

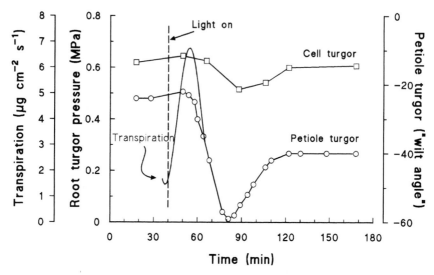

Figure 4.8. *Propagation of water potential signal to root of sugarbeet on sudden strong illumination. Transpiration increases immediately. The petiole turgor responds after some 10 min and taproot turgor after 30 min. The subsequent recoveries are due to stomatal closure induced by the strong illumination. Transpiration data points not shown, and entire trace not included for the sake of clarity.*

compensates for nitrate in wheat leaves (Richardson *et al.*, 1992; Tomos *et al.*, 1992b). In the case of sucrose, osmotic pressure simply increases. Constancy of turgor is not regulated by balancing protoplastic osmotic pressure, but by altering apoplast solute levels. We suggest that K^+ is involved, but do not know how widespread is its use in this respect. However, by analogy with the situation in red-beet taproot tissue, and by the fact that wheat epidermal cells have access to sufficient K^+ and Cl^- to adjust their protoplast concentration, there is circumstantial evidence that apoplast K^+ concentrations may be involved in the turgor regulation of both sugarbeet and wheat leaf mesophyll.

These two scenarios are consistent with sucrose fluxes being largely independent of water relations control. Sucrose is synthesized, moved and stored independently of the water environmental conditions. However, presumably this can only occur because of extensive and efficient turgor regulatory processes based on other solutes. There seems no theoretical limit on the ability of tissues to keep turgor pressure down. In *S. maritima*, even under non-saline conditions, leaf water potential is around −1.5 MPa. Turgor pressure, however, is only 0.1–0.3 MPa and remains unchanged on salt 'stress' when cell water potential drops to −3 MPa (Clipson *et al.*, 1985). Apparently turgor is kept low by high concentrations of solutes in the wall. Presumably this strategy is developed to avoid damage to sensitive tissue architecture whenever rain, or another temporary source of high water potential, would raise turgor pressure. A corollary is that the apoplast of these tissues must be inaccessible to rain. In *Su-*

aeda, wheat leaf mesophyll and beet taproot this must be achieved by a waxy cuticle — in position as much to stop water getting in and washing out wall solutes as it is to prevent tissue dehydration by evaporation.

Recent experiments in our laboratory have shown that wheat epidermal cell walls do have a measurable lowered water potential which is not altered by vacuum infiltration of the leaves with distilled water, but is brought up to zero when a trace of non-ionic detergent is added to the water (Arif *et al.*, unpublished data). This provides further evidence for hydrophobic barriers within the leaf architecture.

Of the three responses available to cells involved in carbon flux, wall rheology plays at most only a minor role. Membrane transport, on the other hand, plays a key role. The evidence for a role for polymerization remains equivocal. In the leaf, drastic measures to alter turgor pressure with mannitol do influence the relationship between sucrose and fructan, suggesting that this may contribute to turgor homeostasis. The elegant work of Oparka and Wright (1988a,b) has shown that starch synthesis in potato tubers is influenced by turgor pressure, although the regulation of turgor pressure of wheat leaves in the absence of fructan synthesis indicates that this is not a prerequisite of turgor maintenance.

Acknowledgements

We would like to acknowledge Drs John Farrar and Chris Pollock for their support and interest during much of this work, and their encouragement to get it published. We also acknowledge the role of Prof. Roger Wyse and the technical assistance of Mr Eirion Owen in some of the work on sugarbeet. Most of the work was supported by grants from the AFRC.

References

Boehringer (1989) *Methods of Biochemical Analysis and Food Analysis.* Boehringer Mannheim GmbH, Mannheim, Germany.

Clipson, N.J.W., Tomos, A.D., Flowers, T.J. and Wyn Jones R.G. (1985) Salt tolerance in the halophyte *Suaeda maritima* L. Dum. The maintenance of turgor pressure and water-potential gradients. *Planta*, 165, 392–396.

Cosgrove, D.J. (1986) Biophysical control of plant cell growth. *Ann. Rev. Plant Physiol.* 37, 377–405.

Erwee, M.G., Goodwin, P.B. and van Bel, A.J.E (1985) Cell-cell communication in the leaves of *Commelina cyanea* and other plants. *Plant Cell Environ.* 8, 173–178.

Hall, J.L. (1977) Fine structure and cytochemical changes occurring in beet discs in response to washing. *New Phytol.* 79, 559–566.

Hempling, H.G. (1982) Osmosis: the push and pull of life. In: *Biophysics of Water* (eds F. Franks and S.F. Mathias). John Wiley, Chichester, pp. 205–214.

Hill, B.S. and Findlay, G.P. (1981) The power of movement in plants: the role of osmotic machines. *Q. Rev. Biophys.* 14, 173–222.

Hinde, P., Tomos, A.D. and Leigh, R.A. (1992) Ion concentrations in individual epidermal cells of *Hordeum vulgare. J. Exp. Bot.* 43 (Suppl.), 27.

Hüsken, D., Steudle, E. and Zimmermann, U. (1978) Pressure probe technique for measuring water relations of cells in higher plants. *Plant Physiol.* 61, 158–163.

Leigh, R.A. and Tomos, A.D. (1983) An attempt to use isolated vacuoles to determine the distribution of sodium and potassium in cells of storage roots of red beet. *Planta*, 159, 469–475.

Malone, M. and Stankovic, B. (1991) Surface-potentials and hydraulic signals in wheat leaves following localised wounding by heat. *Plant Cell Environ.* 14, 431–436.

Malone, M., Leigh, R.A. and Tomos, A.D (1989) Extraction and analysis of sap from individual wheat leaf cells: the effect of sampling speed on the osmotic pressure of extracted sap. *Plant, Cell Environ.* 12, 919–926.

Malone, M., Leigh, R.A. and Tomos, A.D (1991) Concentrations of vacuolar inorganic ions in individual cells of intact wheat leaf epidermis. *J. Exp. Bot.* 42, 305–309.

Maynard, J.W. and Lucas, W.J. (1982) Sucrose and glucose uptake into *Beta vulgaris* leaf tissue. A case for general (apoplastic) retrieval systems. *Plant Physiol.* 70, 1436–1443.

Morgan, J.M. (1984) Osmoregulation and water stress in higher plants. *Ann. Rev. Plant Physiol.* 35, 299–319.

Nobel, P.S. (1991) *Physicochemical and Environmental Plant Physiology.* Academic Press, San Diego.

Oparka, K.J. and Wright, K.M. (1988a) Osmotic regulation of starch synthesis in potato tubers. *Planta*, 174, 123–126.

Oparka, K.J. and Wright, K.M (1988b) Influence of cell turgor on sucrose partitioning in potato tuber storage tissues. *Planta*, 175, 520–526.

Palta, J., Wyn Jones, R.G. and Tomos, A.D. (1987) Leaf diffusive conductance and tap root cell turgor pressure in sugarbeet. *Plant Cell Environ.* 10, 735–740.

Perry, C.A., Leigh, R.A., Tomos, A.D., Wyse, R.A. and Hall, J.L. (1987) The regulation of turgor pressure during sucrose mobilisation and salt accumulation by excised storage-root tissue of red beet. *Planta*, 170, 353–361.

Philip, J.R. (1958) The osmotic cell, solute diffusibility and the plant water economy. *Plant Physiol.* 33, 264–271.

Poole, R.J. (1988) Plasma membrane and tonoplast. In: *Solute Transport in Plant Cells and Tissues.* (eds D.A. Baker and J.L. Hall). Longman, Harlow, pp. 83–105.

Richardson, P., Tomos, A.D. and Leigh, R.A. (1992) Anion concentrations and osmotic pressure in wheat leaf epidermal cells. *J. Exp. Bot.* 43 (Suppl.), 28.

Steudle, E., Smith, J.A.C. and Lüttge, U. (1980) Water-relation parameters of individual mesophyll cells of the crassulacean acid metabolism plant *Kalanchoe daigremontiana*. *Plant Physiol.* 66, 1155–1163.

Tomos, A.D (1988) Cellular water relations in plants. In: *Water Science Reviews*, Volume 3 (ed. F. Franks). Cambridge University Press, Cambridge, pp. 186–277.

Tomos, A.D. (1989) Turgor pressure and membrane transport. In: *Plant Membrane Transport: The Current Position* (eds J. Dainty, M.I. de Michelis, E. Marré and F. Rasi-Caldogno). Elsevier, Amsterdam, pp. 559–562.

Tomos, A.D., Leigh, R.A., Shaw, C.A. and Wyn Jones, R.G. (1984) A comparison of methods for measuring turgor pressures and osmotic pressures of red beet storage tissue. *J. Exp. Bot.* 35, 1675–1683.

Tomos, A.D., Pritchard, J., Thomas, A. and Arif, H. (1989) Using the pressure probe to study salt, water and cold stress. In: *Plant Water Relations and Growth Under Stress* (eds M. Tazawa, Y. Katsumi, Y. Masuda and H. Okamoto). MYU K.K., Tokyo, pp. 245–252.

Tomos, A.D., Hinde, P., Richardson, P., Pritchard, J. and Fricke, W. (1992a). Micro-sampling and measurements of solutes in single cells. In: *Plant Cell Biology — A Practical Approach* (eds N. Harris and K.J. Oparka). IRL Press, Oxford, in press.

Tomos, A.D., Leigh, R.A., Hinde, P., Richardson, P. and Williams, J.H.H. (1992b) Measuring water and solute relations in single cells *in situ*. In: *Current Topics in Plant Biochemistry and Physiology*, Volume 11 Interdisciplinary Plant Biochemistry and Physiology Program, University of Missouri, Columbia, in press.

Trydan Niwclear (1991) *Trawsfynydd. Yr Orsaf Bwer. Gwybodaeth am Orsaf Bwer Niwclear.* (Information booklet). Trydan Niwclear ccc. Trawsfynydd, Blaenau Ffestiniog, Gwynedd.

Wiencke, C., Gorham, J., Tomos, A.D. and Davenport, J. (1992) Incomplete turgor adjustment in *Cladophora rupestris* under fluctuating salinity regimes. *Estuarine, Coast. Shelf Sci.* 34, 413–427.

Wyse, R.E., Zamski, E. and Tomos, A.D. (1986) Effect of turgor on the kinetics of sucrose uptake. *Plant Physiol.* 81, 478–481.

Zhen, R.-G., Koyro, H.-W., Leigh, R.A., Tomos, A.D. and Miller, A.J. (1991) Compartmental nitrate concentrations in barley root cells measured with nitrate-selective microelectrodes and by single-cell sap sampling. *Planta,* 185, 356–361.

Zimmermann, U. and Hüsken, D. (1980) Turgor pressure and cell volume relaxation in *Halicystis parvula. J. Memb. Biol.* 56, 55–64.

5

Sugar transport and metabolism in the potato tuber

K.J. Oparka, R. Viola, K.M. Wright and D.A.M. Prior

5.1 Introduction

Translocated carbon compounds leave the phloem at several sites within the plant and are subsequently allocated to a wide variety of cellular processes. Newly imported sugars may be utilized immediately for respiration or for structural growth or they may be stored for longer periods of time as insoluble polymeric forms such as starch. Several fates may befall imported sugars simultaneously (see, for example, Chapter 8). However, these two major processes, that is, utilization and storage, have been used to define sink organs based on the predominant allocation of assimilates in them (Ho, 1988).

This article will consider the potato tuber as an example of a storage sink in which starch is the major reserve. We will emphasize the role played by the symplast in phloem unloading and present new evidence that plasmodesmata may be regulated by turgor pressure gradients. The potential pathways and mechanisms available for sugar transport between the phloem and storage cells of the tuber will be considered, as will the metabolic fate of sucrose entering the storage cell. Finally, we will consider the consequences arising from the reversal of carbon fluxes during a sink to source transition.

5.2 General concepts

Before considering the pathways and mechanisms of sugar transport in the potato tuber, it is necessary to introduce a number of general concepts which are common to a wide range of storage sinks. Several of these will play a central role in the arguments we will subsequently develop.

5.2.1 *Phloem unloading: definitions and pathways*

The exit of assimilates from the sieve element (SE) is the first step in a complex series of short-distance transport events and is clearly a potential site at which solute fluxes might be regulated in a storage sink. There has been considerable confusion in the literature as to the precise nature of phloem unloading events, based largely on the use of inappropriate or ambiguous terminology (Oparka, 1990). In an attempt to rectify this matter we have suggested a clarification of existing terminology which we hope will provide an accurate structural basis for future physiological studies of phloem loading and unloading (see Chapter 12). Note that a major distinction is drawn between SE unloading and phloem unloading.

5.2.2 Symplastic sieve-element unloading

The exit of assimilates from the sieve element–companion cell (SE–CC) complex by the symplast is the pathway of least resistance and is prevalent in a number of sink tissues. Ho (1988) has suggested that symplastic SE unloading is a general feature of utilization sinks, for example, developing leaves (Gougler Schmalstig and Geiger, 1985) and root tips (Chapleo and Hall, 1989; Dick and ap Rees, 1975). However, it is now clear that the symplast plays a major role both in SE unloading and subsequent short-distance transport in several storage sinks (Patrick, 1990).

5.2.3 Apoplastic sieve-element unloading

To date there has been no direct evidence put forward to support apoplastic SE unloading in a storage sink. It is conceivable that sugars are unloaded from the SE–CC complex across the plasmalemma but such a hypothesis has proved difficult to substantiate. Neither the ability of sink cells to take up exogenous sugars supplied to the apoplast (see Ho, 1988, and references therein) nor the presence of apoplastic invertase (Eschrich, 1989) have proved to be convincing arguments for apoplastic SE unloading (Oparka, 1990). However, it should be stressed that apoplastic SE unloading of sugars is likely to occur in stem tissues in which the pathway SE–CC complexes are virtually isolated symplastically. In such circumstances, lateral leakage of sugars might occur by localized inhibition of the sucrose retrieval mechanism at the SE–CC complex (Patrick, 1990). There seems little doubt that, in several storage sinks, transport to the apoplast occurs at some point along the phloem unloading pathway. However, the case for apoplastic SE unloading has so far proved difficult to substantiate. Transgenic plants in which yeast-derived invertase has been targeted to different compartments (apoplast, cytosol and vacuole) are likely to prove invaluable tools in elucidating the pathways of phloem unloading in storage sinks in the future (Sonnewald et al., 1991).

5.3 Maintenance of a turgor pressure gradient

The pressure-flow hypothesis of phloem transport predicts that bulk solute flow along the phloem will be driven by a pressure gradient between source and sink, this gradient being maintained by the continued loading of solutes into the SE–CC complex in source leaves and their continued exit from the phloem in sink regions (Geiger and Fondy, 1980). A number of 'strategies' are available to storage sinks to allow them to maintain a low sucrose concentration external to the SE–CC complex and thereby ensure the maintenance of a turgor gradient sufficient to drive the continued transport of sucrose into the sink. These will be considered briefly.

5.3.1 *Apoplastic invertase*

In the case of apoplastic SE unloading, leakage to the apoplast could potentially be brought about by inhibition of sugar retrieval mechanisms on the plasma membrane of the SE–CC complex (Patrick, 1990). With sucrose as the trans-located substrate, apoplastic inversion to hexoses could enhance SE unloading by steepening the turgor gradient along the pathway (Eschrich, 1989). It is clear that in several sinks apoplastic invertase is present (Ho, 1988). However, the evidence that apoplastic inversion of sucrose is a prerequisite for the maintenance of SE unloading appears to be losing favour (Gougler Schmalstig and Hitz, 1987; Lingle, 1989).

5.3.2 *Vacuolar compartmentation*

A second means of maintaining a turgor gradient along the phloem is by compartmentation of the newly imported sugars into the vacuoles of storage cells. This mechanism is likely to be of primary importance in sugar-accumulat-ing sinks, such as sugar beet (Leigh *et al.*, 1979) but may also be of significance in starch-storing sinks (Oparka *et al.*, 1990).

5.3.3 *Storage product synthesis*

It has been suggested that the import of assimilate into sink cells may be controlled by the predominant storage process (Ho, 1988). For example, the continued conversion of imported sucrose to starch may provide a means by which a sucrose concentration gradient could be maintained between the SE–CC complex and the storage cells, allowing 'downhill' symplastic transport to occur directly into the cytosol of the storage cells. In this respect, some of the enzymes involved in sucrose breakdown in storage sinks, for example sucrose synthase, may represent rate-limiting steps in the subsequent conversion of imported sucrose to starch (Oparka *et al.*, 1990; Sung *et al.*, 1989).

5.4 Turgor-dependent transport events

As well as its central role as the driving force for translocation, turgor pressure has also begun to emerge as an important regulator of a number of sink-related transport events. Traditionally, plant physiologists have considered turgor to be a consequence of a number of physiological events occurring in sink cells. Here we will develop the concept that turgor is a cause, rather than a consequence, of several transport events.

5.4.1 Turgor-regulated sucrose transport

The lowering of cell turgor of sink cells has been shown to stimulate sugar uptake at the plasma membrane, the enhanced uptake being brought about by stimulation of H^+-ATPase activity (Wyse et al., 1986). The capacity of cells to take up exogenously supplied sugars appears to be a ubiquitous feature of parenchyma elements from both leaves (Daie and Wyse, 1985) and storage tissues (Oparka and Wright, 1988; Wyse et al., 1986). There have been a considerable number of studies on the uptake of sugars by isolated cells and protoplasts and clearly much has been learnt about the control of transmembrane transport and its reponsiveness to turgor. In several studies, the ability of cells to take up sugars from the apoplast has been used as evidence that the phloem unloading pathway is apoplastic (see references in Ho, 1988). Several of these studies have taken little account of the high frequencies of plasmodesmata which often connect individual storage cells. For example, Mierzwa and Evert (1984) showed that there may be sufficient plasmodesmata between the conducting tissues and the storage parenchyma of sugar beet storage roots to support a symplastic phloem unloading pathway. The storage cells, however, can accumulate sucrose actively from the apoplast (Wyse et al.,1986). Consequently, it has been difficult to assess the relative roles of the apoplast versus the symplast in solute transport from the phloem. This problem will be exemplified when considering the phloem unloading pathway in the potato tuber.

5.4.2 Turgor regulation of symplastic transport?

The concept of a turgor 'homeostat', which functions to determine and integrate the rate of phloem import into a sink, has been discussed by Patrick (1990). The homeostat may regulate compartmentation and, through control of solute fluxes, solute metabolism to a turgor 'set point'. This in turn sets the turgor gradient necessary to drive bulk flow of phloem sap through the interconnected symplast (Patrick, 1990).

Symplastic transport is clearly a key feature in solute transport in several sinks and considerable research attention has recently been focused on plasmodesmata, the narrow cytoplasmic channels which functionally interconnect the cytoplasm of individual cells. Plasmodesmata have a complex substructure and major recent advances have been made in elucidating their function through the use of micro-injection techniques (for review see Robards and Lucas, 1990). It is now clear that plasmodesmata are not static pores in the walls between cells but controllable 'valves', opening and closing in response to a number of internal stimuli. Plasmodesmata share a number of interesting properties with animal gap junctions, including a size exclusion limit of approximately 800 Da (Goodwin,

1983; Terry and Robards, 1987), the presence of connexin-like proteins (Yahalom *et al.*, 1991) and closure by elevation of cytosolic Ca^{2+} levels (Tucker, 1990).

A recent hypothesis has been that plasmodesmata might operate as pressure-sensitive valves, opening or closing in response to turgor pressure gradients between adjacent cells (Barclay and Fensom, 1984). Evidence for this proposal has come from experiments with the giant alga *Chara* (Côte *et al.*, 1987) and the liverwort *Conocephalum* (Trebacz and Fensom, 1989) in which mannitol-induced pressure gradients were applied between cells. In *Chara* the applied pressure gradient caused an increase in electrical resistance across the node which was interpreted as evidence for the partial closure of the plasmodesmata (Côte *et al.*, 1987).

5.4.3 *Direct evidence for pressure regulation of plasmodesmata*

We have recently examined the pressure dependency of plasmodesmata directly using a modified pressure probe/micro-injection system. With this technique the turgor of a cell can be manipulated relative to its neighbours and the effects on the intercellular transport of fluorescent probes subsequently examined (Oparka *et al.*, 1991). We chose the leaf trichome of *Nicotiana clevelandii* as a simplified model experimental system to study intercellular transport. The trichome consists of about four to five large elongate cells with a terminal secretory cell. The size exclusion limit of the plasmodesmata connecting the cells is about 750 Da (Derrick *et al.*, 1990). The bulbous tip cell is relatively easy to micro-inject due to the high ratio of cytoplasm to vacuole present and Lucifer yellow CH (LYCH) injected into this cell moves rapidly and basipetally into the larger vacuolate cells.

5.4.4 *Puncturing experiments*

In one set of experiments we punctured different cells in the trichome using the pressure probe. The effects of this treatment on the turgor pressures of cells on either side of the originally impaled cell were then examined. The results are shown in Figure 5.1(a–d). A consistent feature of puncturing was the generation of large turgor differentials between different cells in the trichome, with cells apical to the punctured cell showing greater turgor losses than those below. Thus, when cell 2 was punctured (Figure 5.1c) the turgor of cell 3 fell from approximately 400 kPa to 70 kPa while the turgor of cell 1 only fell to 350 kPa. The rapid drop in turgor of cells apical to the punctured cell can be explained by their hydraulic isolation from the main plant axis. In contrast, the turgor of cells basal to the punctured cell was restored due to a hydraulic flux into them from the leaf surface.

The above puncturing treatments were subsequently repeated, followed by introduction of LYCH into the tip cell. The dye moved rapidly out of the tip cell and was unimpeded in its movement into the punctured cell. However, dye

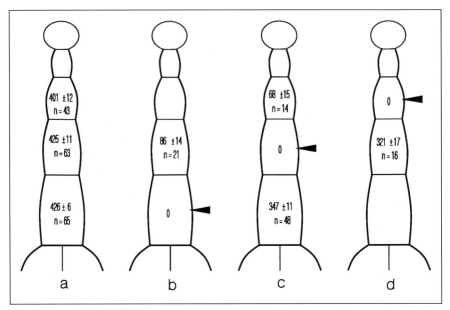

Figure 5.1. *Induction of pressure differentials in a leaf trichome. Cells 1–3 (from base upwards) were punctured by removal of the pressure probe (arrowheads) and the turgor of cells adjacent to the punctured cells recorded. (a) Control. Without puncturing there was little variation in the turgor of cells 1–3. (b–d) Effects of puncturing cells 1, 2 and 3, respectively. Large turgor differentials were generated at the base of the punctured cell, irrespective of its position in the trichome. Reproduced with permission from Oparka and Prior (1992).*

movement out of the punctured cell into the next basal cell was severely impeded. In all cases dye movement was blocked exactly at the sites predicted by the location of the first major turgor differential induced by puncturing (Figure 5.2).

5.4.5 *Pressure 'clamp' experiments*

In a second set of experiments we used the pressure probe to raise the turgor pressure of a single cell by varying degrees above its measured value, thereby creating a pressure differential (ΔP) at its upper, as well as lower, wall. The magnitude of ΔP was clearly important in determining the degree of symplastic impedence. When cells were clamped at 300 kPa above their initial pressure, dye was completely excluded from the clamped cell (this value coincides with the same differential induced at the base of punctured cells cf. Figures 5.1 and 5.2). The same effect could be reproduced regardless of the position of the clamped cell in the trichome (Figure 5.3a,b). A ΔP of 200 kPa allowed trace movement of dye into the clamped cell (Figure 5.3c), while a ΔP of 100 kPa was ineffective at preventing dye movement into the clamped cell (Figure 5.3d). Experiments with flaccid trichomes, in which the overall turgor pressure was lowered,

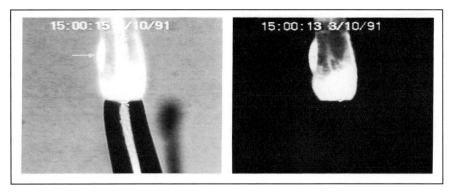

Figure 5.2. *Blockage of intercellular transport of LYCH by a pressure differential. Cell 2 was punctured with the pressure probe (arrow) prior to injection of the tip cell with LYCH. A strong symplastic barrier (corresponding to the site of the largest pressure differential; cf. Figure 5.1) occurs at the base of cell 2. Left, combined bright field/fluorescence. Right, fluorescence only. Reproduced with permission from Oparka and Prior (1992).*

demonstrated that absolute turgor did not influence intercellular transport; a differential in excess of 200 kPa was required to impede symplastic transport.

These experiments provide direct evidence that plasmodesmata may function as pressure-sensitive valves. The results have obvious implications for wounding and suggest that one of the first responses when a cell is cut or pierced is a pressure-generated closure of plasmodesmata, symplastically isolating the damaged cell from its neighbours. As yet we have no evidence for the mechanism of plasmodesmatal closure in response to a generated pressure differential.

5.4.6 *Pressure regulation of symplastic sieve element unloading?*

It is clear from the above observations that the presence of plasmodesmata between cells cannot be taken as unequivocal evidence that symplastic transport is occurring. Within the phloem there is normally an enormous pressure difference between the SE–CC complex and adjoining cells (Warmbrodt, 1987) and plasmodesmata of varying ultrastructure are present throughout this tissue (Beebe and Turgeon, 1991; Robards and Lucas, 1990). An entirely symplastic phloem loading pathway has been suggested in some species (van Bel and Gamelei, 1990) and the possibility that symplastic phloem unloading might be under the control of a turgor homeostat has already been raised (Patrick, 1990).

If symplastic SE unloading is occurring then turgor control of this process at the SE–CC complex invokes a regulatory role of the plasmodesmata connecting the SE–CC complex with adjacent cells. It is difficult to reconcile the presence of plasmodesmata between the SE–CC complex and adjoining cells with the presence of a steep turgor differential at this site (Warmbrodt, 1987). However, these two features are compatible if one considers the possibility that the plasmodesmata might be closed, symplastically isolating the SE–CC complex.

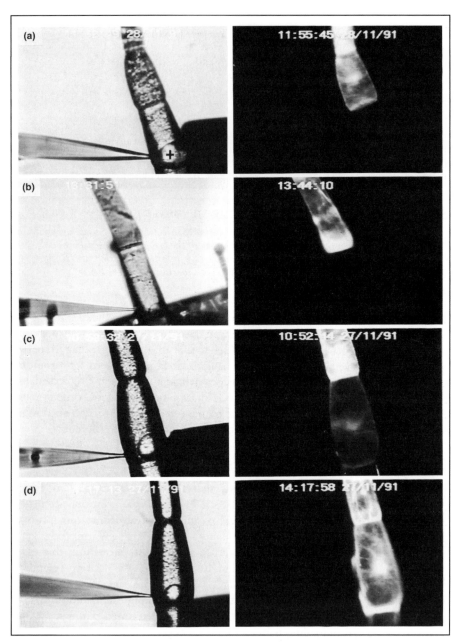

Figure 5.3. *Pressure-generated closure of plasmodesmata. The tip cell was injected with LYCH. Individual cells in the trichome were then 'clamped' with the pressure probe at values (ΔP) higher than the original measured turgor by introducing an oil droplet into the vacuole (+). A ΔP of 300 kPa (a and b) completely inhibited transport into the clamped cell. A ΔP of 200 kPa was partially effective at inhibiting transport (c), while a ΔP of 100 kPa was ineffective at preventing dye movement into the clamped cell (d). Left, bright field. Right, fluorescence. Reproduced with permission from Oparka and Prior (1992).*

Patrick (1990) has drawn attention to seasonal shifts in the pathway of SE unloading in bean stems, suggesting that these may result from a change in the hydrodynamic properties of the plasmodesmata interconnecting the SE–CC complexes with the phloem parenchyma. Other source/sink ratio effects were consistent with the operation of pressure-regulated plasmodesmatal valves (Patrick, 1990).

The presssure dependency of SE loading/unloading should not be beyond the realms of experimental validation in the near future using micro-injection techniques in which pressure measurement forms an integral part of the micro-injection procedure (see Oparka *et al.*, 1991).

5.5 The sink to source transition in the potato tuber — a case history

In the remainder of this article we will turn our attention to the potato tuber, a starch storage organ which has formed the basis of several studies in our laboratory.

The sink to source transition has been studied extensively in leaves (Turgeon, 1989) but there has been little information available on sugar fluxes in storage organs which undergo a sink to source transition. In the potato tuber a sink to source transition occurs during sprouting as reserves are mobilized from the 'parent' tuber to the developing shoots. This transition occurs with no apparent change in internal anatomy and the major direction of carbon flux within the tuber completely reverses. The potato tuber therefore provides a useful system with which to study some of the basic regulatory processes involved in sugar transport.

5.5.1 *The tuber as a sink*

In rapidly growing potato plants, sucrose is transferred continuously from the shoots to the growing tubers. In the storage parenchyma cells of the central perimedulla, the sucrose is converted predominantly to starch which eventually represents about 70% of the final tuber dry matter (Figure 5.4a).

5.5.2 *The tuber as a source*

Following shoot senescence (sometimes induced prematurely in 'early' crops by burning down the shoots) the tubers cease growth and a period of dormancy ensues. Under appropriate conditions sprouting may follow (Figure 5.4b). Anatomical evidence indicates that starch mobilization initially occurs around existing phloem regions of the perimedulla (Ross and Davies, 1985), the same tissue in which phloem unloading of sucrose predominated in the growing sink tuber.

We will consider first the flux of sucrose from the phloem to the cytosol of the storage parenchyma cells.

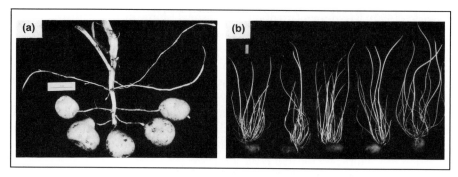

Figure 5.4. *The sink to source transition in the potato tuber. In the sink tuber (a) sucrose is unloaded from the internal phloem of the perimedulla and transported to the storage cells. In the source tuber (b) starch degradation occurs and sucrose is mobilized via the existing internal phloem network to the growing shoots.*

5.5.3 *Sieve element unloading*

As pointed out in previous reviews (Oparka, 1990; Patrick, 1990) SE unloading is extremely difficult to study due to both the technical and conceptual difficulties of isolating the SE–CC complex from the surrounding cells. It can be inferred from plasmodesmatal frequencies, although it should be stressed that the occurrence of plasmodesmata does not in itself demonstrate symplastic continuity and, as will be demonstrated, may lead to confusion in interpreting the pathway of solute transport when the direction of sugar flux in the sink is reversed during the sink to source transition.

5.5.4 *Demonstration of symplastic sieve element unloading*

In phloem-rich regions of the potato tuber plasmodesmata occur between the SE and companion cell and also on all faces connecting the SE–CC complex with adjacent parenchyma). 'Plasmodesmagrams' (e.g. Figure 5.5) indicate the potential for symplastic SE unloading; direct physiological demonstration of SE unloading is more difficult. Plasmolysis of phloem-associated tissues has been used as a technique to demonstrate the presence of a symplastic transport step in the phloem unloading pathway, but not necessarily at the SE–CC complex. For example, in the potato tuber plasmolysis of the storage parenchyma adjacent to the internal phloem severely inhibits assimilate transfer to solute-collecting wells made in the perimedulla of the tuber while transport to control wells containing buffered osmotica is unimpeded (Oparka and Prior, 1987).

5.5.5 *Micro-injection studies*

Recently, more direct inroads to studying SE unloading have come through the use of micro-injection techniques in which membrane-impermeant fluorescent probes have been introduced directly into SEs, or other cells of the phloem. van

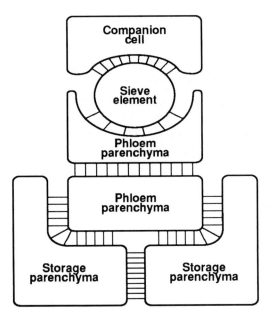

Figure 5.5. *'Plasmodesmagram' of the phloem unloading pathway in the potato tuber. Reproduced from van Bel (1992) with permission from Elsevier.*

Bel and Kempers (1990) were able to demonstrate symplastic isolation of the SE–CC complex in the phloem pathway of a number of species, a finding consistent with the observed paucity of plasmodesmata between the SE–CC complex and surrounding cells.

In the stem of the potato plant we have made similar observations. Micro-injection of the membrane-impermeant fluorescent probe Cascade blue into a single SE resulted in rapid movement along the sieve tube and lateral spread only into the companion cell (Figure 5.6a). However, in the tuber we were able to identify two types of sieve tube, those which transported fluorescent probes only in a longitudinal direction (as in Figure 5.6a) and those which also allowed lateral spread of the probe to adjacent parenchyma elements. An example of the latter is shown in Figure 5.6b. In some instances lateral spread to other internal phloem bundles was observed. It is conceivable, therefore, that some sieve tubes within a sink organ might be involved predominantly in long distance transport (this would seem inevitable if sugars are to be transported through mature regions of sink tissue to those accumulating solutes) while others may be specialized for lateral transport. Conceivably, the proportions of 'pathway' and 'unloading' SEs may change spatially within the sink. Micro-injection techniques could be used to examine this further.

We have also shown that storage parenchyma cells are symplastically coupled (Oparka and Prior, 1988). Thus, in the sink potato tuber the phloem unloading pathway appears to be predominantly symplastic (Oparka, 1986). It is assumed

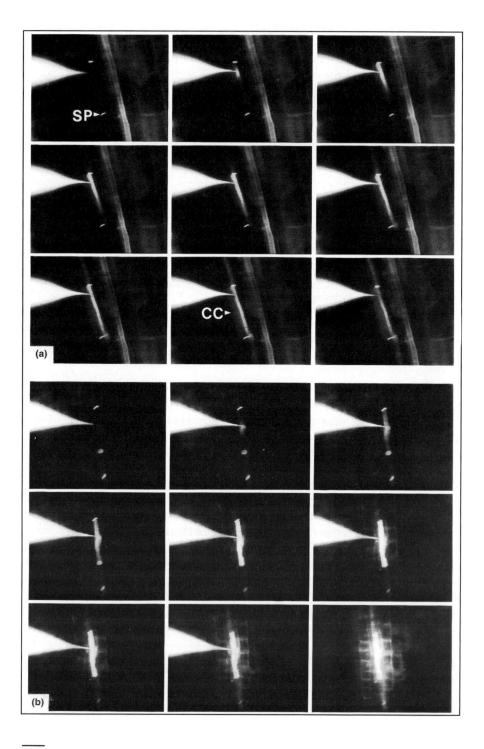

that *in vivo* the turgor differential between the SE–CC complex and the adjoining cells is sufficiently small to permit movement through the plasmodesmata at this site, due to the maintenance of a sucrose concentration gradient between the SE–CC complex and the cytosol of the storage cells. In the storage cells sucrose is allocated both to starch synthesis and to a vacuolar pool which may be constantly turning over.

5.5.6 *Turgor-regulated starch synthesis*

Although the *in vivo* pathway of sucrose movement between storage cells would appear to be symplastic, we have shown previously that when sucrose is supplied exogenously to storage parenchyma discs both the flux of sucrose across the plasma membrane and also the conversion of sucrose to starch were sensitive to the turgor pressure of the storage cells. Starch synthesis showed a distinct optimum at an external mannitol concentration of 300 mM, corresponding to a turgor of approximately 80 kPa (Oparka and Wright, 1988). Thus, in common with several other types of parenchyma element, the storage cells of sink potato tubers display an effective turgor-sensitive mechanism for taking up sucrose from the apoplast. What is the significance of this observation if sucrose is moving between storage cells symplastically? One possibility is that the plasma membrane transport system functions not to take up sucrose which is moving apoplastically but rather to retrieve sucrose which has escaped from the symplast. Madore and Lucas (1989) have discussed in detail the concept of apoplastic retrieval of sugars in leaves and it would appear that solute retrieval may also be a general feature of parenchyma cells in storage sinks (Oparka, 1990).

An additional factor which may influence the levels of cytosolic sucrose is that passive, non-specific leakage to the apoplast is increased at high turgors (Oparka *et al.*, 1989). Thus, at high cell turgors, passive solute leakage to the apoplast is enhanced while the retrieval mechanism is effectively eliminated. Conversely, as the turgor of the storage cell is lowered, the retrieval pump for sucrose is activated and passive solute leakage is reduced. We have suggested that sucrose leakage to the apoplast (and consequently cytosolic sucrose levels) might be regulated by both pump and leak mechanisms in response to fluctuations in cell turgor (Oparka *et al.*, 1989).

Vegetative storage sinks such as the potato tuber can undergo considerable diurnal fluctuations in turgor pressure, turgor being lowest during the day and highest at night (Gandar and Tanner, 1976). These fluctuations are mirrored by visible diurnal growth fluctuations (Stark and Halderson, 1987). If carbon partitioning in the tuber is influenced by turgor, then during the day (low turgor)

Figure 5.6. *Microinjection of the membrane-impermeant fluorescent probe Cascade Blue into pathway (a) and sink (b) SEs. Note that in (a) the dye moved longitudinally into other SEs and laterally only as far as the companion cell (CC). In (b) the dye spread laterally via the symplast to adjoining parenchyma elements. The tissue was stained with a dilute aniline blue solution in order to locate the sieve plates (SP).*

allocation to starch would be enhanced, while at night (high turgor) a larger fraction of the imported sucrose would be utilized in growth processes. Similarly, prolonged periods of water stress would shift the partitioning of carbon towards starch synthesis. This hypothesis is supported by observations that in continuously drought-stressed tubers starch levels may be as high as 90% of tuber dry matter, compared to only 60% in continuously irrigated plants (Oparka and Wright, 1988).

5.5.7 *The metabolic fate of sucrose in the storage cell*

Some of the photosynthate unloaded from the phloem of the tuber is utilized to support cell division and growth. Approximately 25% of the total sucrose metabolized in growing tubers is utilized for respiratory metabolism, organic acids, amino acids and protein biosynthesis (Morrell and ap Rees, 1986). Synthesis of cell wall constituents (4% of tuber dry weight at maturity) represents a minor pathway for the products of sucrose catabolism (Hoff and Castro, 1969). Thus, during tuber growth, the conversion of sucrose into starch represents the major metabolic event in the tuber cell and, as a result, starch represents over 65–70% of the tuber dry matter at maturity.

Although developing potato tubers contain both sucrose synthase and acid or alkaline invertase, measurements of maximum catalytic activities of these enzymes show that only sucrose synthase activity can sustain a rate of sucrose breakdown comparable to that observed *in vivo* (Morrell and ap Rees, 1986). Thus it is likely that the first step in the conversion of sucrose into starch is largely catalysed by sucrose synthase (Figure 5.7). This enzyme is abundant in developing tubers and in other starch synthesizing tissues (Stitt and Steup, 1985). However, the enzyme displays a rather low affinity for sucrose (the K_m value for the potato tuber enzyme in the presence of uridine diphosphate [UDP] is 55 mM [Pressey, 1969; see also Avigad, 1982 and references therein]). Thus, it is possible that sucrose breakdown may represent a rate limiting step during starch biosynthesis, especially in tissues with a low sucrose content.

Black and co-workers (Sung *et al.*, 1989; Xu *et al.*, 1989) proposed that sucrose synthase (SS) activity could be taken as a marker for sink strength in potato tubers and other actively filling sucrose sinks. Although the *in vitro* activity of SS extracted from developing tubers greatly exceeds the rate of sucrose breakdown (Morrell and ap Rees, 1986) or starch biosynthesis (Morrell and ap Rees, 1986; Sowokinos, 1973; Sung *et al.*, 1989; Xu *et al.*, 1989), this may not be the case *in vivo*. The sucrose:hexoses ratio increases in tuberizing stolon tips, but the actual sucrose content declines in potato tubers as starch accumulation commences. Morrell and ap Rees (1986) reported a decrease in sucrose content from approximately 50 mM to 10 mM or less during tuber growth (tubers up to 20 g fresh weight). Similar values can be extrapolated from data provided by Burton and Wilson (1970). It is uncertain whether the sucrose concentration in the cells of growing tubers is high enough to allow SS activity to be at, or near, V_{max}. However, the enzyme could be working at V_{max} if much of

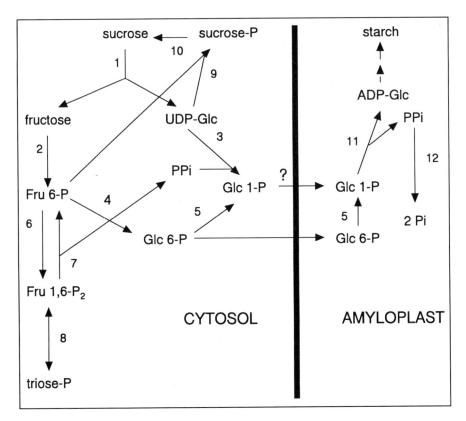

Figure 5.7. *Proposed pathways of sucrose metabolism and starch biosynthesis in potato tubers. For clarity, phosphorylated nucleotides have been omitted. Key to enzymes: (1) sucrose synthase; (2) fructokinase; (3) UDPGlc pyrophosphorylase; (4) hexose-P isomerase; (5) phosphoglucomutase; (6) phosphofructokinase; (7) pyrophosphate:fructose-6-phosphate:phosphotransferase; (8) aldolase; (9) sucrose-P synthase; (10) sucrose-P phosphatase; (11) ADPGlc pyrophosphorylase; (12) inorganic pyrophosphatase.*

the incoming sucrose is not compartmentalized in the vacuole but is readily metabolized in the cytosol, where SS is located.

SS catalyses a reversible reaction *in vitro*. However, in growing tubers, the irreversible conversion of fructose into fructose-6-phosphate (Fru6P) catalysed by a specific fructokinase is likely to drive net flux in the direction of sucrose breakdown *in vivo*. The rapid increase in fructokinase activity observed at the onset of starch deposition in potato tubers and other starch-storing organs (e.g. faba beans) coincides with that of SS (A. Gardner, personal communication). The rapid decline of fructose content at the early stages of tuberization can also be attributed to the activation of a specific fructokinase (Davies and Oparka, 1985). The presence of pyrophosphate (PPi) pools in plant tissues (including potato tubers) allows conversion of UDPglucose (UDPGlc) into glucose-1-phosphate

(Glc1P) via UDPGlc pyrophosphorylase (Figure 5.7). This ensures rapid equilibration between the products of sucrose catabolism in the cytosol.

Carbon flux into the amyloplast to sustain starch biosynthesis is likely to occur in the form of hexose-P (Hatzfeld and Stitt, 1990; Hill and Smith, 1991; Keeling et al., 1988; Viola et al., 1991). It remains to be addressed as to whether the rate of carbon flux through the amyloplast membrane, or the rate of energy supply to the amyloplast, represent limiting steps during starch synthesis. The available evidence strongly suggests that starch synthesis is primarily regulated by the activity of adenosine diphosphate glucose (ADPGlc) pyrophosphorylase in the amyloplast. A good correlation between extractable ADPGlc pyrophosphorylase activity and the rate of starch synthesis has been provided for many starch-storing tissues, including potato tubers (see Preiss, 1988 for a review). However, it is difficult to extrapolate the activity of this enzyme *in vivo* from *in vitro* measurements. The activity of ADPGlc pyrophosphorylase from potato tubers is known to be strongly regulated by the concentration ratio 3-phosphoglycerate (3PGA)/Pi and the concentration of these metabolites in the amyloplast compartment is not known.

The reaction catalysed by ADPGlc pyrophosphorylase is reversible *in vitro*. However, the net flux of carbon in the direction of ADPGlc formation (and of starch synthesis) is driven by the rapid hydrolysis of PPi in the amyloplast, catalysed by a specific inorganic pyrophosphatase (Figure 5.7). The necessity for PPi hydrolysis in the amyloplast during starch synthesis has recently been demonstrated (Viola and Davies, 1991). When sodium fluoride is supplied to discs excised from developing tubers, PPi accumulates as a result of the inhibition of inorganic pyrophosphatase and this leads to an almost complete block in the incorporation of a range of labelled precursors into starch.

The close association between ADPGlc pyrophosphorylase activity and the rate of starch synthesis is also demonstrated by tuber removal experiments (Oparka et al., 1990). In these experiments, the ability of storage parenchyma discs from growing tubers, detached from their stolon, to convert labelled sugars into starch was compared with attached tubers from the same plant. In attached tubers, 54.4% of [^{14}C]glucose was converted into starch, while in excised tubers this proportion decreased to 27.3%. The proportion of [^{14}C]sucrose converted into starch also declined markedly (19.7% in attached tubers, 6.6% in excised tubers). This phenomenon was associated with a 50% decline in ADPGlc pyrophosphorylase activity. The marked decrease in the capacity for incorporation of label into starch from [^{14}C]sucrose in excised tubers, compared to attached tubers, was also probably explained by the sevenfold drop of extractable SS activity following tuber detachment. Thus, the termination of sucrose import in developing tubers has a marked effect on the capacity of the tissue to metabolize sucrose and sustain starch synthesis.

The hypothesis that starch synthesis is under coarse control regulation in developing sinks is confirmed by studies on the expression of genes coding for key enzymes involved in the conversion of sucrose to starch. The expression of genes encoding SS (Salanoubat and Belliard, 1989) and ADPGlc pyrophosphorylase (Müller-Röber *et al.*, 1990) in potato is stimulated by sucrose. There-fore, the rate of sucrose influx into the tuber may regulate directly the capacity for its conversion into starch. This idea is supported by the observed positive correlation between the sucrose concentration of growing tubers and their ^{14}C content following $^{14}CO_2$ assimilation by the shoots (Engels and Marschner, 1986; Oparka, 1985).

The rate of sucrose conversion into starch, however, is not the only factor determining overall sink strength. When labelled precursors are supplied to storage parenchyma discs, rapid conversion into sucrose is usually observed. ^{13}C-NMR analysis of label distribution indicates that both starch and sucrose are synthesized from the hexose-P pool (Viola *et al.*, 1991). The similar degree of label redistribution observed in the glucosyl and fructosyl moieties of sucrose synthesized from specifically labelled [^{13}C]glucose suggests that sucrose synthesis takes place most likely via the sucrose-P synthase/sucrose phosphatase pathway (Figure 5.7). The isotopic equilibrium observed in the two hexosyl moieties of newly synthesized sucrose is explained if one assumes that Fru6P and UDPGlc are the precursors of sucrose synthesis (via sucrose-P synthase) rather than fructose and UDPGlc (via SS). Equilibration between UDPGlc and fructose is unlikely to take place during the course of the experiment.

The relevance of a sucrose synthesizing pathway in sink tissues involved in the active conversion of sucrose into starch is unclear. There is increasing evidence that the partitioning of incoming carbon between sucrose and starch in the tuber cell is strictly regulated. When the capacity for starch synthesis in the tuber tissue is reduced experimentally, an increase in the partitioning of incoming carbon into sucrose is observed (Oparka *et al.*, 1990; Viola and Davies, 1991). However, this does not always result in a decrease in the capacity for carbon import into the tuber cells. For example, the ability to take up [^{14}C]sucrose from the apoplast was not reduced in storage parenchyma discs isolated from tubers detached for 7 days, compared to attached tubers, in spite of the much reduced starch synthesis in the detached tubers (Oparka *et al.*, 1990). When [^{14}C]glucose was used, 80% of the radioactivity was recovered as sucrose in discs from detached tubers, compared to 28% in control discs (attached tubers). Similarly, conversion of a range of labelled precursors into starch in discs from growing tubers was almost completely inhibited by pre-incubating the tissue in 10 mM sodium fluoride. With all precursors, the decrease in labelling of starch was associated with an almost proportional increase in the labelling of sucrose, while total uptake of precursor was not affected (Viola and Davies, 1991; see also Figure 5.8). These data provide evidence that starch synthesis is not necessary to drive the continued uptake of sugars into the storage cell.

In the above experiments the precursors (including sucrose) were supplied to the apoplast, that is, not by the demonstrated *in vivo* route which is symplastic. However, recent interesting observations have been made on potato plants genetically manipulated to express antisense mRNA encoding ADPGlc pyrophosphorylase in the tubers. The extractable ADPGlc pyrophosphorylase activity in the tuber was dramatically reduced. Transgenic plants produced more

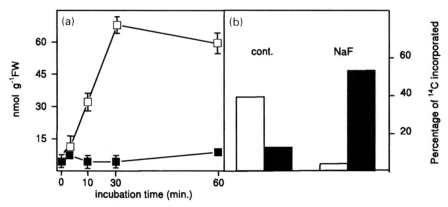

Figure 5.8. *(a) Effect of 1 h pre-incubation in buffer plus 10 mM sodium fluoride (open symbols) versus buffer alone (closed symbols) on PPi concentration in storage parenchyma discs excised from growing tubers. (b) Partitioning of [U-¹⁴C]glucose between starch (open bars) and sucrose (closed bars) as influenced by 1 h pre-incubation in 10 mM sodium fluoride. Redrawn from Viola and Davies (1991) with permission from ASPP.*

numerous tubers compared to the wild-type. Although starch synthesis was effectively eliminated in the transgenic plants, the phloem clearly continued to unload successfully and sucrose concentrations reached levels as high as 30% of the tuber dry weight (Müller-Röber *et al.*, 1992). Thus, elimination of starch synthesis, either by sink removal or by inhibition or elimination of a starch synthesizing enzyme, does not prevent phloem unloading in the potato tuber.

In the long term, starch synthesis may be necessary to maintain sugar import into the sink. However, in the short term, we suggest that fluctuations in starch synthesis are likely to have little effect on phloem unloading. During periods of fluctuating import, carbon partitioning into sucrose (temporarily held in the vacuole?) could act as a buffer, preventing drops in the import capacity of the sink.

An important role in regulating net flux from the hexose-P pool toward starch or sucrose biosynthesis in the tuber cell is probably played by the relative activities of cytosolic sucrose-P synthase and plastidic ADPGlc pyrophosphorylase. Both these enzymes are known to be regulated by coarse and fine control mechanisms which modulate activities *in vivo*. In this respect, fluctuations in the subcellular distribution of Pi (a strong inhibitor of both enzymes) might be important for the overall regulation of carbon partitioning in the tuber cell. (For a more general discussion of starch synthesis in storage sinks, see Chapter 6.)

5.5.8 *The phloem loading pathway during sprouting*

In the source (sprouting) tuber, starch breakdown occurs and sucrose is mobilized towards the internal phloem strands. Sucrose may move symplastically as the storage cells remain symplastically coupled (unpublished data). However, a major change occurs at the plasma membrane of the storage cells. Sugar uptake

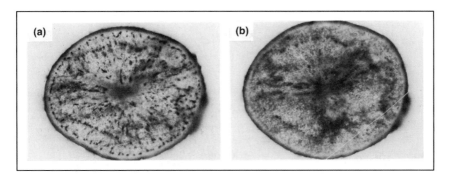

Figure 5.9. *Inhibition of phloem loading by PCMBS in a source potato tuber. Adjacent tuber slices were incubated in the absence (a) or presence (b) of the inhibitor. Note the strong labelling of the internal phloem bundles in (a).*

studies have shown that uptake into storage cells is essentially diffusional and unaffected by *p*-chloromercuribenzene sulphonic acid (PCMBS, an inhibitor of the sucrose carrier) or the protonophore carbonyl cyanide *m*-chlorophenyl hydrazone (CCCP). Furthermore, the sensitivity of sucrose uptake to cell turgor is lost (Wright and Oparka, 1990). These results suggest that in the source tuber sucrose is free to diffuse to the apoplast since the retrieval mechanism is inactivated or no longer present. A major change also occurs at the SE–CC complex of the source tuber. We have shown recently that phloem loading, but not phloem unloading, of sucrose is sensitive to PCMBS (Wright and Oparka, 1991). When slices of source tubers were incubated in [^{14}C]sucrose intense labelling of the internal phloem strands was observed (Figure 5.9a). Labelling of the phloem was not observed when slices of sink tubers were incubated in sucrose, indicating a lack of retrieval into the phloem (data not shown). However, when slices of source tuber were incubated in the presence of PCMBS, phloem loading was eliminated and only background diffusional uptake was observed (Figure 5.9b). These results suggest that SE loading, in contrast to SE unloading, is carrier-mediated and occurs across the plasma membrane of the SE–CC complex, that is, SE unloading is symplastic while SE loading is apoplastic.

How can the anatomy of the transport pathway be compatible with both the above transport mechanisms? One possibility is that different SEs are involved in SE unloading (plasmodesmata present) and SE loading (plasmodesmata absent). A second possibility is that the plasmodesmata present during SE unloading are non-functional (occluded) during SE loading. We have found no obvious structural inclusions in the plasmodesmata of source tubers and suggest instead that the plasmodesmata may be closed by the high turgor differential generated between the SE–CC complex and adjoining cells during SE loading. Such pressure-generated closure of plasmodesmata may not be apparent when examining electron micrographs of plasmodesmata prepared by conventional fix-

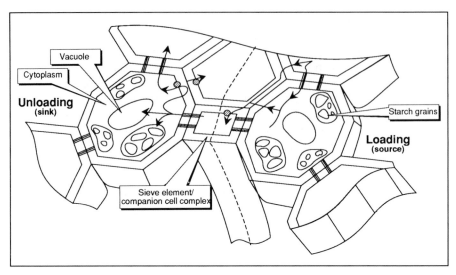

Figure 5.10. *Schematic representation of the sink to source transition in the potato tuber. Note that unloading in the sink tissue occurs via plasmodesmata, followed by sucrose storage in the vacuole and starch synthesis in the amyloplast. The active, turgor-sensitive sucrose transport system on the plasma membrane functions as a retrieval mechanism for sucrose escaping to the apoplast. SE loading in the source tissue occurs via the apoplast, that is, across the plasmalemma of the SE–CC complex. Inactivation of the retrieval mechanism of the storage cells allows diffusion to the apoplast. It is postulated that the plasmodesmata are closed during SE loading by generation of a turgor differential between the SE–CC complex and adjoining cells, the creation of this differential being brought about by activation of a phloem-specific H^+-ATPase. Circles represent membrane transport events.*

ative procedures as the turgor pressure within cells would rapidly be dissipated during tissue preparation. Activation of a phloem-specific H^+-ATPase, coupled to the sucrose carrier, would drive uptake against a sucrose concentration gradient, generating the turgor differential necessary to close the plasmodesmata. SE loading could then proceed apoplastically. Techniques such as high-pressure freezing (Dahl and Staehelin, 1989) are likely to prove invaluable in future studies of *in situ* plasmodesmatal structure (see also Ding *et al.*, 1991).

According to the above hypothesis, SE loading is brought about by a loss of active, carrier-mediated transport at the plasma membrane of the storage cells and its 'switching on' at the SE–CC complex. The recent demonstration of the localization of a phloem-specific ATPase (de Wit *et al.*, 1991) is likely to make such a hypothesis amenable to experimentation in the near future.

A schematic model for the sink to source transition in the potato tuber, summarizing a number of the above features, is shown in Figure 5.10.

5.6 Concluding remarks

Storage sinks appear to utilize a variety of pathways and mechanisms in transporting carbon compounds from the phloem to their eventual sites of accumulation. No unifying hypothesis has emerged for the control of phloem unloading and it would appear that several 'strategies' have been developed among different types of storage sinks for regulating the flux of carbon in them. Each case must be treated individually and it may be misleading to assume common ground.

In this chapter we have emphasized the potential role played by plasmo-desmata in determining the pattern of solute fluxes in a storage sink. Traditionally the symplast has been viewed as something of a 'poor relative' when examining solute transport processes; membrane transport events have received the bulk of attention. This is presumably because the symplast has been relatively inaccessible to experimental challenge. As a consequence, plasmo-desmata have been viewed as rather passive structures, exerting little direct control over symplastic solute transport. In this article we have deliberately raised a number of issues which are, at present, speculative but which we hope will form a challenge for future studies on the events which control carbon fluxes in a storage sink. We concur with the view that plasmodesmata are considerably more dynamic than has previously been thought (see reviews by Beebe and Turgeon, 1991 and Robards and Lucas, 1990) and suggest that the regulation of plasmodesmatal conductance during phloem loading/unloading events will prove to be an important area for future research.

References

Avigad, G. (1982) Sucrose and other disaccharides. In: *Encyclopaedia of Plant Physiology*, Volume XIII (eds F.A. Loewus, W. Tanner). Springer-Verlag, Heidelberg, pp. 217–347.

Barclay, G.F. and Fensom, D.S. (1984) Physiological evidence for the existence of pressure-sensitive valves in plasmodesmata between internodes of *Nitella*. In: *Membrane Transport in Plants* (eds W.J. Cram, J. Janácek, R. Rybová and K. Sigler). Academic Press, Praha, pp. 316–317.

Beebe, D.U. and Turgeon, R. (1991) Current perspectives on plasmodesmata: structure and function. *Physiol. Plant.* 83, 194–199.

van Bel, A.J.E. (1992) Mechanisms of sugar translocation. In: *Crop Photosynthesis: Spatial and Temporal Determinants* (eds N.R. Baker and H. Thomas). Elsevier, Amsterdam.

van Bel, A.J.E. and Kempers, R. (1990) Symplastic isolation of the sieve element–companion cell complex in the phloem of *Ricinus communis* and *Salix alba* stems. *Planta*, 183, 69–76.

van Bel, A.J.E. and Gamelei, Y. V. (1991) Multiprogrammed phloem loading. In: *Recent Advances in Phloem Transport and Assimilate Compartmentation* (eds J.L. Bonnemain, S. Delrot, W.J. Lucas and J. Dainty). Ouest Editions, Nantes, pp. 128–140.

Burton, W.G. and Wilson, A.R. (1970) The apparent effect of the latitude of the place of cultivation upon the sugar content of potatoes. *Eur. Pot. J.* 2, 105–116.

Chapleo, S. and Hall, J.L. (1989) Sugar unloading in roots of *Ricinus communis* L. III. The extravascular pathway of sugar transport. *New Phytol.* 111, 391–396.

Côte, R., Thain, J.F. and Fensom, D.S. (1987) Increase in electrical resistance of plasmo-desmata of *Chara* induced by an applied pressure gradient across nodes. *Can. J. Bot.* 65, 509–511.

Dahl, R. and Staehelin, L.A. (1989) High-pressure freezing for the preservation of biological structure: theory and practice. *J. Elec. Micros. Tech.* 13, 165–174.

Daie, J. and Wyse, R. (1985) Evidence on the mechanism of enhanced sugar uptake at low cell turgor i leaf discs of *Phaseolus coccinius*. *Physiol. Plant.* 64, 547–552.

Davies, H.V. and Oparka, K.J. (1985) Hexose metabolism in developing tubers of potato (*Solanum tuberosum* L.). *J. Plant Physiol.* 126, 387–396.

Derrick, P.M., Barker, H. and Oparka, K.J. (1990) Effect of virus infection on symplastic transport of fluorescent tracers in *Nicotiana clevelandii* leaf epidermis. *Planta*, 181, 555–559.

Dick, P.S. and ap Rees, T. (1975) The pathway of sugar transport in roots of *Pisum sativum*. *J. Exp. Bot.* 26, 305–314.

Ding, B., Turgeon, R. and Parthasarathy, M.V. (1991) Plasmodesmatal substructure in cryofixed developing tobacco leaf tissue. In: *Recent Advances in Phloem Transport and Assimilate Compartmentation* (eds J.L. Bonnemain, S. Delrot, J. Dainty, W.J. Lucas). Ouest Editions, Nantes, pp. 317–323.

Engels, C.H. and Marschner, H. (1986) Allocation of photosynthate to individual tubers of *Solanum tuberosum* L. *J. Exp. Bot.* 37, 1804–1812.

Eschrich, W. (1989) Phloem unloading of photoassimilates. In: *Transport of Photoassimilates* (eds D.A. Baker and J.A. Milburn). John Wiley and Sons, New York, pp. 206–263.

Gandar, P.W. and Tanner, C.B. (1976) Potato tuber leaf and tuber water potential measurements within a pressure chamber. *Am. Pot. J.* 53, 1–14.

Geiger, D.R. and Fondy, B.R. (1980) Phloem loading and unloading: pathways and mechanisms. *What's New Plant Physiol.* 11, 25–28.

Goodwin, P.B. (1983) Molecular size limit for movement in the symplast of the *Elodea* leaf. *Planta*, 157, 124–130.

Gougler Schmalstig, J. and Geiger, D.R. (1985) Phloem unloading in developing leaves of sugar beet 1. Evidence for pathway through the symplast. *Plant Physiol.* 79, 237–241.

Gougler Schmalstig, J. and Hitz, W.D. (1987) Transport and metabolism of a sucrose analog (1'-fluorosucrose) into *Zea mays* L. endosperm without sucrose hydrolysis. *Plant Physiol.* 85, 902–905.

Hatzfeld, W.D. and Stitt. M. (1990) A study of the rate of recycling of triose phosphates in heterotrophic *Chenopodium rubrum* cells, potato tubers, and maize endosperm. *Planta*, 180, 198–204.

Hill, L.M. and Smith, A.M. (1991) Evidence that glucose-6-phosphate is imported as the substrate for starch synthesis by the plastids of developing pea embryos. *Planta*, 185, 91–96.

Ho, L.C. (1988) Metabolism and compartmentation of imported sugars in sink organs in relation to sink strength. *Ann. Rev. Plant Physiol. Plant Mol. Biol.* 39, 355–378.

Hoff, J.E. and Castro, M.D. (1969) Chemical composition of potato cell wall. *J. Agric. Food Chem.* 17, 1328–1331.

Keeling, P.L., Wood, J.R., Tyson, R.H. and Bridges, I.G. (1988) Starch biosynthesis in developing wheat grain: evidence against the direct involvement of triose phosphates in the metabolic pathway. *Plant Physiol.* 87, 311–319.

Leigh, R.A., ap Rees, T., Fuller, W.A. and Banfield, J. (1979) The location of acid invertase activity and sucrose in vacuoles isolated from storage roots of red beet (*Beta vulgaris* L.). *Biochem. J.* 178, 539–547.

Lingle, S.E. (1989). Evidence for uptake of sucrose intact into sugarcane internodes. *Plant Physiol.* 90, 6–8.

Madore, M.A. and Lucas, W.J. (1989) Transport of photoassimilates between leaf cells. In: *Transport of Photoassimilates* (eds D.A. Baker and J.A. Milburn). John Wiley and

Sons, New York, pp. 49–78.

Mierzwa, R.J. and Evert, R.F. (1984) Plasmodesmatal frequency in the root of sugar beet. *Am. J. Bot.* 71(Suppl.), 39.

Morrell, S. and ap Rees, T. (1986) Sugar metabolism in developing tubers of *Solanum tuberosum*. *Phytochemistry*, 25, 1579–1586.

Müller-Röber, B.T., Koßmann, J., Hannah, L.C., Willmitzer, L. and Sonnewald, U. (1990) One of two different ADP-glucose pyrophosphorylase genes from potato responds strongly to elevated levels of sucrose. *Mol. Gen. Genet.* 224, 134–146.

Müller-Röber, B.T., Sonnewald, U. and Willmitzer, L. (1992) Inhibition of the ADP-glucose pyrophosphorylase in transgenic potatoes leads to sugar-storing tubers and influences tuber formation and expression of tuber storage protein genes. *EMBO J.* in press.

Oparka, K.J. (1985) Changes in partitioning of current assimilate during tuber bulking in potato (*Solanum tuberosum* L.) cv. Maris Piper. *Ann. Bot.* 55, 705–707.

Oparka, K.J. (1986) Phloem unloading in the potato tuber, pathways and sites of ATPase. *Protoplasma*, 131, 201–210.

Oparka, K.J. (1990) What is phloem unloading? *Plant Physiol.* 94, 393–396.

Oparka, K.J. and Prior, D.A.M. (1987) ^{14}C sucrose efflux from the perimedulla of growing potato tubers. *Plant Cell Environ.* 10, 667–675.

Oparka, K.J. and Prior, D.A.M. (1988) Movement of Lucifer Yellow in potato tuber storage tissues. A comparison of apoplastic and symplastic transport. *Planta*, 176, 533–540.

Oparka, K.J. and Wright, K.M. (1988) Influence of cell turgor on sucrose partitioning in potato tuber storage tissues. *Planta*, 175, 520–526.

Oparka, K.J. and Prior, D.A.M. (1992) Direct evidence for pressure-generated closure of plasmodesmata. *Plant J.* 2, 741–750.

Oparka, K.J., Wright, K.M. and Prior, D.A.M. (1989) Control of starch synthesis in potato tubers by turgor-sensitive sucrose transport at the plasmalemma. In: *Plant Membrane Transport: The Current Position* (eds J. Dainty, M.I. de Michelis, E. Marre and F. Rasi-Coldogno). Elsevier, Amsterdam, pp. 629–632.

Oparka, K.J., Davies, H.V., Wright, K.M., Viola, R. and Prior, D.A.M. (1990) Effect of sink removal on sugar uptake and starch synthesis by potato-tuber storage parenchyma. *Planta*, 182, 113–117.

Oparka, K.J., Murphy, R., Derrick, P.M., Prior, D.A.M. and Smith, J.A.C. (1991) Modification of the pressure probe technique permits controlled intracellular microinjection of fluorescent probes. *J. Cell Sci.* 98, 539–544.

Patrick, J.W. (1990). Sieve element unloading: cellular pathway, mechanism and control. *Physiol. Plant.* 78, 298–308.

Preiss, J. (1988) Biosynthesis of starch and its regulation. In: *The Biochemistry of Plants*, Volume XIV (ed. J. Preiss). Academic Press, London, pp. 181–254.

Pressey, R. (1969) Potato sucrose synthetase — purification, properties and changes associated with maturation. *Plant Physiol.* 44, 759–764.

Robards, A.W. and Lucas, W.J. (1990) Plasmodesmata. *Ann. Rev. Plant Physiol. Plant Mol. Biol.* 41, 369–419.

Ross, H.A. and Davies, H.V. (1985) A microscopic examination of starch depletion in tubers of cv. Maris Piper during sprouting. *Pot. Res.* 28, 113–118.

Salanoubat, M. and Belliard, G. (1989). The steady-state level of potato sucrose synthase mRNA is dependent on wounding, anaerobiosis and sucrose concentration. *Gene*, 84, 181–185

Sonnewald, U., Brauer, M., von Schaewen, A., Stitt, M. and Willmitzer, L. (1991) Transgenic tobacco plants expressing yeast-derived invertase in either the cytosol, vacuole or apoplast: a powerful tool for studying sucrose metabolism and sink/source interactions. *Plant J.* 1, 95-106.

Sowokinos, J.R. (1973) Maturation of *Solanum tuberosum*. I. Comparative sucrose and sucrose synthetase levels between several good and poor processing varieties. *Am. Pot. J.* 50, 234-247.

Stark, J.C. and Halderson, J.L. (1987) Measurement of diurnal changes in potato tuber growth. *Am. Pot. J.* 64, 245-248.

Stitt. M. and Steup, M. (1985) Starch and sucrose degradation. In: *Encyclopaedia of Plant Physiology*, Volume XVIII (eds R. Douce and D.A. Day). Springer-Verlag, Heidelberg, pp. 347-390.

Sung, S.-J., Xu, D.-P. and Black, C.C. (1989). Identification of actively filling sucrose sinks. *Plant Physiol.* 89, 1117-1121.

Terry, B.R. and Robards, A.W. (1987) Hydrodynamic radius alone governs the mobility of molecules through plasmodesmata. *Planta*, 171, 145-157.

Trebacz, K. and Fensom, D.S. (1989) The uptake and transport of ^{14}C in cells of *Conocephalum conicum* L. in light. *J. Exp. Bot.* 40, 1089-1092.

Tucker, E.B. (1990) Calcium-loaded 1,2 bis(2-aminophenoxy)ethane-N,N,N,N-tetraacetic acid blocks cell-to-cell diffusion of carboxyfluorescein in staminal hairs of *Setcreasea purpurea*. *Planta*, 182, 34-38.

Turgeon, R. (1989) The sink-source transition in leaves. *Ann. Rev. Plant Physiol. Plant Mol. Biol.* 40, 119-138.

Viola, R. and Davies, H.V. (1991) Fluoride-induced inhibition of starch biosynthesis in developing potato, *Solanum tuberosum* L., tubers is associated with pyrophosphate accumulation. *Plant Physiol.* 97, 638-643.

Viola, R., Davies, H.V. and Chudeck, A.R. (1991) Pathways of starch and sucrose biosynthesis in developing tubers of potato (*Solanum tuberosum* L.) and seeds of faba bean (*Vicia faba* L.). *Planta*, 183, 203-208.

Warmbrodt, R.D. (1987) Solute concentrations in the phloem and apex of the root of *Zea mays*. *Am. J. Bot.* 74, 394-402.

de Wit, N.D., Harper, J.F. and Sussman, M.R. (1991) Evidence for a plasma membrane proton pump in phloem cells of higher plants. *Plant J.* 1, 121-128.

Wright, K.M. and Oparka, K.J. (1990) Hexose accumulation and turgor-sensitive starch synthesis in discs derived from source versus sink potato tubers. *J. Exp. Bot.* 41, 1355-1360.

Wright, K.M. and Oparka, K.J. (1991) Sugar uptake and metabolism in sink and source potato tubers. In: *Recent Advances in Phloem Transport and Assimilate Compartmentation* (eds J.L. Bonnemain, S. Delrot, W.J. Lucas and J. Dainty). Ouest Editions, Nantes, pp. 258-264.

Wyse, R.E., Zamski, E. and Tomos, A.D. (1986) Turgor regulation of sucrose transport in sugar beet taproot tissue. *Plant Physiol.* 81, 478-481.

Xu, D.P., Sung, S.S., Loboda. T., Kormanik, P.P. and Black, C.C. (1989) Characterization of sucrolysis via the uridine diphosphate and pyrophosphate-dependent sucrose synthase pathway. *Plant Physiol.* 90, 635-642.

Yahalom, A., Warmbrodt, R.D., Laird, D., Traub, O., Revel, J.-P., Willecke, K. and Epel, B. (1991) Maize mesocotyl plasmodesmata proteins cross-react with connexin gap junction protein antibodies. *Plant Cell*, 3, 407-417.

6

Synthesis of storage starch

T. ap Rees

I intend to discuss our present understanding of the mechanism of storage starch synthesis in the non-photosynthetic cells of higher plants and to emphasize those developments that have occurred since I last reviewed this topic (ap Rees, 1988; ap Rees and Entwistle, 1989). The way in which carbon for starch synthesis is delivered to the non-photosynthetic cells may affect the pathway whereby this carbon is metabolized, and the control of that pathway. Thus my starting point is the carbon in the phloem of the starch storing tissue. Processes prior to this are covered in chapters 3, 5 and 11 on translocation and on phloem loading.

6.1 Delivery of carbon for starch synthesis

Sucrose is by far the dominant form in which carbon is translocated to starchy tissues. In certain families sugar alcohols, and oligosaccharides related to raffinose, are also transported (Ziegler, 1975). Our knowledge of how the latter compounds are converted to starch is restricted so I confine myself to sucrose. There are two basic routes from phloem to the starch storing cell. The first is via the symplasm and entails movement from cell to cell via the plasmodesmata. The second or apoplastic route is characterized by the fact that at some stage in the transfer carbon moves via the apoplast. Where plasmodesmata are not continuous, transport must be apoplastic: where there are plasmodesmata the two routes are not necessarily mutually exclusive.

6.1.1 Symplastic transport

Convincing evidence of symplastic transport of carbon for starch synthesis was provided initially by showing that flooding the apoplast of pea roots with non-radioactive sugars did not affect the movement of [^{14}C]sucrose from the stele to the cortex (Dick and ap Rees, 1975). The same work showed that the sucrose moved without inversion and strongly suggested that movement of carbon was as sucrose itself. Subsequent studies have shown that the symplastic pathway operates widely but not necessarily universally (Patrick, 1991). Present evidence suggests that where the product of translocation is metabolized rapidly to insoluble compounds in the sink, then transport is symplastic. However, where the product of translocation is stored as soluble compounds which attain a high osmotic potential, for example, the storage root of sugar beet, there is a significant apoplastic component. In developing seeds, where there are no plasmodesmata between the embryo and the maternal tissues, an apoplastic step is obligatory.

6.1.2 *Apoplastic transport*

Where transport is apoplastic a further complication arises. This is the extent to which uptake from the apoplast occurs as sucrose or as hexoses produced by hydrolysis of the sucrose by an apoplastic invertase. Proton-coupled transport of sucrose has been demonstrated across the plasma membrane of a wide range of plants (Bush and Li, 1991). Appreciable progress is being made towards the recognition and resolution of the proteins involved in this transport (Gallet *et al.*, 1992).Good evidence that a number of different developing seeds absorb carbon for starch synthesis from the apoplast is available for tissues as taxonomically diverse as wheat endosperm and the embryos of soybean (Ho *et al.*, 1991).

In many plant tissues an acid invertase is present in the cell wall. This is indicated by the observation that differential centrifugation of plant extracts often yields a cell wall fraction containing tightly bound acid invertase. This evidence is not necessarily conclusive because vacuolar acid invertase is readily bound to cell wall fragments. However, complementary evidence is provided by the demonstration of extracellular hydrolysis of sucrose. The enzyme is a glycoprotein and the gene for this glycoprotein in carrots has been cloned and sequenced (Sturm and Chrispeels, 1990). The protein has a signal sequence for entry into the endoplasmic reticulum, and a propeptide at the N-terminus which is not present in the mature protein. The question of whether cell wall acid invertase is distinct from the vacuolar acid invertase is not resolved. Studies of the extracted proteins, summarized by Sturm and Chrispeels (1990), suggest that the two are distinct. Work with the *miniature*-1 seed mutation of maize has led to the suggestion that both invertases may be coded for by the same gene (Miller and Chourey, 1992). This work provides evidence that there is a cell wall invertase in maize endosperm and that it is essential for the normal development of the seed. The argument that this enzyme is specified by the same gene as that responsible for vacuolar acid invertase is not so conclusive. Less than 10% of the total acid invertase activity in the wild-type endosperm was recovered in the soluble fraction. It is conceivable that this represents cell wall invertase that was solubilized during extraction of the tissue, and that the gene for vacuolar acid invertase was not expressed in the tissues studied.

If there is significant acid invertase on the cell wall of tissues in which apoplastic transport of sucrose is occurring, then at least some of that sucrose is likely to be hydrolysed to hexoses in the apoplast, and these may be absorbed directly by the storage cell. The ability to take up glucose and fructose is widely distributed in plant tissues and is catalysed by a proton/hexose transporter in the plasma membrane (Rausch, 1991). The gene for the glucose transporter has been cloned and sequenced from *Arabidopsis thaliana*. The sequence predicts a protein of molecular weight of 57 kDa with 12 transmembrane segments (Sauer *et al.*, 1990).Whether fructose is absorbed by the same or a related transporter is not yet resolved.

Two of the most quoted examples of apoplastic transport involving hydrolysis of sucrose in the apoplast are in the developing endosperm of maize and the storage tissue of sugar cane. In both there is compelling evidence of substantial uptake of hexose from the apoplast. Although previous views that uptake into these tissues is dependent upon hydrolysis in the apoplast have been weakened by evidence that both types of cells can take up sucrose (Ho *et al.*, 1991); the recent work of Miller and Chourey (1992) makes it very likely that the original hypothesis is correct. It is likely that the relative contributions of sucrose and hexose to carbon transport via the apoplast will be governed by the activity of the cell wall invertase and the capacity of the storage cells to take up hexoses as opposed to sucrose.

It is clear that there is considerable flexibility in the pathway of carbon movement from the phloem to the starch storing cell. Three mechanisms have been identified: sucrose via the symplast, sucrose via the apoplast, and hexose via the apoplast. The relative importance of the different routes is not known. No study comparable in its exemplary breadth and thoroughness to that carried out on leaves by Gamalei (1991) has been reported for sink tissues. Thus any generalization as to the more common route of transport is vulnerable. From the species examined to date it is likely that most of the carbon for starch synthesis in non-photosynthetic cells is delivered to those cells as sucrose via either the symplast or the apoplast. However, we need to remember that more than one mechanism may operate in the same tissue at any one time, and that the relative activities of the different routes may change during development and also in response to changes in the plant's environment.

6.2 Immediate metabolism of incoming sugar

6.2.1 *Hexose metabolism*

The simplest situation is where the incoming sugars are hexoses. The speed with which labelled glucose and fructose are metabolized when supplied to plant tissues in general strongly suggests that the immediate fate of these hexoses is phosphorylation to give their 6-phosphate esters. Plants contain a range of hexokinases and two general types may be recognized, those with a preference for glucose and those that are almost exclusive for fructose. For example, a recent study of the developing kernels of maize demonstrated two kinases with a preference for glucose, and two specific fructokinases (Doehlert, 1989). At least one of the two main types of hexokinase appears to be in the cytosol (Copeland and Morell, 1985; Schnarrenberger, 1990). Hexose arriving in the cytosol could be transported directly into the vacuole (Rausch, 1990). Comparison of the K_m values reported for the kinases (Doehlert, 1989) with those reported for the transporter (Rausch, 1990) suggests that most of the incoming hexose will be converted to hexose 6-phosphate. The fact that there is little evidence that hexose 6-phosphates can regulate plant hexokinases favours this view. There is sound evidence that the reactions catalysed by phosphoglucomutase and phosphohexoisomerase are close to equilibrium in plants (ap Rees et al., 1991). Thus, in general, the immediate fate of glucose and fructose absorbed into the cytosol from the apoplast will be conversion to an equilibrium mixture of glucose 6-phosphate (Glc6P), fructose 6-phosphate (Fru6P) and glucose 1-phosphate (Glc1P).

6.2.2 Sucrose metabolism

Where carbon is delivered to the cell as sucrose its immediate metabolism will be determined by the only two cytosolic enzymes known to be capable of metabolizing sucrose: alkaline invertase and sucrose synthase.

Alkaline invertase catalyses the physiologically irreversible hydrolysis of sucrose:

$$\text{sucrose} + H_2O \longrightarrow \text{glucose} + \text{fructose}$$

$$\Delta G' = -27.6 \text{ kJ}.$$

This enzyme is widely distributed, distinct from the acidic cell wall and vacuolar invertases, and shows optimum activity over the range pH 7.0–7.8. There have been relatively few detailed studies of the pure enzyme. Estimates of the K_m for sucrose vary: 10 mM, soybean root nodules (Morell and Copeland, 1984); 27 mM, carrot root; 22 mM, turnip root (Ricardo, 1974); 33 mM, sugar beet suspension cells (Masuda et al., 1988); 65 mM, sycamore cells (Huber and Akazawa, 1986). There is evidence of inhibition by fructose, but only at high concentrations, 15–30 mM. If the sucrose arriving in the starch storing cell is cleaved by alkaline invertase then it would be expected that the resulting hexoses would be metabolized via hexokinases as described above for the metabolism of exogenous hexoses.

The other cytosolic enzyme capable of metabolizing sucrose is sucrose synthase (SS). This catalyses the readily reversible reaction:

$$\text{sucrose} + \text{UDP} \rightleftharpoons \text{UDPGlc} + \text{fructose}$$

$$\Delta G' = -4.2 \text{ kJ}.$$

Two isozymes of SS have been demonstrated (Echt and Chourey, 1985) and their activities vary independently (Heinlein and Starlinger, 1989). The kinetic properties of the two isozymes have not been shown to differ markedly: in maize K_m for sucrose was 52 mM for one and 63 mM for the other. Values from preparations from other tissues are: sycamore cells, 15 mM (Huber and Akazawa, 1986); soybean root nodules, 31 mM (Morell and Copeland, 1985).

The enzyme shows activity with ADP as well as UDP. The K_m for UDP is generally higher than that for ADP (Pontis, 1977). The enzymes from soybean (Morell and Copeland, 1985) and maize endosperm (Echt and Chourey, 1985) showed a V_{max} with ADP that was much lower than that found with UDP. Furthermore, the studies of Murata *et al.* (1966) on the partially purified enzyme from rice suggested that UDP inhibited activity with ADP. Recently, Pozueta-Romero *et al.* (1991b) have argued that there is an ADP-specific SS. This claim cannot be assessed because it was not accompanied by any description of the methods used nor by any experimental data. Present evidence strongly suggests that the product of SS *in vivo* is predominantly UDPGlc. This view is supported by the evidence that UDPGlc is largely confined to the cytosol and by its rapid labelling when [14C]sucrose is supplied to pea embryos (ap Rees, 1988). However, the evidence for UDPGlc as the product of SS is not strong enough to exclude the possibility that in some tissues at least, significant amounts of ADPGlc may be found. A previously unrecognized ADP-linked SS may exist, but, as yet, there is no convincing published evidence that it does.

6.2.3 *UDPGlc metabolism*

If incoming sucrose is metabolized to UDPGlc by SS then the question arises as to how the glucosyl group of that UDPGlc is made available for general metabolism. Clearly some of this UDPGlc is used directly for the synthesis of structural polysaccharide. However, the breakdown of sucrose via SS does not necessarily correlate with the demand for UDPGlc for the synthesis of structural polymers. There are clear instances where far more UDPGlc is formed by SS than can be used for the synthesis of cell wall materials. Thus there must be a route from UDPGlc to starch and respiration (ap Rees, 1988).

The possibility that UDPGlc acts as a direct precursor of starch should be considered. The evidence for this view is threefold. First, unfractionated extracts of starchy tissues will often incorporate label from preparations of UDP[14C]Glc into material that is insoluble in methanol–KCl but soluble in boiling water. Much of this is likely to be due to α-glucan phosphorylase using [14C]Glc1P that contaminates most preparations of UDP[14C]Glc (ap Rees *et al.*, 1984). Furthermore, no-one has yet succeeded in purifying a protein that will synthesize starch from UDPGlc. Secondly, there is often a very close correlation between starch synthesis and the maximum catalytic activity of UDP-linked SS (Doehlert and Chourey, 1991). This may reflect nothing more than the fact that in such tissues SS is the principal route of sucrose breakdown and starch synthesis is by far the dominant compound formed from the products of sucrose breakdown. Thirdly, where UDP[14C]Glc has been supplied to isolated amyloplasts there is evidence of incorporation into insoluble material. However, such incorporation has not been shown to reflect starch synthesis by intact plastids. In wheat amyloplasts the incorporation was largely resistant to amyloglucosidase (Tyson and ap Rees, 1988). The incorporation by maize amyloplasts was not shown to be dependent upon the integrity of the plastids (Echeverria *et al.*, 1988). When labelled

UDPGlc was supplied to intact plastids from sugar beet storage tissue, label was incorporated into material that was immobile in the solvent used to isolate sugars by paper chromatography (Ivanov *et al.*, 1991). About 25% of this incorporation was released by treatment with amyloglucosidase or amylase (Semenov and Ivanov, 1991). This indicates that some of the incorporation represented the formation of $\alpha(1 \rightarrow 4)$ linkages. However, the labelled material was not shown to be insoluble in methanol–KCl and could represent the synthesis of soluble glycans, or a heteroglycan such as that described by Steup *et al.* (1991).

The evidence that UDPGlc is a direct precursor of starch is not strong enough to establish this view. If we consider the positive evidence that ADPGlc is the sole precursor of storage starch then the hypothesis becomes even weaker. This positive evidence has been discussed (ap Rees and Entwistle, 1989) and is now summarized. First, starch synthase and ADPGlc pyrophosphorylase have been identified, purified, shown to be specific for ADPGlc, and to be confined to the plastid in non-photosynthetic and photosynthetic tissues. Secondly, the activities of these enzymes have been shown to correlate with, and to be higher than, the rates of starch accumulation found *in vivo*. Finally, there is a great deal of evidence from mutants that ADPGlc is the sole precursor of storage starch. In this respect the argument is clinched by the recent work of Müller-Röber *et al.* (1992); they produced transgenic potato plants with tubers from which the ability to produce ADPGlc had been specifically eliminated: such tubers did not synthesize starch.

My argument is that in tissues where sucrose breakdown is predominantly via SS the UDPGlc so formed may exceed that required for the synthesis of structural polymers and cannot act as a direct precursor of starch. Under these conditions the UDPGlc will be converted to hexose monophosphates that can be used for the synthesis of ADPGlc or act as respiratory substrate. The issue is how such a conversion occurs. Evidence that it is catalysed by a UDPGlc phosphorylase

$$\text{UDPGlc} + \text{Pi} \longrightarrow \text{Glc1P} + \text{UDP}$$

has not been sustained (ap Rees, 1988; Tyson and ap Rees, 1989). There is now considerable evidence that the key step is catalysed by UDPGlc pyrophosphorylase, which is held close to equilibrium by the presence of appreciable amounts of PPi in the cytosol of higher plant cells. This evidence has been reviewed (ap Rees, 1988) and the proposed pathway is

$$\text{sucrose} + \text{UDP} \rightleftharpoons \text{fructose} + \text{UDPGlc}$$

$$\text{fructose} + \text{UTP} \longrightarrow \text{Fru6P} + \text{UDP}$$

$$\text{UDPGlc} + \text{PPi} \rightleftharpoons \text{Glu1P} + \text{UTP}.$$

A key feature of this proposal is the need for one mole of PPi for each molecule of sucrose metabolized. Striking confirmation of this scheme has been provided by the production of transgenic potato plants in which a bacterial inorganic pyrophosphatase was introduced into the cytosol of the tubers. The consequent disturbance of carbohydrate metabolism is consistent with the need for PPi for sucrose breakdown (Sonnewald, 1992). We initially suggested that this PPi was produced by pyrophosphate: Fru6P 1-phosphotransferase [PFK(PPi)]:

$$Fru1,6P_2 + Pi \rightleftharpoons PPi + Fru6P.$$

Subsequent studies require this view to be modified. First, it is now clear that PFK(PPi) is not specifically associated with sucrose breakdown by SS. In the endosperm of the *shrunken* mutant of maize, sucrose breakdown is largely via alkaline invertase as opposed to SS in the wild-type. Wild-type and mutant do not differ in respect of their maximum catalytic activities of PFK(PPi). In addition, appreciable activities of PFK(PPi) have been demonstrated in red algae, which lack sucrose (Dancer and ap Rees, 1989). Secondly, transgenic plants with little or no PFK(PPi) do not seem impaired. Thus PFK(PPi) is almost certainly not the only source of PPi used in sucrose breakdown. The most likely additional source is the PPi released during the synthesis of proteins and nucleic acids. However, the fact that a transgenic plant may be able to do without PFK(PPi), does not show that PFK(PPi) does not function as a source of PPi for sucrose breakdown in the normal plant. The extensive randomization between carbons 1 and 6 of the hexosyl units of sucrose and starch in tissues fed specifically labelled glucose is clear evidence of recycling between $Fru1,6P_2$ and Fru6P. In non-gluconeogenic tissues such recycling is almost certainly a reflection of the activity of PFK(PPi) (Hatzfeld and Stitt, 1990). PFK(PPi) remains a strong candidate for the formation of PPi required for sucrose breakdown, but it is clearly not the only source and possibly not even a universally essential source.

Measurements of the maximum catalytic activities of alkaline invertase and SS provide compelling evidence that the relative activities of the two pathways of sucrose breakdown vary widely with the tissue (ap Rees, 1988). Comparisons of these activities with estimates of the rates of sucrose breakdown reveals a close correlation between SS and sucrose breakdown and suggests that this is the main pathway. Certainly it must be so in tissues such as the developing tubers of potato, where the activity of alkaline invertase is too low to make a quantitatively significant contribution to sucrose breakdown. The activity of alkaline invertase varies widely and in some tissues, such as the spadix of *Arum maculatum*, its activity is high enough to mediate the whole of sucrose breakdown. Where we have data, tissues with high alkaline invertase also have sufficient SS to mediate sucrose breakdown. In such instances it is difficult to assess the relative activities of the two pathways. As discussed earlier, the affinities of alkaline invertase and SS for sucrose have not been shown to differ significantly in plants as a whole. Thus it seems likely that the relative activities of the two pathways will be determined primarily by the relative maximum catalytic activities of alkaline invertase and SS.

The significance of there being two pathways of sucrose breakdown is not known. Suggestions that one pathway operates at low concentrations of sucrose, and the other at high concentrations are not really borne out by the available K_m values for sucrose of SS and alkaline invertase. As yet it has not proved possible to correlate either pathway with any particular aspect of sink metabolism (ap Rees, 1988). The co-existence of two pathways does offer another piece of striking evidence of the plasticity of plant metabolism where alternative routes exist for essentially the same conversion, that is, sucrose to hexose monophosphates. The significance of such alternatives may be the increased flexibility. That one pathway can substitute for the other is indicated by two recent studies. In the *shrunken* mutant of maize, SS activity remains low but there is a marked compensatory increase in the maximum catalytic activity of alkaline invertase (Dancer and ap Rees, 1989). Conversely, there is good evidence that SS can substitute for alkaline invertase in mutants of maize that lack invertase in their primary roots (Duke *et al.*, 1991).

6.3 Entry of carbon for starch synthesis into the amyloplast

The conclusive argument that ADPGlc is the precursor of storage starch raises the question of how it is synthesized. The simplest hypothesis is that it is made in the plastid by ADPGlc pyrophosphorylase. The latter is known to be confined to the plastid and there is now unequivocal proof that this enzyme is essential for the synthesis of storage starch. Müller-Röber *et al.* (1992) transformed potatoes with an antisense gene for the β-subunit of ADPGlc pyrophosphorylase. The levels of mRNA, and protein as revealed by immunoblotting, in the transformed tubers were below the limit of detection. The measured activity of ADPGlc pyrophosphorylase in the transformed tubers was less than 2% of that in the untransformed tubers. There can be no question from this evidence that the activity of the enzyme had been almost completely eliminated. This elimination resulted in the cessation of starch synthesis in the tuber. The expression of other genes involved in starch synthesis was not affected. The genetic evidence from studies of maize, pea and *Arabidopsis* (mentioned by Müller-Röber *et al.*, 1992) suggests that the essential requirement of ADPGlc pyrophosphorylase for starch synthesis is general.

The question of how starch synthesis is supplied with carbon now resolves itself into how does carbon get from hexose monophosphates in the cytosol to Glc1P in the plastid. The initial question was whether it was necessary to convert the products of sucrose breakdown to triose phosphate in the cytosol to allow the carbon for starch synthesis to enter via the phosphate translocator. The evidence against the general operation of this C-3 pathway is now decisive. As it has previously been reviewed (ap Rees and Entwistle, 1989) it need only be summarized. First, of a wide range of amyloplast-containing tissues examined, only those from the developing embryos of peas contained any detectable fructose-1,6-bisphosphatase. As PFK(PPi) is confined to the cytosol this lack of fructose-1,6-bisphosphatase leaves the amyloplast with no known route from triose phosphate to hexose monophosphate. Secondly, the C-3 pathway would involve extensive exchange between carbons 1 and 6 of the cytosolic hexose monophosphates when they were converted to triose phosphates, which would be equilibrated by triose phosphate isomerase in the cytosol and the plastid. Such extensive equilibration does not occur. Finally, there is no conclusive evidence that amyloplasts can convert triose phosphates into starch. There is evidence that amyloplasts can convert hexose monophosphates into starch. At present, the only proven exception to these genealizations is the rather special case of etioplasts. Here there is clear evidence that, in etiolated barley leaves, triose phosphate enters the plastid and provides carbon for starch synthesis (Batz *et al.*, 1992).

The key question at the moment is which hexose monophosphate enters the amyloplast to support starch synthesis. In the following discussion bear in mind that our techniques for isolating intact functional amyloplasts are still rudimentary. Thus negative results, that is, instances where amyloplasts failed to incorporate a given substrate into starch, must be viewed with the utmost caution. The preparative procedures may easily have damaged the transport system in question.

Amyloplasts from protoplasts from wheat endosperm incorporated label from [^{14}C]Glc1P into material that was insoluble in methanol–KCl (Tyson and ap Rees, 1988). Treatment of this insoluble material with amyloglucosidase released the label as [^{14}C]glucose. The incorporation was dependent upon the intactness of the amyloplasts and was stimulated by ATP. The labelling of starch was not due to prior conversion of G1P to hexose-6-phosphates outside the amyloplasts: neither of the hexose-6-phosphates labelled starch in these preparations.

The next advance was made by Denyer, Hill and Smith. First, Denyer and Smith (1988) developed a rapid mechanical method for isolating amyloplasts from the developing cotyledons of peas. This method gives good yields of uncontaminated and intact plastids. Above all it does so without the lengthy and potentially damaging procedures associated with the use of protoplasts. Hill and Smith (1991) have shown that amyloplasts from pea embryos will incorporate Glc6P into starch when ATP is provided. The incorporation depended upon the intactness of the plastids and occurred at a rate and at a concentration of Glc6P that were close to those found in the intact embryos. The incorporation was specific for Glc6P: neither Glc1P nor Fru6P, with or without ATP, were incorporated at a significant rate. There can be little doubt that in this tissue Glc6P is the form in which carbon for starch synthesis enters the amyloplast.

There is some additional data on the relative importance of Glc1P and Glc6P as transport compounds for starch synthesis, but not enough to resolve the question. The observation that there is little starch in root plastids which lack phosphoglucomutase favours Glc6P as the transported compound (Kiss *et al.*, 1989). If Glc1P could support starch synthesis then an ability to interconvert Glc6P and Glc1P should not reduce starch synthesis unless the absorbed Glc1P also served as the source of ATP via plastidic glycolysis. Amyloplasts from wheat endosperm possess such a glycolytic capacity (Entwistle and ap Rees, 1988). Further support for Glc6P as the transport compound is provided by the demonstration that Glc6P can be taken up and metabolized by plastids from pea roots (Emes and Bowsher, 1991). These observations are balanced by two in favour of Glc1P. Coates (1990) showed that amyloplasts from soybean protoplasts incorporated labelled Glc1P but not Glc6P into starch. ATP did not increase incorporation from Glc6P. Finally, Tetlow *et al.* (1992) have provided evidence that amyloplasts isolated mechanically from wheat endosperm will incorporate Glc1P into starch in the presence of ATP. This suggests that the original demonstration of incorporation of Glc1P into starch by wheat amyloplasts is a true reflection of their *in vivo* activity and not an artefact arising from the use of protoplasts. The possibility exists that different plastids take up different hexose monophosphates to support starch synthesis. Further work on a wider range of species should answer this question.

The most recent development in the study of what enters the amyloplast is a series of substantial papers from Akazawa's laboratory that argue against central features of the hypotheses that I have just championed (Akazawa *et al.*, 1991; Pozueta-Romero *et al.*, 1991a,b). The essence of Akazawa's hypothesis is that sucrose is metabolized to ADPGlc in the cytosol by an ADP-specific SS, and that this ADPGlc then enters the plastid via an adenylate transporter and is promptly used by starch synthase. The evidence for this hypothesis is as follows:

(a) There is an extra-plastidic SS specific for ADP (Pozueta-Romero *et al.*, 1991b). As I have said already this claim cannot be assessed, and the evidence that is available suggests that SS will form UDPGlc, not ADPGlc, in the cytosol.

(b) Both amyloplasts (Pozueta-Romero *et al.*, 1991a) and chloroplasts (Pozueta-Romero *et al.*, 1991b) contain an adenylate transporter that can transport ADPGlc as well as ATP, ADP and AMP. This point is substantiated but does not prove that the transporter catalyses the movement of ADPGlc into the plastid in *vivo*. As Akazawa *et al.* (1991) point out, the mitochondrial adenylate transporter will also transport ADPGlc. Thus the latter property may be an inherent inability of the adenylate transporter to distinguish between ATP, ADP and ADPGlc.

(c) Uptake of hexose monophosphates by sycamore amyloplasts is marginally small (Pozueta-Romero *et al.*, 1991a). I have already commented on the frailty of such negative evidence.

(d) 'Glucose 1-phosphate is not in equilibrium with ADPglucose, eliminating the predominant role of ADPglucose in sucrose–starch transformation' (Pozueta-Romero *et al.*, 1991a). I am not aware of any authenticated measurements of ADPGlc, ATP, Glc1P and PP_i in amyloplasts that prove this point. The evidence that PPi is largely confined to the cytosol and that a very active pyrophosphatase is confined to the amyloplast (ap Rees, 1988) make it very likely that the instant removal of PPi keeps the reaction catalysed by ADPGlc pyrophosphorylase very far from equilibrium *in vivo*.

(e) The Gibbs Effect, the assymetric labelling of the two halves of the hexosyl units of starch during short-term photosynthesis in $^{14}CO_2$, is more easily explained by arguing that chloroplast starch is made from ADPGlc generated in the cytosol from sucrose (Pozueta-Romero *et al.*, 1991b). This effect may also be explained by a lag in isotopic equilibration between the two triose phosphates, or a back reaction of transketolase.

(f) The activities of key enzymes of the classical pathway are marginally small (Frehner *et al.*, 1990; Pozueta-Romero *et al.*, 1991a). Even for the sycamore cells used by Akazawa and his colleagues, this argument is not convincing if we accept that it is a C-6, not a C-3, compound that enters the amyloplast. No data are presented for ADPGlc pyrophosphorylase. The activities of phosophoglucomutase and phosphohexoisomerase in the amyloplasts were not shown to be incapable of supporting the observed rate of starch synthesis. In soybean and wheat amyloplasts the activities of all three enzymes have been shown to be appreciable (ap Rees and Entwistle, 1989).

I think that, even in respect of the cultures of sycamore cells used in Akazawa's studies, the evidence that starch is made from cytosolically synthesized ADPGlc is not conclusive. By the same token, for these cultures we do not have positive evidence that the hypothesis is wrong. However, when the evidence from plants as a whole is considered the hypothesis fails because it does not explain the absolute dependence of starch synthesis on the activity of plastidically located ADPGlc pyrophosphorylase that I have already discussed. Suggestions (Akazawa *et al.*, 1991) that the importance of ADPGlc pyrophosphorylase lies in a role in starch breakdown in that it converts Glc1P, formed during starch breakdown, to ADPGlc for export from amyloplast to cytosol are not convincing. There is overwhelming evidence from developmental studies (ap Rees *et al.*, 1984), genetic studies, and work with transgenic plants, that ADPGlc pyrophosphorylase is responsible for the synthesis, not the breakdown, of starch.

6.4 Conversion of ADPGlc to starch

The formation of starch grains from ADPGlc is at least, if not more, complex than the conversion of sucrose to ADPGlc. From a single precursor the amyloplast synthesizes two complex polymers in a precise ratio and builds them into a characteristically shaped granule. One polymer, amylose, is essentially linear; the other, amylopectin, is highly branched, yet both are made in the same amyloplast. The usual ratio of amylose to amylopectin is 1:3 by weight, but there is considerable variation in this ratio, in the molecular size of the polymers, and in the degree of branching of the amylopectin. The $\alpha(1 \to 4)$ bonds are made by starch synthase which catalyses the transfer of the glucosyl group from ADPGlc to the non-reducing end of pre-existing glucose chains. The $\alpha(1 \to 6)$ linkages are formed by the branching enzyme which breaks an $\alpha(1 \to 4)$ chain and transfers a portion of it on to another $\alpha(1 \to 4)$ chain via an $\alpha(1 \to 6)$ linkage. Both starch synthase and branching enzyme occur as isoforms. The function and the precise relationships between these isoforms is not yet established (Preiss, 1991).

Starch synthase may be fractionated into soluble and granule-bound forms. The former remain in the supernatant when lysed amyloplasts are centrifuged and are presumed to be located in the amyloplast stroma. The latter sediment with the starch granules and are not removed by repeated washing. The granule-bound starch synthases can be removed by treating homogenates of starch granules with amylase and glucoamylase. There is convincing evidence from a range of plants that there are two soluble starch synthases that differ in their molecular weight and, to some extent, their kinetic properties (Preiss, 1991). For example, the two soluble starch synthases from pea embryos have molecular weights of 60 kDa and 77 kDa; in the presence of citrate their K_m for ADPGlc is 0.29 mM and 0.48 mM, respectively (Denyer and Smith, 1992).

There are also two forms of granule-bound starch synthase. Evidence that one of these is the product of the *waxy* gene, a mutation recognized in maize, is now considerable (Dry *et al.*, 1992; Preiss, 1991). A granule-bound protein of 59–60 kDa has been demonstrated. In maize endosperm there is a close relationship between the amount of this protein, granule-bound starch synthase activity, and the dosage of the *waxy* allele The derived sequence of the protein from maize (Preiss, 1991), and pea embryos (Dry *et al.*, 1992) is similar to that of bacterial glycogen synthase. *Waxy* phenotypes show a parallel decrease in amylose content and granule-bound starch synthase activity. The basic similarity of the data from maize and peas and the occurrence of 'waxy' mutants in other species makes it highly likely that one of the granule-bound starch synthases is the product of the *waxy* gene and plays a key role in the synthesis of amylose in storage tissue. This protein does not appear to be needed for amylopectin synthesis as the total amounts of starch made in *waxy* mutations are comparable to those found in wild-type. The second granule-bound starch synthase from pea embryos has a molecular weight of 77 kDa. The major difference between this protein and the 59 kDa granule-bound starch synthase in peas is that the former has a very hydrophilic serine-rich domain of 203 amino acids at the *N*-terminus. In peas the activity of the larger enzyme is sufficient to account for the rate of starch synthesis observed in the developing embryo. The extent to which the granule-bound and soluble enzymes are related is not established nor do we know what each enzyme does. What is clear is that the different forms are expressed at different stages in development and it is likely that this contributes to the regulation of starch granule formation.

The situation in respect of branching enzyme is only slightly less complex than that found for starch synthase. Three forms of branching enzyme have been reported for maize endosperm, although two of them may be encoded by the same gene (Preiss, 1991). Two forms of branching enzyme, encoded by different genes, are present in pea embryos (Smith, 1988). The latter are expressed at different stages of development and it is the specific lack of one of them that is responsible for the reduced starch synthesis that produces the 'wrinkled' phenotype used by Mendel (Edwards *et al.*, 1988; Smith, 1988).

6.5 Regulation of starch synthesis

We now have sufficient knowledge of the pathway of starch synthesis to be able to study its control. Progress to date has been meagre but the promise for the future is excellent. Present information about control of storage starch synthesis is largely limited to studies of isolated enzymes. No clear picture of how the pathway is controlled has emerged from these studies because, on the whole, the enzymes studied have not consistently shown the properties of enzymes that might be expected to play a major role in control. ADPGlc pyrophosphorylase from maize endosperm has been shown to be inhibited by P_i and activated by 3-PGA in a manner comparable to that found for the enzyme from chloroplasts (Plaxton and Preiss, 1987). The significance of this property of the maize enzyme is difficult to assess if one accepts that carbon enters the amyloplast as hexose monophosphate. The extent to which ADPGlc pyrophosphorylase from all non-photosynthetic tissues shows these regulatory properties is not known. Studies of the enzyme from pea embryos suggest that not all ADPGlc pyrophosphoryl-ases behave as the one from maize embryo (Hylton and Smith, 1992). A vital aspect of the work on the maize enzyme is the demonstration that unless great care is taken to inhibit proteases during the purification of ADPGlc pyrophos-phorylase then a 1 kDa fragment is removed from the native enzyme which greatly decreases the enzyme's sensitivity to its allosteric effectors. This empha-sizes how careful we must be in deciding that an enzyme has no allosteric regula-tors from work *in vitro*. However, as it stands there is little evidence for such effectors for the enzymes responsible for the conversion of sucrose to starch. In assessing the available data it must be borne in mind that the concentration at which putative effectors act must be comparable to that found *in vivo*. The pos-sibility that the activities of some of the enzymes involved is regulated by protein phosphorylation also needs to be considered as does coarse control of enzyme activity. Good evidence of coarse control of SS by sucrose is available. The activity of the enzyme (Claussen *et al.*, 1986) and the steady state level of its mRNA (Salanoubat and Belliard, 1989) have been shown to vary with the sucrose concentration.

A more profitable approach to the study of the control of storage starch synthesis would be to identify which steps make significant contributions to con-trol and then make a careful study of the relevant enzymes. The role of the indi-vidual reactions in overall control of the pathway may be determined by comparing the fractional change in flux with the fractional change in capacity of the individual enzymes when the latter is varied. The wide range of available mutants and the advent of transgenic plants are capable of producing the plants required for this type of quantitative study of control. The feasibility of this approach has already been demonstrated in respect of starch synthesis in leaves by Smith *et al.* (1990).

A point of perhaps even greater importance than the quantitative study of control is that we will be able to use transgenic plants and mutants to assess the significance of storage starch synthesis in the physiology of the plant as a whole. An example of this approach is provided by the recent exciting paper of Müller-Röber *et al.* (1992). Their work shows that removal of the ability to synthesize starch by a single gene change led to potato plants that still produced tubers but produced large numbers of small tubers full of sugar rather than the usual crop of normal sized tubers containing starch. This demonstrates that the developmental control of tuber formation is not directly linked to the synthesis of starch, the main component of the tubers.

References

Akazawa, T., Pozueta-Romero, J. and Fernando, A. (1991) New aspects of starch biosynthesis in the plant cell. In: *Molecular Approaches to Compartmentation and Metabolic Regulation* (eds A.H.C. Huang and L. Taiz). American Society of Plant Physiologists, Rockville, pp. 74–85.

ap Rees, T. (1988) Hexose phosphate metabolism by non-photosynthetic tissues of higher plants. In: *The Biochemistry of Plants*, Volume 14 (ed. J. Preiss). Academic Press Inc., New York, pp. 1–33.

ap Rees, T. and Entwistle, G. (1989) Entry into the amyloplasts of carbon for starch synthesis. In: *Physiology, Biochemistry and Genetics of Nongreen Plastids* (eds C.T. Boyer, J.C. Shannon and R.C. Hardison). American Society of Plant Physiologists, Rockville, pp. 49–62.

ap Rees, T., Leja, M., Macdonald, F.D. and Green, J.H. (1984) Nucleotide sugars and starch synthesis in spadix of *Arum maculatum* and suspension cultures of *Glycine max. Phytochemistry*, 23, 2463–2468.

ap Rees, T., Entwistle, G. and Dancer, J.E. (1991) Interconversion of C-6 and C-3 sugar phosphates in non-photosynthetic cells of plants. In: *Compartmentation of Plant Metabolism in Non-photosynthetic Tissues* (ed. M.J. Emes). Cambridge University Press, Cambridge, pp. 95–110.

Batz, O., Scheibe, R. and Neuhaus, H.E. (1992) Transport processes and corresponding changes in metabolite levels in relation to strach synthesis in isolated barley (*Hordeum vulgare* L.) etioplasts. *Plant Physiol.* in press.

Bush, D.R. and Li, Z.-C. (1991) Proton-coupled sucrose and amino acid transport across the plant plasma membrane. In: *Recent Advances in Phloem Transport and Assimilate Compartmentation* (eds J.L. Bonnemain, S. Delrot, W.J. Lucas and J. Dainty). Ouest Editions, Nantes, pp. 148–153.

Claussen, W., Loveys, B.R. and Hawker, J.S. (1986) Influence of sucrose and hormones on the activity of sucrose synthase and invertase in detached leaves and leaf sections of egg plants (*Solanum melongena*). *J. Plant Physiol.* 124, 345–357.

Coates, S.A. (1990) *The metabolism of hexose phosphates by soybean leucoplasts.* Ph.D. thesis, University of Cambridge, UK.

Copeland, M. and Morell, M. (1985) Hexose kinases from the plant cytosolic fraction of soybean nodules. *Plant Physiol.* 79, 114–117.

Dancer, J.E. and ap Rees, T. (1989) Relationship between pyrophosphate:fructose 6-phosphate 1-phosphotransferase, sucrose breakdown, and respiration. *J. Plant Physiol.* 135, 197–206.

Denyer, K. and Smith, A.M. (1988) The capacity of plastids from developing pea cotyledons to synthesize acetyl CoA. *Planta*, 173, 172–182.

Denyer, K. and Smith, A.M. (1992) The purification and characterisation of the two forms of soluble starch synthase from developing pea embryos. *Planta*, 186, 609–617.

Dick, P.S. and ap Rees, T. (1975) The pathway of sugar transport in roots of *Pisum sativum. J. Exp. Bot.* 26, 305–314.

Doehlert, D.C. (1989) Separation and characterization of four hexose kinases from developing maize kernels. *Plant Physiol.* 89, 1042–1048.

Doehlert, D.C. and Chourey, P.S. (1991) Possible roles of sucrose synthase in sink function. In: *Recent Advances in Phloem Transport and Assimilate Compartmentation* (eds J.L. Bonnemain, S. Delrot, W.J. Lucas and J. Dainty). Ouest Editions, Nantes, pp. 187–195.

Dry, I., Smith, A., Edwards, A., Bhattacharyya, M., Dunn, P. and Martin, C. (1992) Characterization of cDNAs encoding two isoforms of granule-bound starch synthase which show differential expression in developing storage organs of pea and potato. *Plant J.* 2, 193–202.

Duke, E.R., McCarty, D.R. and Koch, K.E. (1991) Organ-specific invertase deficiency in the primary root of an inbred maize line. *Plant Physiol.* 97, 523–527.

Echeverria, E., Boyer, C.D., Thomas, P.A., Liu, K.-C. and Shannon, J.C. (1988) Enzyme activities associated with maize kernel amyloplasts. *Plant Physiol.* 86, 786–792.

Echt, C.S. and Chourey, P.S. (1985) A comparison of two sucrose synthetase isozymes from normal and *shrunken*-1 maize. *Plant Physiol.* 79, 530–536.

Edwards, J., Green, J.H. and ap Rees, T. (1988) Activity of branching enzyme as a cardinal feature of the Ra locus in *Pisum sativum*. *Phytochemistry* 27, 1615–1620.

Emes, M.J. and Bowsher, C.G. (1991) Integration and compartmentation of carbon and nitrogen metabolism in roots. In: *Compartmentation of Plant Metabolism in Non-Photosynthetic Tissues* (ed. M.J. Emes). Cambridge University Press, Cambridge, pp. 147–165.

Entwistle, G. and ap Rees, T. (1988) Enzymic capacities of amyloplasts from wheat (*Triticum aestivum*) endosperm. *Biochem. J.* 295, 391–396.

Frehner, M., Pozueta-Romero, J. and Akazawa, T. (1990) Enzyme sets of glycolysis, gluconeogenesis, and oxidative pentose phosphate pathway are not complete in non-green highly purified amyloplasts of sycamore (*Acer pseudoplatanus* L.) cell suspension cultures. *Plant Physiol.* 94, 538–544.

Gallet, O., Lemoine, R., Gaillard, C., Larsson, C. and Delrot, S. (1992) Selective inhibition of active uptake of sucrose into plasma membrane vesicles by polyclonal sera directed against a 42 kilodalton plasma membrane polypeptide. *Plant Physiol.* 98, 17–23.

Gamalei, Y. (1991) Phloem loading and its development related to plant evolution from trees to herbs. *Trees* 5, 50–64.

Hatzfeld, W.-D. and Stitt, M. (1990) A study of the rate of recycling of triose phosphates in heterotrophic *Chenopodium rubrum* cells, potato tubers, and maize endosperm. *Planta*, 180, 198–204.

Heinlein, M. and Starlinger, P. (1989) Tissue- and cell-specific expression of the two sucrose synthase isoenzymes in developing maize kernels. *Mol. Gen. Genet.* 215, 441–446.

Hill, L.M. and Smith, A.M. (1991) Evidence that glucose-6-phosphate is imported as the substrate for starch synthesis by the plastids of developing pea embryos. *Planta*, 185, 91–96.

Ho, L.C., Lecharny, A. and Willenbrink, J. (1991) Sucrose cleavage in relation to import and metabolism of sugars in sink organs. In: *Recent Advances in Phloem Transport and Assimilate Compartmentation* (eds J.L. Bonnemain, S. Delrot, W.J. Lucas and J. Dainty). Ouest Editions, Nantes, pp. 178–186.

Huber, S.C. and Akazawa, T. (1986) A novel sucrose synthase pathway for sucrose degradation in cultured sycamore cells. *Plant Physiol.* 81, 1008–1013.

Hylton, C.M. and Smith, A.M. (1992) The Rb mutation of peas causes structural and regulatory changes in ADPglucose pyrophosphorylase from developing embryos. *Plant Physiol.* in press.

Ivanov, A.A., Semenov, I.L. and Timonina, V.N. (1991) Isolation and some characteristics of plastids from the storage root of sugar beet. *Fiziol. Rast.* 38, 864–873.

Kiss, J.Z., Hertel, R. and Sack, F.D. (1989) Amyloplasts are necessary for full gravitropic sensitivity in roots of *Arabidopsis thalliana. Planta,* 177, 198–206.

Masuda, H., Takakashi, T. and Sugawara, S. (1988) Acid and alkaline invertases in suspension cultures of sugar beet cells. *Plant Physiol.* 86, 312–317.

Miller, M.E. and Chourey, P.S. (1992) The maize invertase-deficient *miniature*-1 seed mutation is associated with aberrant pedical and endosperm development. *Plant Cell,* 4, 297–305.

Morell, M. and Copeland, L. (1984) Enzymes of sucrose breakdown in soybean nodules. *Plant Physiol.* 74, 1030–1034.

Morell, M. and Copeland, L. (1985) Sucrose synthase of soybean nodules. *Plant Physiol.* 78, 149–154.

Murata, T., Sugiyama, T., Minamikawa, T. and Akazawa, T. (1966) Enzymic mechanisms of starch synthesis in ripening rice grains. *Arch. Biochem. Biophys.* 113, 34–44.

Müller-Röber, B., Sonnewald, U. and Willmitzer, L. (1992) Inhibition of the ADP-glucose pyrophosphorylase in transgenic potatoes leads to sugar-storing tubers and influences tuber formation and expression of tuber storage protein genes. *EMBO J.* 11, 1229–1238.

Patrick, J.W. (1991) Control of phloem transport to and short distance transfer in sink regions: an overview. In: *Recent Advances in Phloem Transport and Assimilate Compartmentation* (eds J.L. Bonnemain, S. Delrot, W.J. Lucas and J. Dainty). Ouest Edition, Nantes, pp. 167–177.

Plaxton, W.C. and Preiss, J. (1987) Purification and properties of nonproteolytic degraded ADPglucose pyrophosphorylase from maize endosperm. *Plant Physiol.* 83, 105–112.

Pontis, H. (1977) Riddle of sucrose. In: *International Review of Biochemistry,* Volume 13 (ed. D.H. Northcote). University Park Press, Baltimore, pp. 79–117.

Pozueta-Romero, J., Frehner, M., Viale, A.M. and Akazawa, T. (1991a) Direct transport of ADPglucose by an adenylate translocator is linked to starch biosynthesis in amyloplasts. *Proc. Natl Acad. Sci. USA,* 88, 5769–5773.

Pozueta-Romero, J., Ardila, E. and Akazawa, T. (1991b) ADPglucose transport by the chloroplast adenylate translocator is linked to starch biosynthesis. *Plant Physiol.* 97, 1565–1572.

Preiss, J. (1991) Biology and molecular biology of starch synthesis and its regulation. *Oxford Surv. Plant Mol. Cell Biol.* 7, 59–114.

Rausch, T. (1991) The hexose transporters at the plasma membrane and the tonoplast of higher plants. *Physiol. Plant.* 82, 134–142.

Ricardo, C.P.P. (1974) Alkaline β-fructofuranosidases of tuberous roots: possible physiological functions. *Planta,* 118, 333–343.

Salanoubat, M. and Belliard, G. (1989) The steady state level of potato sucrose synthase mRNA is dependent upon wounding, anaerobiosis and sucrose concentration. *Gene* 84, 181–185.

Sauer, N., Friedlander, K. and Gräml-Wicke, U. (1990) Primary structure, genomic organization and heterologous expression of a glucose transporter from *Arabidopsis thalliana. EMBO J.* 9, 3045–3050.

Schnarrenberger, C. (1990) Characterization and compartmentation, in green leaves, of hexokinases with different specificities for glucose, fructose and mannose and for nucleoside triphosphates. *Planta*, 181, 249–255.

Semenov, I.L. and Ivanov, A.A. (1991) Metabolism of UDPG in the vacuolar fraction from the storage root of sugar beet. *Fiziol. Rast.* 38, 126–133.

Smith, A.M. (1988) Major differences in isoforms of starch-branching enzyme between developing embryos of round- and wrinkled-seeded peas (*Pisum sativum* L.). *Planta*, 175, 270–279.

Smith, A.M., Neuhaus, H.E. and Stitt, M. (1990) The impact of decreased activity of starch-branching enzyme on photosynthetic starch synthesis in leaves of wrinkled-seeded peas. *Planta*, 181, 310–315.

Sonnewald, U. (1992) Expression of *E. coli* inorganic pyrophosphatase in transgenic plants alters photoassimilate partitioning. *Plant J.* 2, 571–581.

Steup, M., Yang, Y., Greve, B. and Weiler, E.W. (1991) Characterization of a heteroglycan from higher plants. In: *Recent Advances in Phloem Transport and Assimilate Compartmentation* (eds J.L. Bonnemain, S. Delrot, W.J. Lucas and J. Dainty). Ouest Editions, Nantes, pp. 224–232.

Sturm, A. and Chrispeels, M.J. (1990) cDNA cloning of carrot extracellular β-fructosidase and its expression in response to wounding and bacterial infection. *Plant Cell*, 2, 1107–1119.

Tetlow, I.J., Blisset, K.J. and Emes, M.J. (1992) Starch synthesis in amyloplasts isolated from developing wheat endosperm. *J. Exp. Bot.* 43 (Suppl), 69.

Tyson, R.H. and ap Rees, T. (1988) Starch synthesis by isolated amyloplasts from wheat endosperm. *Planta*, 175, 33–38.

Tyson, R.H. and ap Rees, T. (1989) Failure to detect UDPglucose phosphorylase in the developing endosperm of maize and wheat. *Plant Sci.* 59, 71–76.

Ziegler, H. (1975) Nature of transported substances. *Encycl. Plant Physiol.* 1, 59–136.

Carbon metabolism in the legume nodule

A.J. Gordon

7.1 Introduction

The legume root nodule represents a symbiotic association between a colony of bacteria ([Brady] rhizobia) and the plant. The interaction between bacterial cells and the plant root appears to be triggered or stimulated by compounds exuded and perceived by both symbionts (e.g. de Bruijn and Downie, 1991; Long, 1990; Philips *et al.*, 1990). The bacteria invade the root enclosed within an infection thread, meristematic growth is induced in the root and new cells are infected with bacteria. The resulting structure is known as a root nodule. Within the infected cells the bacteria differentiate into bacteroids and are enclosed within a host plant membrane — the peribacteroid membrane. In terms of agricultural significance, the essential feature of this putative symbiotic arrangement is that the bacteroids assimilate atmospheric dinitrogen in return for supplies of nutrients from the plant. The plant benefits because the fixed (reduced) nitrogen can be transported from the nodule and then utilized in plant biomass production. The bacteria ultimately benefit from the supply of nutrients resulting in the proliferation of bacterial cells which are eventually released as the nodule senesces and decays.

Energetically, N_2 fixation is an extremely expensive process requiring both reductant and ATP (at least $8e^- + 8H^+ + 16$ ATP per N_2) to which must be added the cost of formation of amino acids, amides and ureides (the export products) and the cost of building and maintaining the complex machinery which constitutes the nodule (Schubert, 1986; Schubert and Ryle, 1980). Nodule metabolism is entirely dependent on either direct or indirect photosynthetic products transported from the shoots. Since the principal compound transported in the phloem is sucrose, this paper will centre around the supply and metabolism of this compound, with particular reference to nodule structure and to the location of metabolic events within the functional nodule. Recent important reviews in this subject area include: Day and Copeland, 1991; Streeter, 1991; Vance and Heichel, 1991.

7.2 Nodule structure

7.2.1 *Determinate and indeterminate nodules*

I will focus mainly on the nodules of the grain legume, soybean, but occasionally will contrast this type with that of the temperate pasture legume, white clover. Soybean nodules can be described as determinate in that they lack a persistent meristem. The main nitrogenous export products are the ureides, allantoin and allantoic acid. In contrast, indeterminate nodules maintain a persistent meristem where new cells are continually infected with rhizobia. Indeterminate nodules (of, for example, white clover and pea) export mainly amides and amino acids. Based on light microscopic observations (Figure 7.1) nodules can be loosely divided into two regions: the peripheral cortex and the central infected region (Bergersen, 1982).

7.2.2 *The central region*

As the nodule grows following infection the infected cells are confined to the central region (an area surrounded by cortical cells — see below). As nodule

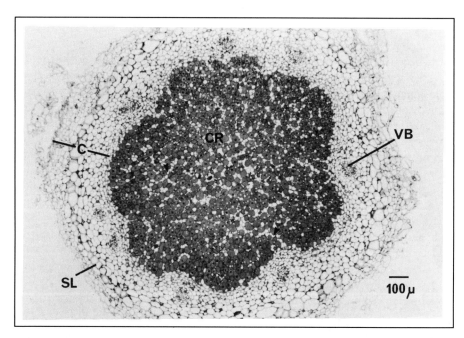

Figure 7.1. *Transverse section of a soybean root nodule showing the cortex (C) and central region (CR). Densely stained cells are infected cells. Lightly stained cells in the central region are uninfected interstitial cells — note that these cells are often in files (arrows) sometimes radiating from the cortex. VB, vascular bundle; SL, scleroid layer. Bar= 100 μm.*

tissue matures, the bacteria, enclosed within the peribacteroid membrane, differentiate both physically (in some cases enlarging) and biochemically (for example, developing the ability to fix N_2). Simultaneous changes in the host plant result in the synthesis of a range of nodule specific (or enhanced) proteins which have been termed 'nodulins' (Verma, 1989). The central region is mostly composed of infected cells which are distinguishable by their dense cytoplasm packed with symbiosomes, the peribacteroid membrane enclosing one (as in white clover) or many (e.g. soybean) bacteroids (Figure 7.2). Mitochondria and plastids are generally found at the periphery of these cells, often adjacent to intercellular airspaces.

Interspersed among the infected cells are a smaller number of uninfected interstitial cells which, in appropriately cut sections, appear as continuous files of cells which radiate from the cortex and possibly form a matrix of interconnected cells within the central zone (Gordon et al., 1992; Newcomb et al., 1979; Selker, 1988) (Figure 7.1). A large proportion of the surface area of uninfected cells probably interfaces with infected cells (Bergersen, 1982; Selker and Newcombe, 1985) and there may be active metabolic exchange through the numerous symplastic connections (Bergersen, 1982; Selker, 1988). In the central region of soybean nodules, Bergersen and Goodchild (1973) estimated that infected cells occupy approximately 80% of the volume.

7.2.3 The cortex

Surrounding the central region and occupying approximately 35% of the nodule volume is the nodule cortex. Although these spongy cells are not directly concerned with nitrogen fixation, the nodule cortex is the site of the oxygen diffusion barrier (Section 7.4.1) and also of the vascular bundles upon which the nodule depends for import of sucrose and export of nitrogenous products. In addition, some reports have suggested that the cortex may be the site of a number of glycolytic enzymes (Copeland et al., 1989b) and also may be involved in ureide synthesis (Newcomb et al., 1985). The cortex has a prominent endodermis with thickened cell walls — the scleroid layer — inside which (in soybean) a series of cell layers have been identified (Parsons and Day, 1990; James et al., 1991). In order, from the scleroid layer to the infected cell region these are: the mid-cortex, the boundary layer and the distribution zone (Figure 7.3, cf. James et al., 1991). The mid-cortex is composed of cells with intercellular air spaces containing a variable amount of a matrix glycoprotein which may be involved in regulating gaseous diffusion. The boundary layer is a band of tightly packed cells with no spaces between the adjoining cell walls. In the distribution zone, air spaces between cells are again evident. The hypothesis relating these structural features to the control of oxygen flux is described later (Section 7.4.1).

Also beneath the scleroid layer are the vascular bundles which, at the root junction, are in direct contact with the root vasculature. The vascular strands branch, surround the nodule, and end in closed loops (Walsh et al., 1989a,b). Vascular morphology has been described in detail elsewhere (Bergersen, 1982;

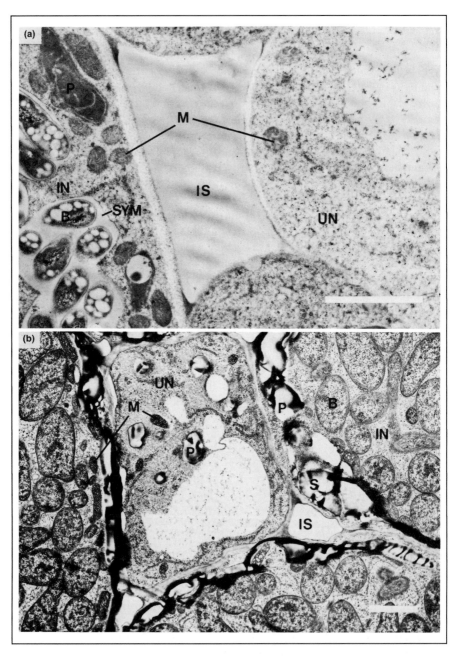

Figure 7.2. *Transmission electron micrographs of infected (IN) and uninfected cells (UN) of soybean (a) and white clover (b) nodules. SYM, symbiosome; B, bacteroid; IS, intercellular space; M, mitochondria; P, plastid; S, starch grain. Bar = 2.5 μm.*

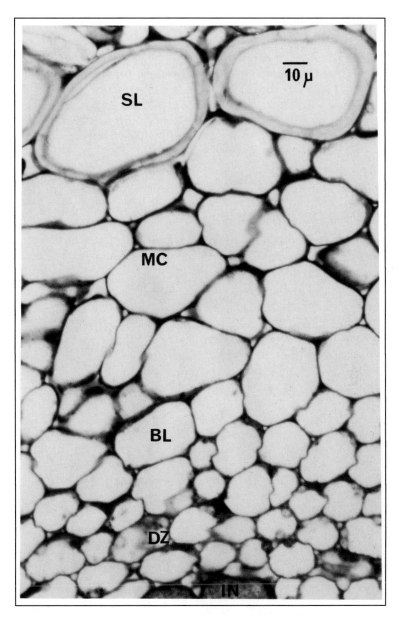

Figure 7.3. *Zones of the soybean nodule cortex (after James et al., 1991). SL, sclereid layer; MC, mid-cortex; BL, boundary layer; DZ, distribution zone; IN, infected cell. Bar= 10 μm.*

Pate *et al.*, 1969; Walsh *et al.*, 1989 a,b). Common features of nodule vascular bundles (Figure 7.4) are a bundle endodermis enclosing a series of pericycle cell layers which may (clover — Figure 7.5) or may not (soybean) have wall protuberances. Both types of pericycle cell are likely to be involved in the loading and unloading of materials either from the phloem sieve elements (SEs) or into the xylem vessels. Outside the bundle endodermis, towards the infected zone, are a number of layers of cortical cells which in some cases can merge with the uninfected cells of the central region (Figures 7.1 and 7.6). It seems likely that this arrangement of cells may constitute transport channels between the active central region of the nodule and the vascular bundles. However, this aspect is still being studied (Gordon *et al.*, 1992; Selker, 1988).

The supply of water to the nodules and the involvement of water in the export of N products has been the subject of recent debate (Raven *et al.*, 1989; Sprent, 1980; Walsh, 1990; Walsh *et al.*, 1989a) which has resulted in a number of proposals (Walsh, 1990). The main points are as follows.

(i) The xylem connections do not provide a continuous stream of water from the lower root, through the nodule and so to the plant shoot; rather the nodule xylem represents a cul-de-sac.

(ii) Water is supplied to the nodule, along with sucrose, via the phloem SEs.

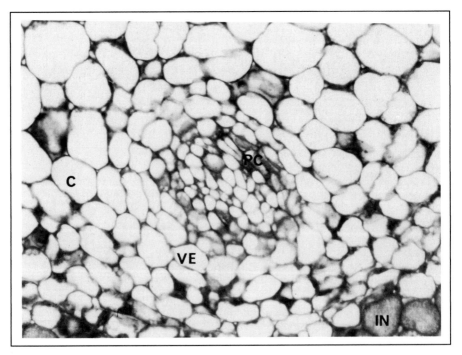

Figure 7.4. *Light micrograph of soybean nodule vascular bundle. VE, vascular endo-dermis; PC, pericycle cell; C, cortex; IN, infected cell. Bar = 10 μm.*

(iii) Export of nitrogenous products in the xylem is thought to be mediated by active unloading of solutes into the apoplast from the bundle pericycle cells. 'A positive xylem pressure is generated osmotically by solutes enclosed by the semi-permeable endodermis, driving water flow and exporting products from the nodule to the root' (Walsh *et al.*, 1989b).

(iv) Water is thought to move directly from the phloem to the xylem.

However, one aspect of vascular function which has not been addressed is that of sucrose unloading. From the arguments advanced by Walsh and co-workers (Walsh, 1990; Walsh *et al.*, 1989) it is assumed that sucrose unloaded from the phloem proceeds down a concentration gradient towards the infected cells. However, within the bundle endodermis, unless sucrose is moved symplastically into the pericycle cells and out into the nodule cortex, then any sucrose leaking into the apoplast would be re-exported in the flow of water (and nitrogenous products) via the xylem. This implies either that the pericycle cells have selective loading characteristics or that sucrose cannot enter the apoplast and thus the

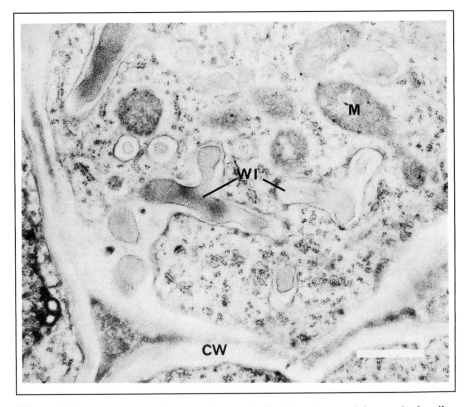

Figure 7.5. *Transmission electron micrograph of a white clover nodule vascular bundle pericycle cell showing wall in-growths (WI), mitochondria (M), and cell wall (CW). Bar = 1 μm.*

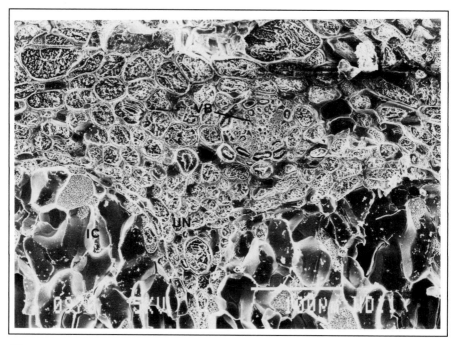

Figure 7.6. *Scanning electron micrograph of a freeze-fractured soybean nodule showing a vascular bundle (VB) and cortical/uninfected cells (UN) funnelling into the central region infected cells (IC). Bar = 100 μm.*

xylem vessels. No information is available on these specific points. However, there is evidence to suggest that selected compounds are exported from nodules (Gunning *et al.*, 1974; Pate *et al* 1969).

7.3 Supply of photosynthetic products

7.3.1 *Carbon flux into and out of nodules*

An obvious experimental approach in the study of carbon supply to nodules has been to use stable or radioactive isotopes of carbon, supplied as CO_2 to the leaves, in order to follow the short- and long-term fate of fixed carbon (e.g. Bach *et al.*, 1958; Gordon *et al.*, 1985, 1987; Kouchi and Yoneyama, 1984a,b; Kouchi *et al.*, 1986; Lawrie and Wheeler, 1975; Romanov *et al.*, 1985). Since legume nodules are organs which process raw materials and then export the products, it is not a simple matter to estimate how much photosynthate is directed to the nodules. The amount of tracer found in nodules at any instant can only represent the balance between a number of interacting processes — namely, import, respiration, re-export of nitrogenous or other compounds, incorporation into storage pools and growth and turnover of nodule components. None the less, these types of experiment, in addition to those designed to interrupt the flow of

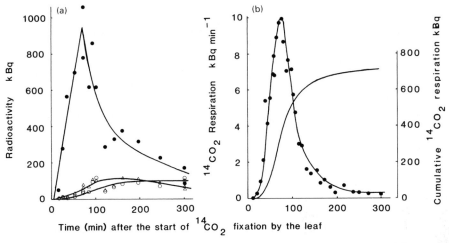

Figure 7.7. (a) Accumulation and depletion of ^{14}C in fractions of soybean nodules following $^{14}CO_2$ fixation for 10 min by the first trifoliate leaf (which was excised at 60 min). Ethanol soluble (80% sucrose) (●); starch (△); structural material (○). (b) Cumulative (continuous line) and rate (●) of $^{14}CO_2$ respiration from the nodulated root of a plant supplied with $^{14}CO_2$ as in (a). Reproduced from Gordon et al. (1985) with permission from the Society for Experimental Biology.

current photosynthate, confirm that nodule activity is dependent on export from the leaves and that sucrose is the primary compound appearing in the nodules (Gordon et al., 1985; Kouchi and Yoneyama, 1984b, Kouchi et al., 1985; Reibach and Streeter, 1983; Walsh et al., 1987). Overall it has been calculated that 40–50% of each day's photosynthetic products are processed by the nodules (Gordon et al., 1987) and that approximately 50% of this is lost as respired CO_2 (Gordon et al., 1987; Ryle et al., 1985, 1986). Pulse-labelling experiments have demonstrated that sucrose synthesized in the leaves is rapidly exported to and metabolized by nodules (Gordon et al., 1985; Kouchi et al., 1985). By terminating the supply of ^{14}C-labelled photosynthate, Gordon et al. (1985) also demonstrated that the pulse of [^{14}C] sucrose is rapidly flushed out of the nodule (Figure 7.7). The decline in the ^{14}C content of the previously labelled nodules was due both to rapid respiratory loss as $^{14}CO_2$ and to export of nitrogenous products. Approximately 60–70% of the loss from nodules was due to respiration while the remaining 30–40% was re-exported. Similar values were compiled by Walsh (1990) from a number of reports and estimates of the relative amounts of respiration and export can also be derived from various other sources of information. For example, in soybean nodules, Ryle et al. (1984) found that the respiratory cost of N_2 fixation was approximately 8.4 mol C per mol N_2. The main export compounds from these nodules are the ureides which have a C:N ratio of unity. Layzell and LaRue (1982), however, found that the xylem exudate had a C:N ratio of 1.6 (due to the additional presence of organic acids). Thus for each mole of N_2 fixed, 8.4 mol C were respired (72%) and 3.2 mol C used in the export products (28%). For white clover the proportions were calculated to be 65% respired versus 35% re-exported, assuming a C:N ratio in export products of 2.5 and respiratory costs of 9.1 mol C mol^{-1} N_2 (Ryle et al., 1986). Thus these estimates are close to those derived from direct experimental observation.

7.3.2 *Carbohydrate storage and diurnal supply for nodule metabolism*

In addition to the large amounts of current photosynthate which are used directly in nodule metabolism, a portion is also used to build up or maintain a carbohydrate reserve (Gordon et al., 1986, 1987; Kouchi et al., 1984b, 1986; Minchin and Pate, 1974; Rainbird et al., 1983; Walsh et al., 1987). Sucrose and starch are the principal storage carbohydrates in nodules (Gordon et al., 1985; Kouchi and Yoneyama, 1984b; Walsh et al., 1987) although small amounts of hexoses and cyclitols are also present (Kouchi and Yoneyama, 1984b; Gordon et al., 1985; Streeter, 1980). In soybean there is very little diurnal variation in nodule carbohydrate levels (Walsh et al., 1987), whilst in nodules of other species carbohydrate levels fluctuate during light and dark periods (e.g. pea: Minchin and Pate, 1974; cowpea: Rainbird et al., 1983; white clover: Gordon et al., 1986, 1987). In white clover, in particular, both sucrose and starch increased during the 12 h light period and declined during subsequent darkness. Sucrose accumulation and depletion was completely in phase with the light and dark periods whereas that of starch was out of phase by 3–4 h (Gordon et al., 1986, 1987). In contrast with soybean, where 80% of the total carbon stored during the light period was as starch in the leaves (Kerr et al., 1985), 50% of carbon accumulated by white clover plants was in organs other than leaves and contributed to continued growth, metabolism and nitrogen fixation during darkness (Gordon et al., 1987). It appears that in soybean nodules, which maintain a low but fairly constant level of reserve carbohydrate (approximately 40 mg g^{-1} dry weight, Walsh et al., 1987), continued metabolism is entirely dependent on export from leaves being maintained during darkness. White clover nodules, in contrast, accumulate both sucrose and starch during the light period and these would be sufficient to maintain N_2 fixation at temperature-adjusted daytime rates for 75% of the dark period (Gordon et al., 1987). Continued export of stored carbon from other parts of the plant ensures uninterrupted fixation during darkness. But as Ryle et al. (1988) have shown, reducing the light period by only a few hours (in plants acclimated to 12 h light periods) can cause precipitous declines in nodule metabolism towards the end of the normal dark period. This is probably because the plant was unable to accumulate sufficient reserves in either the nodules or the other organs. These results support the contention that reserve carbohydrate storage should be considered a necessary and carefully programmed component of day to day plant metabolism (Gordon, 1986) and not simply as excess photosynthate (Vance and Heichel, 1991). However, it is clear from the examples described above that different legume species have adopted different means to ensure continuous nitrogen fixation during the day and night. In this regard, there is increasing interest in the possibility that, in some species, poly-β-

hydroxybutyrate accumulated in bacteroids may serve as a source of reduced carbon during darkness (e.g. Bergersen *et al.*, 1991), although often there is an inverse relationship between poly-β-hydroxybutyrate content and nitrogenase activity (Romanov *et al.*, 1980).

7.4 Control of oxygen flux

Oxygen is essential to drive the respiratory processes of the nodule bacteroids in order to produce ATP and reductant for N_2 fixation; yet the enzyme nitrogenase is inactivated by oxygen. Oxygen flux into the nodule, therefore, must be carefully controlled to supply an adequate amount at a low concentration. This is achieved by the combination of a variable gaseous diffusion barrier and by the presence in the central region cells of the haemoglobin-like protein, leghaemoglobin (Lb).

7.4.1 *The gaseous diffusion barrier*

The gaseous diffusion barrier is located in the inner cortex, where, over a distance of approximately three cell layers, there is a remarkable drop in the concentration of free O_2 (Tjepkema and Yokum, 1974; Witty *et al.*, 1986). It is thought that the boundary layer of cortical cells, which have closely abutted cell walls and no air spaces, offers a permanent resistance to gaseous diffusion (since O_2 flux through water is approximately 10^4 times slower than through air) while the air spaces between the cells of the mid-cortex may be the site of the variable diffusion barrier (James *et al.*, 1991; Parsons and Day, 1990). Changes in the rate of diffusion may be brought about by alterations in the size of the intercellular air spaces either by osmotic adjustments (Hunt, *et al.*, 1990; Layzell *et al.*, 1990) and/or by a mechanism involving a matrix glycoprotein found in the intercellular spaces of the mid cortex (James *et al.*, 1991). The net result is that plants under stress can increase their resistance to O_2 diffusion into the nodule and so protect nitrogenase from inactivation (see Layzell *et al.*, 1990; Vance and Heichel, 1991; Witty and Minchin, 1990 for extensive reviews). The distribution of O_2 from the inner cortex and throughout the central region is thought to occur by diffusion through intercellular air spaces.

7.4.2 *The role of Lb*

Leghaemoglobin, the most abundant protein in legume nodules, is synthesized only in nodules (Appleby, 1984), and has a high affinity for O_2. Soybean Lbs a and c, for example, are 50% oxygenated at 37 and 51 nM O_2, respectively (Bergersen, 1982), which results in the interior of the nodule having a free O_2 concentration in the range 10–20 nM (King *et al.*, 1988). Lb therefore is thought to act as a buffer for O_2 and to facilitate the uniform diffusion of O_2 from the air spaces of the central region to the bacteroids while maintaining a low concentration of free O_2. An important adjunct to these host-derived mechanisms to maintain a low concentration of free O_2, is that the bacteroids produce a terminal oxidase with a high affinity for O_2 such that they can operate effectively at O_2 concentrations in the range 5–10 nM (Bergersen and Turner, 1975).

7.4.3 *Mitochondrial function in nodules*

Studies with mitochondria isolated from nodules suggest that, except for malate and succinate dehydrogenases, tricarboxylic acid (TCA) cycle enzymes are present only at low activities. Malate oxidation and oxaloacetate (OAA) efflux is rapid (Bryce and Day, 1990; Day and Mannix, 1988). These authors suggest that nodule mitochondria operate a 'truncated TCA cycle' and primarily oxidise malate to provide OAA and ATP for NH_3 assimilation (Bryce and Day, 1990). Additionally, Streeter (1991) has proposed that an important function of mitochondria may be to provide the 2-oxoglutarate required in ammonia assimilation. An important point is whether mitochondria could function at the low O_2 concentrations found in infected cells. Rawsthorne and LaRue (1986) found that cowpea nodule mitochondria had a K_m for O_2 of approximately 100 nM. This suggests that in the infected region of the nodule, mitochondrial respiration may be severely restricted. However, mitochondria in infected cells are commonly found close to the intercellular air spaces (where the O_2 concentration may be greater) and there has also been a suggestion that Lb may facilitate O_2 diffusion to mitochondria as well as to the symbiosomes (Suganuma *et al.*, 1987). Lb is located primarily in the cytosol of infected cells but also at a lower concentration in the cytosol of the uninfected interstitial cells (VandenBosch and Newcomb, 1988). Thus oxyLb may also aid mitochondrial function in both infected and uninfected cells.

A question which has not been addressed is whether the free O_2 concentration in the intercellular air spaces and in uninfected cells is significantly higher than that in the infected cells. A low O_2 concentration is maintained in infected cells by rapid bacteroid respiration, the presence of Lb and an appropriate resistance to gaseous diffusion. There must, however, be a concentration gradient from the intercellular spaces to the bacteroids located in the cytosol of the infected cells (cf. Sheehy *et al.*, 1984). In the uninfected cells, which have a much lower Lb concentration and, possibly, lower respiration rates, the free O_2 concentration may be considerably higher. In this situation mitochondria of uninfected cells may behave differently and be involved in different functions to those of infected cells. Any difference in mitochondrial performance from one cell type to another may have important consequences for the integration of metabolic function between different nodule cell types (see Section 7.5.5). One indication that the O_2 concentration in uninfected cells may be higher than that in the infected cells is that the enzyme uricase is located exclusively in peroxisomes of uninfected cells (VandenBosch and Newcomb, 1986). This enzyme has a K_m for O_2 of approximately 30 μM, i.e. nearly 1000 times higher than the estimated free O_2 concentration in infected cells (King *et al.*, 1988; Lucas *et al.*, 1983; Rainbird and Atkins, 1981).

7.5 Nodule metabolism

7.5.1 *Metabolism of sucrose to malate*

As discussed earlier, carbon for nodule metabolism is supplied as sucrose from the leaves. Until fairly recently it was considered that sucrose was hydrolysed by alkaline invertase (INV) although acid invertase is found in developing nodules (Robertson and Taylor, 1973). However, Morell and Copeland (1984, 1985) demonstrated that soybean nodules also develop substantial activity of sucrose synthase (SS). Anthon and Emerich (1990) showed that SS activity increased significantly coincident with the onset of nitrogen fixation and the appearance of Lb. Another important finding, from studies on the molecular biology of legume nodules, was that one of the nodule-specific (or nodule-enhanced) genes (nodulin 100, Thummler and Verma, 1987) encoded SS. These findings suggest that SS has an important role to play in nodule metabolism. Both alkaline INV and SS activities have been found in nodules from a range of legume species (white clover, mung bean, *Lotus corniculatus*, lupin: Gordon, unpublished data; Gordon and Kessler, 1990) and, as shown in Figure 7.8, both enzymes increase in specific activity as N_2 fixation begins in the amide exporting nodules of lupin (Gordon and Reynolds, unpublished data).

During rapid N_2 fixation in nodules either alkaline INV or SS may be involved in the hydrolysis of sucrose (see Figure 7.9). Both enzymes have low affinities for sucrose (10–30 mM), though SS from soybean and white clover has a high affinity for uridine diphosphate (UDP) (soybean — Morell and Copeland 1984, 1985; white clover — Harrison and Gordon, unpublished data). Neither enzyme appears to be subject to fine metabolic control although it has been suggested by Thummler and Verma (1987) that SS activity may be regulated by free haem in nodules. In addition, the level of mRNA coding for SS in nodules responds rapidly to changes in the availability of photosynthetic products (Gordon and Ougham, unpublished data) which may imply some measure of 'coarse control'. Flux through one route or the other (see Figure 7.9) may be determined more by the activities of other enzymes and the availability of other metabolites. Fructo-kinase activity in nodule cytosol is much higher than that of glucose kinase (Copeland and Morell, 1985, also see Figure 7.8) and there is considerable UDPGlc pyrophosphorylase (UDPGlcPP) activity (Copeland *et al.*, 1989b; Gordon, 1991; Figure 7.8). Thus there is sufficient enzymic potential to metabolize the products of the SS reaction, although phosphorylation of glucose may be limited (Copeland and Morell, 1985). Another constraint related to the hexose kinase reactions may be the availability of ATP, which may be in short supply in cells where mitochondrial activity is limited by the low free O_2 concentration (but see discussion in Section 7.4.3). Fructokinase is able to utilize the uridine triphosphate (UTP) produced in the UDPGlcPP reaction and so regenerate UDP for the SS reaction. This latter reaction itself may be limited by the availability of inorganic pyrophosphate (PPi).

However, although PPi levels in nodules have not been measured, results

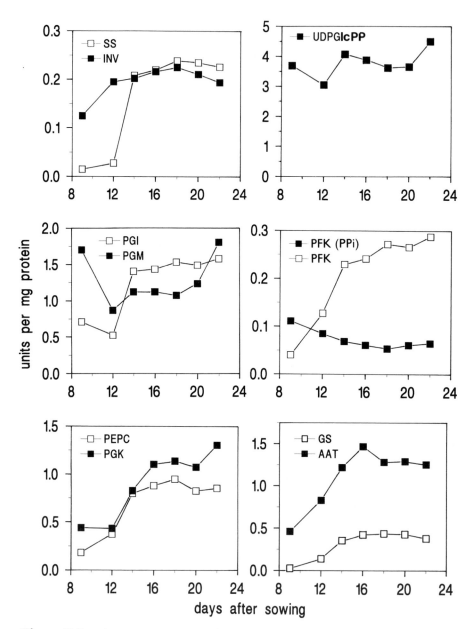

Figure 7.8. *Changes in enzyme specific activity (1 unit = 1 μmol min⁻¹) during the development of lupin nodules. Other enzymes with relatively constant activities were fructokinase (0.052 ± 0.003), hexokinase (0.0035 ± 0.0003) and aldolase (0.026 ± 0.004) (units mg⁻¹ protein). N₂ fixation began 12–14 days after sowing seeds.*

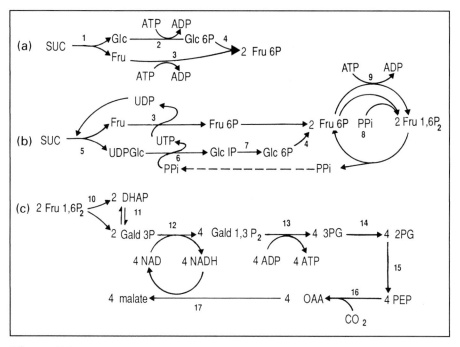

Figure 7.9. *Possible pathways in the metabolism of sucrose to malate in the host plant cells of soybean nodules. (a) The involvement of alkaline invertase in the conversion of sucrose to Fru6P. (b) The involvement of sucrose synthase in the conversion of sucrose to Fru1,6P₂. (c) Metabolism of Fru1,6P₂ to malate. Enzymes involved: 1. Alkaline invertase. 2. Hexokinase. 3. Fructokinase. 4. Phosphoglucose isomerase. 5. Sucrose synthase (SS). 6. UDP glucose pyrophosphorylase (UDPGlcPP). 7. Phosphoglucomutase. 8. Pyrophosphate: fructose-6-phosphate:phosphotransferase (PFK(PPi)). 9. Phosphofructokinase. 10. Aldolase. 11. Triose phosphate isomerase. 12. Glyceraldehyde 3-phosphate dehydrogenase. 13. Phosphoglycerate kinase. 14. Phosphoglycerate mutase. 15. Enolase. 16. Phosphoenol pyruvate carboxylase (PEPC). 17. Malate dehydrogenase.*

from other plant tissues suggest that cytoplasmic concentrations may be in the region 100–300 μM (Edwards *et al*, 1984; Weiner *et al.*, 1987) which is close to the K_m for PPi of the soybean nodule UDPGlcPP (Vella and Copeland, 1990). The production of PPi could arise from a variety of biosynthetic reactions (Taiz, 1986) and if this is the case in nodules then the SS route for sucrose catabolism would not be limited by PPi availability.

Conversion of the glucose phosphates to fructose-6-phosphate (Fru6P) would be catalysed by the substantial activities of phosphoglucomutase and phospho-glucose isomerase found in nodules (Copeland *et al.*, 1989b; Gordon, 1991; Kouchi *et al.*, 1988; Figure 7.8). Phosphorylation of Fru6P to fructose-1,6-bisphosphate (Fru1,6P₂) may involve either phosphofructokinase (PFK) or pyrophosphate:fructose-6-phosphate:phosphotransferase (PFK(PPi)). During nodule development the specific activity of PFK increases while that of PFK(PPi) declines somewhat (see Figure 7.8 for lupin and Anthon and Emerich (1990) for soybean). The same arguments concerning availability of ATP and PPi (described above) also apply here. In addition, the role of fructose-2,6-bisphosphate ($Fru2,6P_2$), the powerful activator of PFK(PPi), in nodules is unknown, but it has been suggested in other systems (Huber and Akazawa, 1986; Morell and ap Rees, 1986) that PFK(PPi) may operate in the direction of Fru6P formation, thus generating PPi which would then be available for the UDPGlcPP reaction. Changes in the cytoplasmic concentrations of $Fru2,6P_2$ may therefore have a bearing on the involvement of PFK(PPi) in glycolysis.

The enzymes responsible for the conversion of $Fru1,6P_2$ to phosphoenol-pyruvate (PEP) were all found to be present in soybean nodules (Copeland *et al.*, 1989b; Kouchi *et al.*, 1988). The lowest activity found was that of aldolase (Anthon and Emerich, 1990; Copeland *et al.*, 1989b) which declined during nodule development (Anthon and Emerich, 1990). However, the data of Kouchi *et al.* (1988) suggest that high activities of aldolase may be present in nodule uninfected cells. Anthon and Emerich (1990) proposed that the aldolase reaction may be partially circumvented by pentose phosphate pathway enzymes. Substantial activities (one to two orders of magnitude greater than that of aldolase) of triose phosphate isomerase, glyceraldehyde-3-phosphate dehydrogenase, phosphoglycerate kinase, phosphoglycerate mutase and enolase were present in soybean nodules (Copeland *et al.*, 1989b; Kouchi *et al.*, 1988).

Phosphoenolpyruvate carboxylase (PEPC) is considered to have a central role in nodule carbon metabolism (see Deroche and Carrayol, 1988; Vance and Heichel, 1991; for reviews). PEPC activity in nodules is 20–100 times greater on a fresh weight basis than that in roots, and (in nodules) 4–10 times greater than that of pyruvate kinase (PK) (Copeland *et al.*, 1989b; Deroche and Carrayol, 1988; Gordon, 1991; Peterson and Evans, 1979) while the K_m values of PEPC and PK for PEP are similar (Peterson and Evans, 1978, 1979). In addition PK may be inhibited by the presence in nodule cells of NH_3 exported from the bacteroids (Peterson and Evans, 1978). This suggests that the greater flux of carbon is likely to be towards OAA production and hence, via malate dehydrogenase (MDH — of which there is substantial activity in legume nodules), to malate. OAA is also likely to be involved in aspartate synthesis (see Section 7.5.4), but may be replenished by limited oxidation of malate by mitochondria (see Section 7.4.3).

Although the foregoing discussion has concentrated mainly on the glycolytic pathway, the enzymes of the pentose phosphate pathway are also present. However, it is still unclear to what degree this pathway operates in legume nodules (cf. Hong and Copeland, 1990; Kohl *et al.*, 1990; Laing *et al*, 1979; Tajima and Yamamoto, 1984).

7.5.2 *Energetics of sucrose utilization*

Since in the presence of OAA, MDH is able to reoxidise cytosolic NADH produced in the glyceraldehyde 3-P dehydrogenase reaction the only energetic considerations relate to ATP/ADP (see Figure 7.9).

Table 7.1. *Potential production of ATP during the metabolism of sucrose to malate (as in Figure 7.9) depending on whether SS or INV is the primary enzyme and on the conditions outlined below*

Primary enzyme	Net ATP production (moles) during the metabolism of 1 mole sucrose → 4 mole malate under conditions*		
	1	2	3
SS	+1	+2	+4
INV	−1	0	+2

*Condition 1. For each mole of sucrose hydrolysed, PFK(PPi) produces 1 mole PPi from Fru1,6P$_2$ (i.e. 3 moles ATP are required to produce 2 mol Fru1,6P$_2$ for further metabolism). NB. This condition would not necessarily apply to the INV route.
Condition 2. Assumes PPi is available from an undefined source for the UDPGlcPP reaction and PFK(PPi) plays no role whatsoever.
Condition 3. Assumes PPi is available in sufficient quantities to be utilized by both UDPGlcPP and PFK(PPi) (to produce UTP + Glc1P, and Fru1,6P$_2$, respectively) and PFK plays no role whatsoever.

Table 7.1 shows the net production or utilization of ATP depending on whether sucrose is metabolized by SS or INV and whether PFK(PPi) is involved or not in the formation of PPi or Fru1,6P$_2$. Although this is a simplistic assessment of the ATP economy of malate synthesis from sucrose, it does illustrate the potential differences of the INV or SS routes under differing circumstances which could be crucial to the overall economy of legume nodules. This may be particularly so if PPi is limiting or if mitochondrial oxidative phosphorylation is low due to limited O$_2$ levels. In the latter case glycolysis, and particularly the phosphoglycerate kinase step could provide ATP for other cytoplasmic reactions (for example, ammonia assimilation by glutamine synthetase). Also if some of the PEP is converted to pyruvate by pyruvate kinase then this would generate a further 4 moles ATP per mole sucrose.

7.5.3 Carbon supply to bacteroids

It is clear from earlier sections that a variety of compounds is likely to be present in nodules engaged in active N$_2$ fixation. However, it is now apparent from elegant work with intact symbiosomes that the bacteroid and peribacteroid membranes permit only selected transport (Day *et al.*, 1989; Herrada *et al.*, 1989; Price *et al.*, 1987; Udvardi *et al.*, 1988a,b). Malate and succinate are taken up and metabolized both by bacteroids and symbiosomes from soybean nodules. Glutamate and the carbohydrates, sucrose and glucose, are transported only slowly across the peribacterial membrane, though glutamate can cross the bacteroid membrane. These observations are summarized in Figure 7.10. Since the nodule cytosol has abundant PEPC and MDH, malate is likely to be synthesized by the sucrose catabolic pathways described previously. In addition, malate is the most abundantly labelled compound if $^{14}CO_2$ is supplied to effective nodules (Rosendahl *et al.*, 1990). On the other hand, since nodule mitochondria are likely to be limited by O_2, lack malic enzyme (Day and Mannix, 1988) and can rapidly oxidise succinate (Bryce and Day, 1990), the nodule system does not appear to have the capacity to produce succinate for use by the bacteroids. Thus the most likely carbon source for bacteroid metabolism in soybeans is malate (Bryce and Day, 1990). It is possible, however, that bacteroids and symbiosomes from nodules of other species may have different uptake properties. For example, Herrada *et al.* (1989) have shown that in addition to efficient uptake of succinate (and malate), bacteroids and symbiosomes from French bean nodules have a low K_m and high V_{max} for glucose uptake.

Day and Copeland (1991) have proposed that since bacteroids contain NAD — and NADP — malic enzymes (Copeland *et al.*, 1989a; McKay *et al.*, 1988) and MDH they have the means to produce OAA and acetyl-CoA from malate and thus drive the TCA cycle. In addition acetyl-CoA and NAD(P)H may be utilized in poly-β-hydroxybutyrate production and thus there may be competition for substrates between poly-β-hydroxybutyrate synthesis and the TCA cycle (Day and Copeland, 1991).

ATP is produced by oxidative phosphorylation and is involved along with reducing equivalents (possibly transferred by ferredoxin or flavodoxin) in the

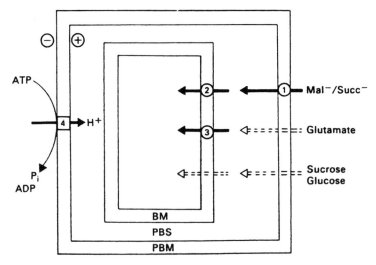

Figure 7.10. Scheme depicting transport properties of the bacteroid (BM) and peribacteroid (PBM) membranes. 1. PBM dicarboxylate anion transporter. 2. BM dicarboxylate transporter. 3. BM glutamate transporter. 4. PBM ATPase. Solid arrows indicate carrier-mediated transport; broken arrows indicate passive diffusion. Reproduced from Day et al. (1989) with permission from CSIRO.

reduction of N_2 to NH_3 by nitrogenase (see Gallon and Chaplin, 1987, for a detailed discussion of the provision of reductant).

7.5.4 *Assimilation of ammonia and the synthesis of export products*

The ammonia generated in the bacteroids diffuses across the bacteroid and peribacteroid membranes into the host plant cytosol where initially, it is probably incorporated into glutamine by the glutamine synthetase (GS), glutamate synthase pathway (Cullimore and Bennett, 1988; Miflin and Lea, 1980; Robertson and Farnden, 1980). In the amide-exporting legumes, transamination of OAA by aspartate amino transferase generates aspartate (Rosendahl *et al.*, 1990) from which asparagine is readily synthesized (Shelp and Atkins, 1984). Glutamine and asparagine are the principal nitrogenous compounds exported from the nodules of temperate legumes. Ureide production in the tropical legumes involves *de novo* purine synthesis followed by nucleotide breakdown and oxidation (Schubert, 1986; Schubert and Boland, 1984; Shelp *et al.*, 1983) as indicated in Figure 7.11).

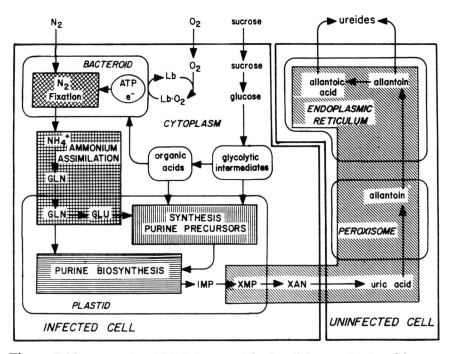

Figure 7.11. *Proposed model (Schubert, 1986) for the cellular organization of the reactions of ammonium assimilation,* de novo *purine synthesis and ureide biogenesis in various compartments and cell types in nodules of ureide producing legumes. Reproduced with permission from the* Annual Review of Plant Physiology, *Volume 37.* © *1986 by Annual Reviews Inc.*

7.5.5 *Integration and localization of events within the nodule*

The central metabolic event in legume nodules is the reduction of N_2 to ammonia by nitrogenase in the bacteroids. Nodule morphology is of vital importance in that it provides an environment which is conducive to this process. This implies (i) a rapid flux of O_2 at a low concentration in order that bacteroid respiration can take place without the danger of inactivating nitrogenase (see earlier sections), (ii) the ability to import sucrose into the nodule and then transport it within the nodule to the appropriate site for catabolism, (iii) the means to metabolize carbohydrate to products required by the bacteroid, for NH_3 assimilation and for the synthesis of export products, and finally (iv) the ability to export these products away from the nodule.

Only 10 years ago, Bergersen (1982) stated that 'in view of the large number of papers about the fluxes of photosynthetic products through nodules (e.g. Pate *et al.*, 1981) it is very surprising that so little is known about the metabolic pathways through which photosynthetic products pass in nodule cells, and of the location of the components of the enzyme systems involved'. Since then much effort has been directed at the location of enzymes involved in ureide metabolism and more recently enzymes in the sucrose → malate pathway have received attention (Copeland *et al.*, 1989b; Gordon, 1991; Kouchi *et al.*, 1988; Streeter, 1982; Suganuma *et al.*, 1987). Three general approaches have been employed, (i) density gradient centrifugation to separate subcellular organelles, (ii) protoplast preparation techniques to separate cell types (particularly infected cells, uninfected interstitial cells and cortex, and (iii) immunogold labelling coupled to electron microscopy which can be used to identify the precise location of an antigen within particular cell types and organelles.

All three approaches have been used successfully to elucidate the location of the enzymes involved in ureide metabolism (Boland *et al.*, 1982; Datta *et al.*, 1991b; Hanks *et al.*, 1981, 1983; Newcomb and Tandon, 1981; Newcomb *et al.*, 1989; Reynolds and Blevins, 1986; Reynolds *et al.*, 1988; Shelp *et al.*, 1983; VandenBosch and Newcomb, 1986). The flow of nitrogen (following fixation of N_2) into the ureides is shown schematically in Figure 7.11 and illustrates the involvement of a number of organelles within at least two cell types. Of particular interest is that the enzyme uricase has been shown to occur only in peroxisomes of the uninfected cells (Hanks *et al.*, 1983; VandenBosch and Newcomb, 1986) and also of the inner cortical and vascular parenchyma cells (Newcomb *et al.*, 1989; Vaughn and Stegink, 1987). The conversion of allantoin to allantoic acid also occurs in the uninfected cells (Hanks *et al.*, 1983). Since these enzymes catalyse the last two steps in ureide synthesis it seems probable that the uninfected cells also constitute the route of ureide export (Selker, 1988), though whether this is entirely symplastic is as yet unclear. As indicated earlier (Section 7.2.1) the uninfected cells radiate from the cortex and in many cases from the vascular bundles (soybean). The finding that inner cortical cells also contain large numbers of peroxisomes (Newcomb *et al.*, 1989) suggests that the route of nitrogen export from infected cells close to the cortex may actually be into the cortex. Transport from these cells to the vascular bundle could be either apoplastic or symplastic, though the final step across the vascular endodermis must be symplastic (Walsh *et al.*, 1989b).

The route of sucrose flux and the location of its subsequent metabolism has not been so well defined but most work has centred on the nodules of soybean. Only three reports have dealt in depth with the distribution of the key enzymes of the sucrose → malate catabolic route (Copeland *et al.*, 1989b; Gordon, 1991; Kouchi *et al.*, 1988). Kouchi *et al.* (1988) and Copeland *et al.* (1989b) used enzymic maceration techniques, the former separating the protoplast suspensions into infected, uninfected and cortical cells while Copeland *et al.* (1989b) separated central region cells from the cortical tissue (which is resistant to attack by cellulytic enzymes). Gordon (1991) separated central region tissue from cortex by manual dissection which avoided the long incubation times involved in protoplast preparation. This may be an important point, since it is clear from the data of Copeland *et al.* (1989b) that their cell preparations may have lost substantial amounts of protein (2 and 1 mg g^{-1} FW for cortical and central zone extracts — compared with 5 and 25 mg g^{-1} FW, respectively, in dissected tissue — Gordon, 1991). Kouchi *et al.* (1988) expressed all their data on a protein basis but did not include information about the protein content of their cell extracts; however, it is clear from the values quoted above that the infected region (and most likely the infected cells) has several times more protein than the cortex. Thus although Kouchi *et al.* (1988) showed that the specific activities of some enzymes were quite high in the cortex (e.g. UDPGlcPP and phosphoglucose isomerase) much more total activity is likely to be found in the central region compared with the cortex. This was demonstrated by Gordon (1991) who calculated that at least 80% of the total activity of all enzymes assayed was located in the central region of the nodule. The enzymes SS, INV and GS were almost exclusively (> 96%) located in the central region. The data of Kouchi *et al.* (1988) none the less indicate that substantial activity of many of the enzymes involved in carbohydrate and amino acid metabolism may reside in the uninfected cells of the central region. This raises questions concerning the integration of metabolic activity between infected and uninfected central region cells. One recent suggestion is that much of the catabolism of sucrose to malate may occur in the uninfected cells (Day and Copeland, 1991). It was envisaged that malate would be supplied to the infected cells to fuel bacteroid respiration and N_2 fixation and to a lesser extent to mitochondria to provide a limited amount of ATP for ammonia assimilation by GS. This suggestion is not entirely supported by published results, since the data of Kouchi *et al.* (1988) indicate the presence of all enzymes in both infected and uninfected cells. In addition, specific activities were not consistently higher in uninfected cells.

Further progress in our understanding of the metabolic communication between different cell types in the nodules is likely to come from *in situ* localization studies using antibodies (e.g. VandenBosch and Newcomb, 1986, 1988) or cDNA probes (Van de Wiel *et al.*, 1990a,b). Antibodies raised to glyceraldehyde 3-P dehydrogenase were used to localize this enzyme in soybean nodules

(Zammit and Copeland, unpublished data — quoted in Day and Copeland, 1991) where it was more prominent in the cortex and uninfected interstitial cells than in the infected cells. In contrast, Kouchi *et al.* (1988), found high and equal specific activity in infected and uninfected cells and very low activity in the cortex. Other studies with PEPC (Vidal *et al.*, 1986) and GS (Datta *et al.*, 1991a) have shown that these enzymes are also present in both infected and uninfected cells.

Recently the location of SS has also been studied using sucrose density gradients and immunogold labelling techniques employing antibodies to soybean nodule SS (Gordon *et al.*, manuscript in press). The protein was found exclusively in the cytoplasm and most prominently in the uninfected interstitial cells where the labelling was 2.5 times greater than in the infected cells. Very little labelling was found in the cortical cells. Again the data of Kouchi *et al.* (1988) are in conflict since they found a twofold greater specific activity of SS in the infected cells compared with uninfected cells. The results do agree to the extent that it appears unlikely that SS is confined exclusively to one particular cell type in the central region.

Another important observation from the immunogold labelling work with SS antibodies (Gordon *et al.*, manuscript in press) was that significant labelling was found in the cytosol of the vascular endodermis and the cortical cells directly between the vascular bundle and the infected cells. Thus it seems possible that the vascular bundle, the intervening cortical cells and the uninfected central region cells may be the direct route of sucrose transport into the metabolically active interior of the nodule — and that SS may somehow be involved in phloem unloading and transport. Other recent findings, that SS activity is associated with vascular bundles in grapefruit (Tomlinson *et al.*, 1991), and that the SS promotor directs phloem-cell specific expression of the reporter gene, β-glucuronidase, in transgenic tobacco plants (Yang and Russell, 1990) support this proposal. However, the role of SS in these processes is not yet clear.

Since in some systems SS genes have been shown to be induced under 'anaerobic' conditions (Ricard *et al.*, 1991; Talierco and Chourey, 1989) and the nodule environment may be considered anaerobic (Section 7.4), it is pertinent to ask whether nodule SS expression is also induced by low O_2 concentration. Two conflicting reports have been published. Xue *et al.* (1991) demonstrated that when soybean stem callus tissue was exposed to low O_2 (4%) the SS mRNA level was approximately fourfold greater than that from tissue grown at 21% O_2. In contrast Govers *et al.* (1986) concluded that genes induced by anaerobiosis in pea roots were not expressed in nodules; nor were nodulins expressed in anaerobic roots. In nodules the situation may be more complex since effective ways have evolved to supply sufficient O_2 at a low concentration (see Section 7.4.2). In addition, it seems likely that other factors (for example, carbohydrate availability) may be involved in the expression of SS (Koch *et al.*, 1992).

Finally, although much effort has been directed towards an understanding of the role of SS it is still not known whether it is this enzyme or alkaline invertase which is the principal enzyme of sucrose hydrolysis in sink tissues. Further metabolic and localization studies are necessary to resolve this issue.

7.6 Summary and conclusions

This article has outlined the morphological complexity of legume nodules and attempted to describe carbon metabolism within the context of structure, the flux and availability of oxygen, the import of sucrose, the requirements of the microsymbiont during N_2 fixation and the export of nitrogenous products. Discussion has centred mostly on the soybean/*Bradyrhizobium japonicum* nodule but it is important to appreciate that metabolism in other legume nodules may differ from that described here. For example, the amide-exporting nodules of temperate legumes do not have the enzymes involved in ureide metabolism. In general, nodules of temperate species have received only limited research attention compared with those of soybean. It is apparent, however, that even with soybean nodules we are still a long way from a full description of the relevant processes and some of the areas of imperfect understanding have been raised in this chapter.

Of particular interest is the relationship between morphology and metabolism and the mode of communication between different cell types with contrasting metabolic functions. I have illustrated this by reference to current views about the localization of ureide synthesis and the recent progress made in elucidating the distribution of enzymes involved in carbohydrate metabolism. I have proposed also (based on immunogold labelling studies with anti-SS antibodies) that the route of sucrose transport into the central region of the nodule may be via the uninfected interstitial cells. However, the limited information we have on the distribution of enzymes involved in carbohydrate and amino acid metabolism suggests that these enzymes are not confined exclusively to one particular cell type (cf. Day and Copeland, 1991). In my view this is a logical strategy since it seems very unlikely that NH_3, exported from the bacteroids into the infected cell cytosol, could be entirely metabolized or contained within those cells. The necessity to assimilate NH_3 to prevent damage to cells was proposed by Givan (1979), and recent work with legumes supplied with NH_3 has demonstrated the synthesis of GS and aspartate amino transferase in roots to levels similar to those found in nodules (Reynolds et al., 1990). Thus it might be expected that uninfected cells and cortical cells, in close proximity to infected cells, will have the necessary enzyme complement to detoxify NH_3 (i.e. to assimilate it) and as a result will also be involved in carbohydrate and (possibly) ureide metabolism in order to maintain this process.

Acknowledgements

I am grateful to Dr B.J. Thomas and Mr P. Roberts for electron microscopy, to Mr R.M. Thomas and Mrs C.L. James for assistance with figures and to Ms M. Hobbelen for typing the manuscript.

References

Anthon, G.E. and Emerich, D.W. (1990) Developmental regulation of enzymes of sucrose and hexose metabolism in effective and ineffective soybean nodules. *Plant Physiol.* 92, 346–351.

Antoniw, L.D. and Sprent, J.I. (1978) Primary metabolites of *Phaseolus vulgaris* nodules. *Phytochemistry*, 17, 675–678.

Appleby, C.A. (1984) Leghaemoglobin and *Rhizobium* respiration. *Ann. Rev. Plant Physiol.* 35, 443–478.

Bach, M.K., Magee, W.E. and Burris, R.H. (1958) Translocation of photosynthetic products to soybean nodules and their role in nitrogen fixation. *Plant Physiol.* 33, 118–124.

Bergersen, F.J. (1982) *Root Nodules of Legumes: Structure and Functions.* Research Studies Press, John Wiley and Sons, Chichester.

Bergersen, F.J. and Goodchild, D.J. (1973) Cellular location and concentration of leghaemoglobin in soybean root nodules. *Aust. J. Biol. Sci.* 26, 741–756.

Bergersen, F.J. and Turner, G.L. (1975) Leghaemoglobin and the supply of O_2 to nitrogen-fixing root nodule bacteroids: presence of two oxidase systems and ATP production at low free O_2 concentration. *J. Gen. Microbiol.* 91, 345–354.

Bergersen, F.J., Peoples, M.B. and Turner, G.L. (1991) A role for poly-β-hydroxybutyrate in bacteroids of soybean root nodules. *Proc. R. Soc., London B*, 245, 59–64.

Boland, M.J., Hanks, J.F., Reynolds, P.H.S., Blevins, D.G., Tolbert, N.E. and Schubert, K.R. (1982) Subcellular organization of ureide biogenesis from glycolytic intermediates and ammonium in nitrogen fixing soybean nodules. *Planta*, 155, 45–51.

deBruijn, F.J. and Downie, J.A. (1991) Biochemical and molecular studies of symbiotic nitrogen fixation. *Curr. Opinions Biotech.* 2, 184–192.

Bryce, J.H. and Day, D.A. (1990) Tricarboxylic acid cycle activity in mitochondria from soybean nodules and cotyledons. *J. Exp. Bot.* 41, 961–967.

Copeland, L. and Morrell, M. (1985) Hexose kinases from the plant cytosolic fraction of soybean nodules. *Plant Physiol.* 79, 114–117.

Copeland, L., Quinnell, R.G. and Day, D.A. (1989a) Malic enzyme activity in bacteroids from soybean nodules. *J. Gen. Microbiol.* 135, 2005–2011.

Copeland, L., Vella, J. and Hong, Z.Q. (1989b) Enzymes of carbohydrate metabolism in soybean nodules. *Phytochemistry*, 28, 57–61.

Cullimore, J.V. and Bennett, M.J. (1988) The molecular biology and biochemistry of plant glutamine synthetase from root nodules of *Phaseolus vulgaris* L. and other legumes. *J. Plant Physiol.* 132, 387–393.

Datta, D.B., Cai, X., Wong, P.P. and Triplett, E.W. (1991a) Immunochemical localization of glutamine synthetase in organs of *Phaseolus vulgaris* L. *Plant Physiol.* 96, 507–512.

Datta, D.B., Triplett, E.W. and Newcomb, E.H. (1991b) Localization of xanthine dehydrogenase in cowpea root nodules: implications for the interaction between cellular compartments during ureide biogenesis. *Proc. Natl Acad. Sci. USA*, 88, 4700–4702.

Day, D.A. and Copeland, L. (1991) Carbon metabolism and compartmentation in nitrogen fixing legumes nodules. *Plant Physiol. Biochem.* 29, 185–201.

Day, D.A. and Mannix, M. (1988) Malate oxidation by soybean nodule mitochondria and the possible consequences for nitrogen fixation. *Plant Physiol. Biochem.* 26, 567–573.

Day, D.A., Price, G.D. and Udvardi, M.K. (1989) Membrane interface of the *Brady-*

rhizobium japonicum — Glycine max. symbiosis: peribacteroid units from soybean nodules. *Aust. J. Plant Physiol.* 16, 69–84.

Deroche, M.E. and Carrayol, E. (1988) Nodule phosphoenolpyruvate carboxylase: a review. *Physiol. Plant.* 74, 775–782.

Edwards, J., apRees, T., Wilson, P.H. and Morrell, S. (1984) Measurement of the inorganic pyrophosphate in tissues of *Pisum sativum* L. *Planta*, 162, 188–191.

Gallon, J.R. and Chaplin, A.E. (1987) *An Introduction to Nitrogen Fixation.* Cassell Education Ltd, London.

Givan, C.V. (1979) Metabolic detoxification of ammonia in tissues of higher plants. *Phytochemistry*, 18, 375–382.

Gordon, A.J. (1986) Diurnal patterns of photosynthate allocation and partitioning among sinks. In: *Phloem Transport* (eds J. Cronshaw, W.J. Lucas and R.T. Giaquinta). Alan R. Liss, Inc., New York, pp. 499–517.

Gordon, A.J. (1991) Enzyme distribution between the cortex and the infected region of soybean nodules. *J. Exp. Bot.* 42, 961–967.

Gordon, A.J. and Kessler, W. (1990) Defoliation-induced stress in nodules of white clover. 2. Immunological and enzymic measurements of key proteins. *J. Exp. Bot.* 41, 1255–1262.

Gordon, A.J., Ryle, G.J.A., Mitchell, D.F. and Powell, C.E. (1985) The flux of ^{14}C-labelled photosynthate through soyabean root nodules during N_2 fixation. *J. Exp. Bot.* 36, 756–769.

Gordon, A.J., Ryle, G.J.A., Mitchell, D.F., Lowry, K.H. and Powell, C.E. (1986) The effect of defoliation on carbohydrate, protein and leghaemoglobin content of white clover nodules. *Ann. Bot.* 58, 141–154.

Gordon, A.J., Mitchell, D.F., Ryle, G.J.A. and Powell, C.E. (1987) Diurnal production and utilization of photosynthate in nodulated white clover. *J. Exp. Bot.* 38, 84–98.

Gordon, A.J., Thomas, B.J. and Reynolds, P.H.S. (1992) Localization of sucrose synthase in soybean root nodules. *New Phytol.* in press.

Govers, F., Moerman, M., Hoymanns, J., van Kammen, A. and Bisseling, T. (1986) Microaerobiosis is not involved in the induction of pea nodulin gene expression. *Planta*, 169, 513–517.

Gunning, B.E.S., Pate, J.S., Minchin, F.R. and Marks, I. (1974) Quantitative aspects of transfer cell structure in relation to vein loading in leaves and solute transport in legume nodules. In: *Transport at the Cellular Level. SEB Symposium* Volume 28. Cambridge University Press, Cambridge, pp. 87–126.

Hanks, J.F., Tolbert, N.E. and Schubert, K. (1981) Localization of enzymes of ureide metabolism in peroxisomes and microsomes of nodules. *Plant Physiol.* 68, 65–69.

Hanks, J.F., Schubert, K. and Tolbert, N.E. (1983) Isolation and characterization of infected and uninfected cells from soybean nodules. *Plant Physiol.* 71, 869–873.

Herrada, G., Puppo, A. and Rigaud, J. (1989) Uptake of metabolites by bacteroid-containing vesicles and free bacteroids from french bean nodules. *J. Gen. Microbiol.* 135, 3165–3177.

Hong, Z.Q. and Copeland, L. (1990) Pentose phosphate pathway enzymes in nitrogen-fixing leguminous root nodules. *Phytochemistry*, 29, 2437–2440.

Huber, S.C. and Akazawa, T. (1986) A novel sucrose synthase pathway for sucrose degradation in cultured sycamore cells. *Plant Physiol.* 81, 1008–1013.

Hunt, S., Denison, R.F., King, B.J., Kouchi, H., Tajima, S. and Layzell, D.B. (1990) An osmotic mechanism for diffusion barrier regulation in soybean nodules. In: *Nitrogen Fixation: Achievements and Objectives* (eds P.M. Gresshoff, E.L. Roth, G. Stacey and

W.E. Newton). Chapman and Hall, London, p. 352.

James, E.K., Sprent, J.I., Minchin, F.R. and Brewin, N.J. (1991) Intercellular location of glycoprotein in soybean nodules: effect of altered rhizosphere oxygen concentration. *Plant Cell Environ.* 14, 467–476.

Kerr, P.S., Rufty, T.W. and Huber, S.C. (1985) Changes in non-structural carbohydrates in different parts of soybean (*Glycine max.* L. Merr.) plants during a light/dark cycle and in extended darkness. *Plant Physiol.* 78, 576–581.

King, B.J., Hunt, S., Weagle, G.E., Walsh, K.B., Pottier, R.H., Canvin, D.T. and Layzell, D.B. (1988) Regulation of O_2 concentration in soybean nodules observed by *in situ* spectroscopic measurement of leghaemoglobin oxygenation. *Plant Physiol.* 87, 296–299.

Koch, K.E., Nolte, K.D., Duke, E.R., McCarhty, D.R. and Avigne, W.T. (1992) Sugar levels modulate differential expression of maize sucrose synthase genes. *Plant Cell,* 4, 59–69.

Kohl, D.H., Lin, J.-J., Shearer, G. and Schubert, K.R. (1990) Activities of the pentose phosphate pathway and enzymes of proline metabolism in legume root nodules. *Plant Physiol.* 94, 1258–1264.

Kouchi, H. and Yoneyama, T. (1984a) Dynamics of carbon photosynthetically assimilated in nodulated soyabean plants under steady-state conditions. 1. Development and application of $^{13}CO_2$ assimilation system at a constant ^{13}C abundance. *Ann. Bot.* 53, 875–882.

Kouchi, H. and Yoneyama, T. (1984b) Dynamics of carbon photosynthetically assimilated in nodulated soyabean plants under steady state conditions. 2. The incorporation of ^{13}C into carbohydrates, organic acids, amino acids and some storage compounds. *Ann. Bot.* 53, 883–896.

Kouchi, H., Nakaji, K., Yoneyama, T. and Ishizuka, J. (1985) Dynamics of carbon photo-synthetically assimilated in nodulated soya bean plants under steady-state conditions. 3. Time-course study on ^{13}C-incorporation into soluble metabolites and respiratory evolution of $^{13}CO_2$ from roots and nodules. *Ann. Bot.* 56, 333–346.

Kouchi, H., Akao, S. and Yoneyama, T. (1986) Respiratory utilization of ^{13}C-labelled photosynthate in nodulated root systems of soybean plants. *J. Exp. Bot.* 37, 985–993.

Kouchi, H., Fukai, K., Katagiri, H., Minamisawa, K. and Tajima, S. (1988) Isolation and enzymological characterisation of infected and uninfected cell protoplasts from root nodules of *Glycine max. Physiol. Plant.* 73, 327–334.

Laing, W.A., Christeller, J.T. and Sutton, W.D. (1979) Carbon dioxide fixation by lupin nodules. II. Studies with ^{14}C-labelled glucose, the pathway of glucose catabolism and the effects of some treatments that inhibit nitrogen fixation. *Plant Physiol.* 63, 450–454.

Lawrie, A.C. and Wheeler, C.T. (1975) Nitrogen fixation in the root nodules of *Vicia faba* L. in relation to the assimilation of carbon. 1. Plant growth and metabolism of photosynthetic assimilates. *New Phytol.* 74, 429–436.

Layzell, D.B. and LaRue, T.A. (1982) Modelling C and N transport to developing soybean fruits. *Plant Physiol.* 53, 96–103.

Layzell, D.B., Hunt, S., Moloney, A.H.M., Fernando, S.M. and Diaz del Castillo, L. (1990) Physiological, metabolic and environmental implications of O_2 regulation in legume nodules. In: *Nitrogen Fixation: Achievements and Objectives* (eds P. Gresshoff, E.L. Roth, G. Stacey and W.E. Newton). Chapman and Hall, London, pp. 21–33.

Long, S.R. (1990) Nodulation genetics: the plant bacterial interface. In: *Nitrogen Fixation: Achievements and Objectives* (eds P.M. Gresshoff, E.L. Roth, G. Stacey and

W.E. Newton). Chapman and Hall, London, pp. 15–19.

Lucas, K., Boland, M.J. and Schubert, K.R. (1983) Uricase from soybean root nodules: Purification, properties and comparison with the enzyme from cowpea. *Arch. Biochem. Biophys.* **226**, 190–197.

McKay, I.A., Dilworth, M.J. and Glenn, A.R. (1988) C_4-dicarboxylate metabolism in free-living and bacteroid forms of *Rhizobium leguminosarum* MNF3481. *J. Gen. Microbiol.* **134**, 1433–1440.

Miflin, B.J. and Lea, P.J. (1980) Ammonia assimilation. In: *The Biochemistry of Plants*, Volume 5 (ed. B.J. Miflin). Academic Press, New York, pp. 169–202

Minchin, F.R. and Pate, J.S. (1974) Diurnal functioning of the legume root nodule. *J. Exp. Bot.* **25**, 295–308.

Morrell, M. and Copeland, L. (1984) Enzymes of sucrose breakdown in soybean nodules. Alkaline invertase. *Plant Physiol.* **74**, 149–154.

Morrell, M. and Copeland, L. (1985) Sucrose synthase of soybean nodules. *Plant Physiol.* **78**, 149–154.

Morrell, S. and ap Rees, T. (1986) Sugar metabolism in developing tubers of *Solarium tuberosum*. *Phytochemistry*, **25**, 1579–1585.

Newcomb, E.H. and Tandon, S.R. (1981) Uninfected cells of soybean root nodules: ultrastructure suggests key role in ureide production. *Science*, **212**, 1394–1396.

Newcomb, W., Sippel, D. and Peterson, R.L. (1979) The early morphogenesis of *Glycine max.* and *Pisum sativum* root nodules. *Can. J. Bot.* **57**, 2603–2616.

Newcomb, E.H., Tandon, S.R. and Kowal, R.R. (1985) Ultrastructural specialization for ureide production in uninfected cells of soybean root nodules. *Protoplasma*, **125**, 1–12.

Newcomb, E.H., Kaneko, Y. and VandenBosch, K.A. (1989) Specialization of the inner cortex for ureide production in soybean root nodules. *Procoplasma*, **150**, 150–159.

Parsons, R. and Day, D.A. (1990) Mechanism of soybean nodule adaptation to different oxygen stresses. *Plant Cell Environ.* **13**, 501–512.

Pate, J.S., Gunning, B.E.S. and Briarty, L.G. (1969) Ultrastructure and functioning of the transport system of the leguminous root nodule. *Planta*, **85**, 11–34.

Pate, J.S., Atkins, C.A. and Rainbird, R.M. (1981) Theoretical and experimental costing of nitrogen fixation and related processes in nodules of legumes. In: *Current Perspectives in Nitrogen Fixation* (eds A.H. Gibson and W.E. Newton). Australian Academy of Science, Canberra, pp. 105–116.

Peterson, J.B. and Evans, H.J. (1978) Properties of pyruvate kinase from soybean nodule cytosol. *Plant Physiol.* **61**, 909–914.

Peterson, J.B. and Evans, H.J. (1979) Phosphoenol-pyruvate carboxylase from soybean nodule cytosol. Evidence for isoenzymes and kinetics of the most active component. *Biochim. Biophys. Acta*, **567**, 445–452.

Phillips, D.A., Hartwig, U.A., Maxwell, C.A., Joseph, C.M., Wery, J., Hungria, M. and Tsai, S.M. (1990) Host legume control of nodulation by flavonoids. In: *Nitrogen Fixation: Achievements and Objectives* (ed. P.M. Gresshoff, E.L. Roth, G. Stacey and W.E. Newton). Chapman and Hall, London, pp. 331–340.

Price, G.D., Day, D.A. and Gresshoff, P.M. (1987) Rapid isolation of intact peribacteroid envelopes from soybean nodules and demonstration of selective permeability to various metabolites. *J. Plant Physiol.* **130**, 157–164.

Rainbird, R.M. and Atkins, C.A. (1981) Purification and some properties of urate oxidase from nitrogen-fixing nodules of cowpea. *Biochem. Biophys. Acta*, **659**, 132–140.

Rainbird, R.M., Atkins, C.A. and Pate, J.S. (1983) Diurnal variation in the functioning of cowpea nodules. *Plant Physiol.* **72**, 308–312.

Raven, J.A., Sprent, J.I., McInvoy, S.G. and Hay, G.T. (1989) Water balance of N_2-fixing root nodules: can phloem and xylem transport explain it? *Plant Cell. Environ.* **12**, 683–688.

Rawsthorne, S. and LaRue, T.A. (1986) Metabolism under microaerobic conditions of mitochondria from cowpea nodules. *Plant Physiol.* **81**, 1097–1102.

Reibach, P.H. and Streeter, J.G. (1983) Metabolism of ^{14}C-labelled photosynthate and distribution of enzymes of glucose metabolism in soybean nodules. *Plant Physiol.* **72**, 634–640.

Reynolds, P.H.S. and Blevins, D.G. (1986) Phosphoserine aminotransferase in soybean root nodules. Demonstration and localization. *Plant Physiol.* **81**, 293–296.

Reynolds, P.H.S., Hine, A. and Rodber, K. (1988) Serine metabolism in legume nodules: purification and properties of phosphoserine aminotransferase. *Physiol. Plant.* **74**, 194–199.

Reynolds, P.H.S., Boland, M.J., McNaughton, G.S., More, R.D. and Jones, W.T. (1990) Induction of ammonium assimilation: leguminous roots compared with nodules using a split root system. *Physiol. Plant.* **79**, 359–367.

Ricard, B., Rivoal, J., Spiteri, A. and Pradet, A. (1991) Anaerobic stress induces the transcription and translation of sucrose synthase in rice. *Plant Physiol.* **95**, 669–674.

Robertson, J.G. and Taylor, M.P. (1973) Acid and alkaline invertases in roots and nodules of *Lupinus angustifolius* infected with *Rhizobium lupini*. *Planta*, **112**, 1–6.

Robertson, J.G. and Farnden, K.J.F. (1980) Ultrastructure and metabolism of the developing legume root nodule. In: *The Biochemistry of Plants*, Volume 5 (ed. B.J. Miflin). Academic Press, New York, pp. 65–113.

Romanov, V.I., Fedulova, N.G., Tchermenskaya, I.E., Shramko, V.I., Molchanov, M.I. and Kretovich, W.L. (1980) Metabolism of poly-β-hydroxybutyric acid in bacteroids of *Rhizobium lupini* in connection with nitrogen fixation and photosynthesis. *Plant Soil*, **56**, 379–390.

Romanov, V.I., Hajy-zadeh, B.R., Ivanov, B.F., Shaposhnikov, G.L. and Kretovich, W.L. 1985. Labelling of lupin nodule metabolites with $^{14}CO_2$ assimilated from the leaves. *Phytochemistry*, **24**, 2157–2160.

Rosendahl, L., Vance, C.P. and Pedersen, W.B. (1990) Products of dark CO_2 fixation in pea root nodules support bacteroid metabolism. *Plant Physiol.* **93**, 12–19.

Ryle, G.J.A., Arnott, R.A., Powell, C.E. and Gordon, A.J. (1984) N_2 fixation and the respiratory costs of nodules, nitrogenase activity and nodule growth and maintenance in Fiskeby soyabean. *J. Exp. Bot.* **35**, 1156–1165.

Ryle, G.J.A., Powell, C.E. and Gordon, A.J. (1985) Defoliation in white clover: regrowth, photosynthesis, and N_2 fixation. *Ann. Bot.* **56**, 9–18.

Ryle, G.J.A., Powell, C.E. and Gordon, A.J. (1986) Defoliation in white clover: nodule metabolism, nodule growth and maintenance, and nitrogenase functioning during growth and regrowth. *Ann. Bot.* **57**, 263–271.

Ryle, G.J.A., Powell, C.E. and Gordon, A.J. (1988) Responses of N_2 fixation-linked respiration to host-plant energy status in white clover acclimated to a controlled environment. *J. Exp. Bot.* **39**, 879–887.

Schubert, K.R. (1986) Products of biological nitrogen fixation in higher plants: synthesis, transport and metabolism. *Ann. Rev. Plant Physiol.* **37**, 539–574.

Schubert, K.R. and Ryle, G.J.A. (1980) The energy requirements for N_2 fixation in nodulated legumes. In: *Advances in Legume Science* (eds R.J.S. Summerfield and A.H. Bunting). Kew Botanical Gardens, London, pp. 85–96.

Schubert, K.R. and Boland, M.J. (1984) The cellular and intracellular organization of ureide biogenesis in nodules of tropical legumes. In: *Advances in Nitrogen Fixation Research* (eds C. Veeger and W.E. Newton). Nijhoff/Junk, The Hague, pp. 445–451.

Selker, J.M.L. (1988) Three-dimensional organization of uninfected tissue in soybean root nodules and its relation to cell specialization in the central region. *Protoplasma* 147, 178–190.

Selker, J.M.L. and Newcomb, E.H. (1985) Spatial relationships between uninfected and infected cells in root nodules of soybean. *Planta*, 165, 446–454.

Sheehy, J.E., Minchin F.R. and Witty, J.F. (1984) Control of nitrogen fixation in a legume nodule: an analysis of the role of oxygen diffusion in relation to nodule structure. *Ann. Bot.* 55, 549–562.

Shelp, B.J. and Atkins, C.A. (1984) Subcellular location of enzymes of ammonia assimilation and asparagine synthesis in root nodules of *Lupinus albus* L. *Plant Sci. Lett.* 36, 225–230.

Shelp, B.J., Atkins, C.A., Storer, P.J. and Canvin, D.T. (1983) Cellular and subcellular organization of pathways of ammonia assimilation and ureide synthesis in nodules of cowpea (*Vigna unguiculata* L. Walp.) *Arch. Biochem. Biophys.* 224, 429–441.

Sprent, J.I. (1980) Root nodule anatomy, type of export product and evolutionary origin in some Leguminosae. *Plant Cell Environ.* 3, 35–43.

Streeter, J.G. (1980) Carbohydrates in soybean nodules. II. Distribution of compounds in seedlings during the onset of nitrogen fixation. *Plant Physiol.* 66, 471–476.

Streeter, J.G. (1982) Enzymes of sucrose, maltose, and α,α-trehalose catabolism in soybean root nodules. *Planta*, 155, 112–115.

Streeter, J.G. (1991) Transport and metabolism of carbon and nitrogen in legume nodules. *Adv. Bot. Res.* 18, 129–187.

Suganuma, N., Kitou, M. and Yamamoto, Y. (1987) Carbon metabolism in relation to cellular organization of soybean root nodules and respiration of mitochondria aided by leghaemoglobin. *Plant Cell Physiol.* 28, 113–122.

Taiz, L. (1986) Are biosynthetic reactions in plant cells thermodynamically coupled to glycolysis and tonoplast proton motive force? *J. Theor. Biol.* 123, 231–238.

Tajima, S. and Yamamoto, Y. (1984) Fluctuation of enzyme activities related to nitrogen fixation, C6/C1 ratio, and nicotinamide nucleotide contents during soybean plant development. *Soil Sci. Plant Nutr.* 30, 85–94.

Talierco, E.W. and Chourey, P.S. (1989) Post-transcriptional control of sucrose expression in anaerobic seedlings of maize. *Plant Physiol.* 90, 1359–1364.

Thummler, F. and Verma, D.P.S. (1987) Nodulin-100 of soybean is the subunit of sucrose synthase regulated by the availability of free heme in nodules. *J. Biol. Chem.* 262, 14730–14736.

Tjepkema, J.D. and Yocum, C.S. (1974) Measurement of oxygen partial pressure within soybean nodules by oxygen microelectrodes. *Planta*, 119, 351–360.

Tomlinson, P.T., Duke, E.R., Nolte, K.D. and Koch, K.E. (1991) Sucrose synthase and invertase in isolated vascular bundles. *Plant Physiol.* 97, 1249–1252.

Udvardi, M.K., Price, G.D., Gresshoff, P.M. and Day, D.A. (1988a) A dicarboxylate transporter on the peribacteroid membrane of soybean nodules. *FEBS Lett.* 231, 36–40.

Udvardi, M.K., Salom, C.L. and Day, D.A. (1988b) Transport of L-glutamate across the bacteroid but not the peribacteroid membrane of soybean nodules. *Mol. Plant Microbe Interact.* 1, 250–254.

Vance, C.P. and Heichel, G.H. (1991) Carbon in N_2 fixation: Limitation and exquisite adaptation. *Ann. Rev. Plant Physiol. Plant Mol. Biol.* 42, 373–392.

VandenBosch, K.A. and Newcomb, E.H. (1986) Immunogold localization of nodule — specific uricase in developing soybean root nodules. *Planta*, 167, 425–436.

VandenBosch, K.A. and Newcomb, E.H. (1988) The occurrence of leghaemoglobin protein in the uninfected interstitial cells of soybean root nodules. *Planta*, 175, 442–451.

Van de Wiel, C., Norris, J.H., Bochenek, B., Dickstein, R., Bisseling, T. and Hirsch, A.M. (1990a) Nodulin gene expression and ENOD2 localization in effective, nitrogen-fixing and ineffective, bacteria-free nodules of alfalfa. *Plant Cell*, 2, 1009–1017.

Van de Wiel, C., Scheres, B., Franssen, H., vanLierop, M.J., vanLammeren, A., vanKammen, A. and Bisseling, T. (1990b) The early nodulin transcript ENOD2 is located in the nodule parenchyma (inner cortex) of pea and soybean root nodules. *EMBO J.* 9, 1–7.

Vaughn, K.C. and Stegink, S.J. (1987) Peroxisomes of soybean (*Glycine max*) root nodule vascular parenchyma cells contain a 'nodule-specific' urate oxidase. *Plant Physiol.* 71, 251–256.

Vella, J. and Copeland, L. (1990) UDP-glucose pyrophosphorylase from the plant fraction of nitrogen fixing soybean nodules. *Physiol. Plant.* 78, 140–147.

Verma, D.P.S. (1989) Plant genes involved in carbon and nitrogen assimilation in root nodules. *Rec. Adv. Phytochem.* 23, 43–63.

Vidal, J., Nguyen, J., Perrot-Rechenmann, C. and Gadal, P. (1986) Phosphoenolpyruvate carboxylase in soybean root nodules: an immunochemical study. *Planta*, 169, 198–201.

Walsh, K.B. (1990) Vascular transport and soybean nodule function. III. Implications of a continual phloem supply of carbon and water. *Plant Cell Environ.* 13, 893–901.

Walsh, K.B., Vessey, J.K. and Layzell, D.B. (1987) Carbohydrate supply and N_2 fixation in soybean. The effect of varied daylength and stem girdling. *Plant Physiol.* 85, 137–144.

Walsh, K.B., Canny, M.J. and Layzell, D.B. (1989a) Vascular transport and soybean nodule function: II. a role for phloem supply in product export. *Plant Cell Environ.* 12, 713–723.

Walsh, K.B., McCully, M.E. and Canny, M.J. (1989b) Vascular transport and soybean nodule function: nodule xylem is a blind alley, not a throughway. *Plant Cell Environ.* 12, 395–405.

Weiner, H., Stitt, M. and Heldt, H.W. (1987) Subcellular compartmentation of pyrophosphate and alkaline pyrophosphatase in leaves. *Biochim. Biophys. Acta*, 893, 13–21.

Witty, J.F. and Minchin, F.R. (1990) Oxygen diffusion in the legume root nodule. In: *Nitrogen Fixation: Achievements and Objectives* (eds P.M. Gresshoff, E.L. Roth, G. Stacey and W.E. Newton). Chapman and Hall, London, pp. 285–292.

Witty, J.F., Minchin, F.R., Skøt, L. and Sheehy, J.E. (1986) Nitrogen fixation and oxygen in legume root nodules. *Oxford Surv. of Plant Mol. Cell Biol.* 3, 275–314.

Xue, Z., Larsen, K. and Jochimsen, B.U. (1991) Oxygen regulation of uricase and sucrose synthase synthesis in soybean callus tissue is exerted at the mRNA level. *Plant Mol. Biol.* 16, 899–906.

Yang, N.S. and Russell, D. (1990) Maize sucrose synthase-1 promoter directs phloem cell-specific expression of *GUS* gene in transgenic tobacco plants. *Proc. Natl Acad. Sci. USA* 87, 4144–4148.

8

The whole plant: carbon partitioning during development

J.F. Farrar

8.1 Introduction

In 1927, W.H. Pearsall showed that plants, like animals, have allometric growth. At intervals he harvested plants from a batch and measured shoot (S) and root (R) weight, and showed that the two were related such that $S = bR^k$, where b and k are constants. Rephrased, the relative growth rates of shoot and root stayed in constant ratio, k, to each other (Pearsall, 1927). Numerous confirmations that the allometric relationship describes plant growth, including leaf and fruit expansion and leaf area:dry weight relationships, lend confidence to its generality, and it is known that k is heritable, and changes with both developmental switches such as flowering and external stimuli such as changes in nutrient supply. In spite of the obvious challenge of explaining this elegant relationship, and in spite of the importance of dry matter partitioning for food production, there is no satisfactory mechanistic explanation for the allometric relationship. The aim of this chapter is to work towards such an explanation, with the premise that progress will only be made by examining the partial processes of carbon partitioning, since growth is essentially the accretion of carbon metabolites. The chapter should be treated as an extended hypothesis. I will concentrate on shoot:root ratio (S:R) for convenience, without wishing to imply that other aspects of whole plant carbon partitioning are less worthy of attention.

8.2 Models for control of S:R

The remarkable regulation of S:R is not imposed by morphology, since there is almost always a large excess of actual or potential meristems in both shoot and root, and much of the control of plant form is exercised by selective suppression of their growth. The relationships between apical dominance and growth analysis have never been adequately explored, but the idea that plant growth regulators (PGRs) regulate both has a long history.

Theories for PGR control of S:R have been described by Thomas (1986), Patrick (1987) and Brenner (1987). They share the premise that a given PGR is produced locally and is transported to affect growth in a remote part of the plant, a premise shared with theories of apical dominance (Clive, 1991). In addition to critiscisms of such thinking (Trewavas, 1991) there is no complete, mechanistic explanation for how PGRs may control S:R. There is also an unsatisfying indirectness about controlling carbon partitioning — essentially the transport of sucrose — by completely unrelated molecules. Here I will develop a simpler model, with the caveat that PGRs may well be involved in some partial processes, for example, the link between S:R and apical dominance.

A second type of explanation for control of S:R is provided by the 'functional equilibrium' model of Brouwer (1983) which has been treated mathematically by Thornley (1977). Root growth is limited by supply of assimilate from the shoot, and shoot growth by supply of nutrient from the root; the mutual dependence of shoot and root results in their balanced growth. Important as this model is, it is marred by the lumping of shoot as photosynthetic organ with shoot as growing meristems: assimilate is produced in the shoot and either leaves it or is retained and contributes to its growth. In the real world, assimilate exported from a mature source leaf is subsequently partitioned between shoot and root meristems. Barnes (1979) has modified Thornley's model to include partitioning of assimilate leaving source leaves, and shows that his model can generate modified allometric growth. Like many other models, however, his partitioning coefficient is not mechanistically determined.

The models of Barnes and Thornley (and many other crop growth models) share one relevant feature: flux of assimilate depends on the size of the assimilate pool in the source. The success of models making this assumption suggests that the relationship may be causal, and as a result it is commonly assumed that substrate drives growth by mass action. I shall show that the true situation is more complex.

I suggest that the existing models which seek to explain the control of S:R are deeply flawed. Clearly there is need for a fresh approach. Here I suggest that sucrose is central to the regulatory process. Not only is it the major product of source and substrate of sink metabolism, but it can regulate (perhaps indirectly) the expression of genes coding for enzymes centrally involved in the carbon fluxes leading to growth. Before these issues can be discussed for phloem pathway, source and sink, a diversion into terminology is necessary.

8.3 Partial processes

Many of the partial processes involved in carbon partitioning are discussed in Baker and Milburn (1989), Bonnemain *et al.* (1991) and Wardlaw (1990). One impediment to progress in whole plant carbon partitioning is the lack of an agreed and precise terminology. Words such as source, sink, activity, and even partitioning itself, are used loosely, with the same word often stretched to describe quite different phenomena. It is not my intention here to suggest a revised terminology, but merely to spotlight some of the partial processes that often hide behind the words in use. I consider three, mutually interacting, ways of classifying partial processes, listed in Table 8.1: time scales, response, and flux.

There are three distinct time scales of response to perturbation, each linked to a different underlying mechanism. Rapid responses will act on the existing machinery (and in the presence of current sizes of pool of storage carbohydrates such as starch and vacuolar sucrose); they operate via fine control, often via mass action alone but including mechanisms such as allostery, over seconds to minutes. Fine control thus includes the essential and well-studied ways in which metabolism is regulated. Conversely, over a few hours, the size of cytosolic, vacuolar and plastidic storage pools can change, altering fluxes where first- or second-order kinetics operate; this might be termed buffering. Examples in leaf tissue are provided by Farrar and Farrar (1986) and in roots by Farrar and Williams (1990). And these two will be joined, over hours to days, by a third type of response: changes in gene expression and thus of the amount of machinery — enzymes and carriers, for example — often termed coarse control.

Wareing (1976) introduced the useful distinction between direct and programmed responses. Direct responses, such as the effect of temperature on enzyme activity, are non-adaptive and contrast with programmed responses such as stomatal closure on droughting which result in an amelioration of the effects of the changed environment. According to this definition, many changes in S:R could be considered as programmed (Wilson, 1989) whilst effects of temperature on starch metabolism (Pollock and Lloyd, 1987) would be direct.

The third type of classification is concerned with whether or not a given process or pathway is operating at its maximum possible rate. Commonly, pathways sustain less than their maximum flux, photosynthesis being a classical example. A pathway may be limited by supply of substrate, by an environmental constraint (low temperature, for example) or by amount of machinery (gene

Table 8.1. *Classification of partial processes of carbon partitioning*

1. Time scales
   Fine (seconds to minutes; mass action, allostery)
   Buffering (hours; pool sizes of storage carbohydrate)
   Coarse (hours to days; gene expression)

2. Response
   Direct
   Programmed

3. Flux
   Activity (actual flux)
   Capacity
       Current environment (no substrate limitation)
       Optimized environment (no substrate limitation and temperature, etc. optimized)
   Potential (relevant genes fully expressed)

See text for details.

product). Thus, to say that a sink has an 'activity' is hopelessly simplistic; if used to mean actual flux, this is better stated directly as flux or growth rate; if not, it is necessary to identify just what is limiting it.

A complete understanding of any partial process of partitioning will include information from each of these categories of time scale, response and flux. As I progress from considering the transport conduit to discussing source leaves, then sinks, and finally an integration of these components into a whole plant, it will become clear that this is not yet possible, although the quality of our information is improving.

8.4 Transport in the phloem

The properties of the transport conduit connecting sources and sinks could have a considerable effect on partitioning. The majority view is currently that phloem transport involves a mass flow driven by a gradient of turgor pressure (the Munch hypothesis; Figure 8.1). The flux of sucrose, J_{suc} through the phloem is then described by

$$J_{suc} = \delta P.C.k,$$

where δP is the turgor pressure difference between source and sink, C the concentration of sucrose in phloem and k a conductivity coefficient. Where transport is described by the Hagen–Poisseuille relationship

$$k = A.r^2/8\eta.l,$$

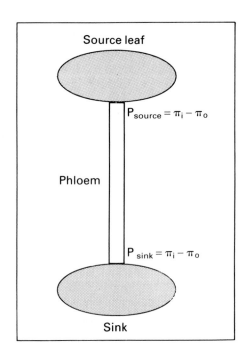

Figure 8.1. *Determinants of transport through the phloem. If the sieve tubes are considered as pressurized vessels connecting a source leaf to a sink, the pressure within sieve tubes in the source leaf, P_{source}, is determined by the difference in osmotic potential $(\pi_i - \pi_0)$ across their boundary membrane. Pressure within the sieve tubes in the sink, P_{sink}, is similarly determined. The difference, $P_{source} - P_{sink}$, gives the driving force for transport through the phloem.*

Source leaf

$P_{source} = \pi_i - \pi_0$

Phloem

$P_{sink} = \pi_i - \pi_0$

Sink

where A is the cross-sectional area of phloem with pore radius r, η is viscosity and l the length of phloem pathway. r is unlikely to vary greatly in different parts of the same plant, and so the variables of most interest are δP, C and l.

The general opinion is that phloem itself does not limit translocatory flux (Milthorpe and Moorby, 1969; Passioura and Ashford, 1974), although transport over distances greater than say 0.5 m may be limited due to the reciprocal relationship between J and l (Canny, 1976; Cook and Evans, 1978). Further, the longer the phloem path, the greater the lateral leakage from it (Minchin and Thorpe, 1987), with a possible reduction in flux. Partitioning between two sinks may thus be determined in part by their relative distances from a common source (Cook and Evans, 1978; Section 8.7).

The turgor pressure in phloem of a source leaf is determined by the difference in solute potential across the phloem bounding membrane (Figure 8.1). The dominant component of turgor will therefore be the concentration of sucrose in the phloem, although in sinks the apoplastic solute potential may have a larger role, as will be discussed below. The extent to which osmoregulation may reduce perturbations in turgor of the phloem is not yet clear (Williams *et al.*, 1991), although there is indirect evidence of changes of pressure gradient in phloem of intact barley following a small osmotic shock (Thorpe *et al.*, unpublished data).

With phloem transport shown to be dependent on water relations of source and sink, first source leaves and then sinks will be discussed with a view to relating their metabolism to regulation of phloem water relations.

8.5 Source leaves

The two questions to be addressed in this section are: what controls the flux of carbon out of a source leaf, and does the source leaf control the direction of travel of assimilate leaving it?

Source leaves often photosynthesize at a rate below that of which they are capable, placing an upper limit on their translocatory flux. The activity or amount of chloroplastic enzymes may be down-regulated from the maximum possible, perhaps as a response to carbohydrate concentration in the source leaf (Pamplin and Chapman, 1978; Stitt, 1991; Stitt *et al.*, 1990; Williams *et al.*, 1992). Feedback inhibition of photosynthesis by sugars, operating on the existing metabolic machinery, is probably less important in long-term regulation of photosynthesis *in vivo* than is control at the gene level. Indeed, genes of both photosynthesis itself and of carbohydrate storage are controlled, possibly indirectly, by sucrose: ADPGlc pyrophosphorylase (starch synthesis) and sucrose–sucrose fructosyl transferase (fructan synthesis) are enhanced by sucrose, and Rubisco and other photosynthetic enzymes repressed by it (Cairns and Pollock, 1988; Krapp *et al.*, 1991; Müller-Röber *et al.*, 1990; Schafer *et al.*, 1992; Sheen, 1990; Stitt *et al.*, 1990). The sucrose status of the source leaf will therefore determine the future potential acquisition of sucrose from photosynthesis by that leaf and the extent of storage.

Sucrose is compartmented between cytosol, vacuole and apoplast (Delrot,

1989; Farrar and Farrar, 1986; Tetlow and Farrar, 1992) and the fluxes between these compartments may place constraints on export; for example, after several hours of darkness, efflux across the tonoplast may rate-limit export in sucrose-storing species. Suggestions that the sucrose content of the source leaf determines the rate of export (Ho and Thornley, 1978; Swanson and Christy, 1976) take no account of compartmentation, although in barley leaves only the content of sucrose in cytosolic and apoplastic pools (and thus readily available for trans-location) correlates with rate of sucrose export (Farrar and Farrar, 1985).

It is unlikely that sucrose concentration alone controls export, since export rate can be varied independently of photosynthesis such that it is partially regulated (Ho, 1976). Furthermore, events outside of the source leaf can influence export rate in the short term: for example, changing the temperature of a fruit results in increased export from a source leaf shortly afterwards (Geiger and Fondy, 1985; Moorby and Jarman, 1975). It may be possible to reconcile these findings if sucrose in the transport pool is the major source of turgor generation when it is loaded into phloem, and it is the turgor gradient between source and sink that determines flux (see Section 8.4). Since potassium can contribute to turgor generation in phloem, and its concentration may vary inversely with that of sucrose, there is the potential for regulation of export flux independently of sucrose (Smith and Milburn, 1980).

There is no evidence for control being exercised during the passage of sucrose from mesophyll cell to phloem, nor in the capacity of the loading system in the leaf, although in barley grown at low light and transferred to high, the resultant increase in photosynthesis was not accompanied by increased translocation, perhaps suggesting a limit on loading (Farrar and Farrar, 1987). Rather, it appears that control is exercised via pool sizes, and more importantly, that there is coarse control of enzymes of photosynthesis and carbohydrate storage.

It is generally held that the direction in which assimilates travel after leaving the source leaf is determined by the sinks. Certainly sinks seem to be supplied largely from source leaves near to them. It has been suggested that the fate of carbon exported during the light period may differ from that exported at night (Huber, 1983). There are sound reasons for believing that the source leaf may have an influence on direction, in particular that altering the flux out of a source leaf may result in a changed distribution of its products; this will be discussed below (Section 8.7).

8.6 Sinks

Two questions are addressed in this section: what are the mechanistic relationships between rate of import into a sink, its sugar status, and its rate of metabolism? and how is coarse control of sink metabolism exercised?

There is substantial evidence that the sugar status of a sink is instrumental in determining import into it. Classically, demonstrations of a whole-tissue gradient in sugar concentration between sources and sinks, although of limited use in determining mechanism, suggest that sink sugar status is part of the control of partitioning (Kursanov, 1984; Mason and Maskell, 1928; Swanson and Christy, 1976). More recently, the role of sink sugar status has been discussed in relation to import (Ho, 1988). Walker and Ho (1977) varied the temperature of tomato fruit and showed that the resultant changes in internal sucrose concentration and rate of import were inversely related, although the localization of the sucrose within and between cells was not determined. A study of partitioning between the two halves of a split root system in barley, where anatomically and metrically identical sinks are compared, thereby eliminating many complicating variables, has shown that sugar status of the roots is central to their rate of import. Feeding sucrose to one root half both increases the size of the cytosolic sugar pool and, after about 1–2 h, decreases dramatically the rate of import of ^{11}C relative to the untreated root half (Farrar and Minchin, 1991). Only metabolized sugars, not sugars such as 3-O-methyl glucose, exert this effect (Farrar et al., unpublished data). Neither metabolic rate (estimated as respiration rate) nor elongation rate were related to import in the short term, so it was suggested that sugar status mediates between import, and growth and metabolism as the ultimate controllers of sink status (Farrar and Minchin, 1991; Farrar et al., unpublished data). In pea fruits the metabolic determinants of sink status are the active uptake of sugar by the embryo and the storage of carbohydrates in the seed coat; again a sugar pool — in the cytosol of the seed coat and in the apoplast between seed coat and embryo — will mediate between these fluxes and import via the phloem, buffering one from the other and acting as an integrated measure of the processes of delivery and consumption.

Changes in apoplastic solute concentration which are needed to alter phloem turgor are also likely to alter metabolism of receiver cells in the sink. In some apoplastically unloading sinks, the uptake of sugars from the apoplast is turgor-dependent, so that sucrose uptake by sugar beet taproot tissue is faster at low turgor (Wyse et al., 1985); and for potato tubers (which unload at least partially symplastically) starch synthesis from exogenous glucose is higher at low turgor (see Chapter 5). In each case, an increase in apoplastic solute concentration will enhance phloem import and metabolism of the sink in tandem.

If the cytosolic sugar pool is indeed important, then there must be a means of communicating the size of this pool to the phloem itself, and translating it into an effect on phloem water relations (Figure 8.1); either the solute potential within, or that immediately outside, the phloem must be modified. That within the phloem is constrained partly by the sugar arriving from source leaves and partly by the rate of unloading of sugar, but it may be that simple mass action cannot control turgor adequately. Apoplastic solute potential is also likely to change and certainly this seems a major driving variable in fruits (Patrick, 1990, 1991; Thorne, 1985). There are problems with applying this model to fibrous roots. Patrick (1990, 1991) has pointed out that the percentage change in cytosolic solutes necessary to change phloem transport is impossibly high; therefore an amplification step of some sort may be necessary. Further, in a fibrous root, the concentration of solutes needed in the apoplast to reduce turgor appreciably is so high that a massive flux of these solutes out into the soil would result. Since it is difficult to design an hypothesis around the idea that mean cytosolic sugar pool size controls import by modifying solute potential in the phloem symplast or apoplast, I wish to consider next a quite different hypothesis to explain the involvement of sugars, which still acknowledges the central importance of water relations.

Elongation growth of cereal roots has been studied from two quite different points of view, and two correspondingly distinct literatures have developed. The more voluminous considers elongation as the flux of water into expanding cells, and contains excellent descriptions of the biophysics of cell expansion (Cosgrove, 1986; Taiz, 1984; Tomos, 1988). Another literature considers the incrementing of dry weight (Farrar and Williams, 1990; Lambers *et al.*, 1991). Clearly each is but a partial description of that process; it is time to integrate the descriptions. In particular, it is desirable to clarify the role of sucrose in elongation since there is no commonly recognized central role for sucrose in elongation *per se*, although it is the main substrate for dry weight increment.

Elongation growth and sucrose import would be directly related if phloem sap flowed directly into expanding cells, providing both the water and the solutes necessary for cell expansion (Figure 8.2). The caveat that both water and solutes can enter expanding cells across the plasma membrane — for example, in excised roots — does not lessen the likelihood of such direct unloading in intact, rapidly photosynthesizing plants. Mature protophloem extends to the distal end of the

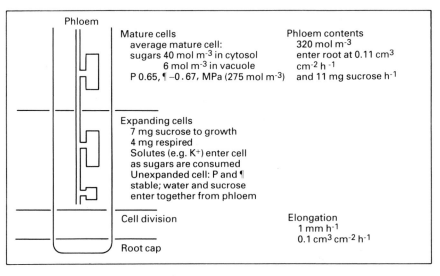

Figure 8.2. A model of cell expansion and phloem import in the apical region of fibrous roots. Calculations apply to barley and are based on data in Farrar and Williams (1990) and Farrar (unpublished data); figures for P and π are taken from measurements on wheat roots (Pritchard et al., 1991). Fluxes of water and sucrose, calculated per cm^2 cross-sectional area of fully expanded root, show a close correspondence between rate of import of phloem contents and rate of accretion of water in expanding cells.

elongation zone in fibrous roots (Esau, 1941; Jensen, 1955; Peterson, 1967) where symplastic unloading occurs (Farrar, 1985; Giaquinta et al., 1983; Murphy, 1989). It may be no coincidence that the net flux of water into growing barley root tips is within 10% of the calculated flux into those root tips via the phloem, and the solute potential in expanding cells is similar to that calculated for the phloem (Figure 8.2; Pritchard et al., 1991). That sucrose would be, for a short period, the dominant osmoticum of expanding cells — although much K^+ would accompany it from the phloem — is consistent with the falling concentration of sugars with increasing distance from root tips (Sutcliffe and Sexton, 1969). Both the high activity of acid invertase (presumably mainly vacuolar) in elongation zones (Brown and Broadbent, 1950; Sutcliffe and Sexton, 1969), and the rapid turnover of sucrose in roots (Farrar and Williams, 1990) suggest that sucrose will be rapidly hydrolysed after unloading, and the hexoses metabolized. Hydrolysis by invertase would sustain osmotic potential, to which there would be a steadily increasing contribution of inorganic ions, taken up from the medium as metabolism removed the sugars.

Such a link between phloem import and expansion growth, similar to that suggested for pea stems (Schmalstig and Cosgrove, 1990) obviates the need for a communication or amplification step between receiver cells and events in the phloem, to relate sucrose status of receiver cells to import. Were cell expansion to be inhibited and sucrose to accumulate in expanding cells, there would be an immediate rise in phloem turgor as unloading reduced. Import to the root would then fall.

A second link between sucrose and elongation may be sought in the equation describing volume growth rate, r:

$$r = m(P - Y),$$

where P is turgor pressure, Y the yield threshold and m a coefficient of volume extensibility; m will incorporate (for example) ability to synthesize new cell wall and other structural material, and thus be dependent on both substrate and energy supply (Green et al., 1971).

The long-term effects of sucrose, as a coarse controller of root growth (Table 8.2; Evans, 1972; Koch et al., 1992; Williams and Farrar, 1990) need a quite different explanation. It is clear that sucrose regulates the activity of a number of key enzymes and pathways in barley roots (Table 8.2) and by analogy with the elegant work of Koch et al. (1992) on sucrose synthase genes in maize roots we assume that it operates at the level of gene expression. Whilst the actual signal molecule is unknown, it is likely to be a metabolite of sucrose as a number of sugars can have a similar effect, and the initial site of action may be on energy-consuming processes rather than on respiratory metabolism (Farrar and Williams, 1990, also unpublished data). It may be that the elongation process itself is sensitive to coarse control by sucrose, but a more interesting possibility exists. The supply of newly-divided cells for elongation may be under the control of sucrose. In roots of Pisum and Vicia sucrose starvation results in the arrest of meristematic cells in either the G_1 or G_2 phase of the cell cycle; resupplying

171

Table 8.2. *The control by sucrose of growth and metabolism of barley roots*

	Shoot-pruned (*shaded)	Control + sucrose	Shoot-pruned + sucrose
Root elongation rate	27	—	57
Flux of sucrose to structure	56	98	126
Respiratory oxygen uptake	46	125	74
Capacity of cytochrome pathway	66	—	109
Respiration of isolated mitochondria			
Using NADH	43*	130	—
Using malate + pyruvate	56*	125	—
Total protein	72	—	83
Activity of acid invertase	33	—	66
Activity of sucrose synthase	63	—	61
Activity of fumarase	65	130	—
Activity of cytochrome oxidase	70	128	—

Young barley plants grown hydroponically were either pruned by removing all leaves (or all leaves were darkened), and 24 h later 20 mol m^{-3} sucrose was added to the root system; alternatively, unpruned plants were supplied with sucrose for 24 h. Data, expressed as a percentage of control values, from Bingham and Farrar (1988), Williams and Farrar (1990), McDonnell and Farrar (unpublished data), and Williams *et al.* (unpublished data). Original measurements were on a weight basis, apart from mitochondrial respiration, fumarase and cytochrome oxidase which were on a mitochondrial protein basis.

sucrose causes the cell cycle to restart but only after a lag of some hours, suggesting that sucrose is more than simply an energy source (van't Hof and Kovacs, 1972; van't Hof *et al.*, 1973). The *cdc2* gene is not expressed in nutrient-deficient cells, and its gene product p34^{cdc2} may be a major controller of the plant cell cycle (Francis, 1992), making it a candidate for a sucrose control site.

An attractive feature of this hypothesis is that, since sucrose will have to diffuse from the site of phloem unloading to the meristem, a rather high sensitivity to sucrose will be achieved. It is likely that flux from phloem to meristem will be largely diffusive as the positive pressure in the phloem will be dissipated in cell expansion.

Import into sinks is thus controlled in the short term by growth-related processes, apparently mediated by cytosolic sugar pool size. This fine control works within a framework of maximum growth set by the sucrose-determined expression of key genes coding for proteins central to the rate control of growth and respiration.

8.7 Generation complexity: multiple sinks

When a single source supplies assimilate to multiple sinks (previously called allocation; Farrar, 1989) it is necessary to determine what controls the proportion of assimilate going to each sink, a proportion which may vary with the flux from the source. The simplest approach is to examine the allocation between two sinks predicted by use of the Poisseuille equation; the fluxes J_1 and J_2 are then related by

$$J_1/J_2 = (\delta P_1.A_1.r^2_1.l_2)/(\delta P_2.A_2.r^2_2.l_1),$$

where δP is pressure difference, A cross-sectional area, r radius and l distance. Concentration of sucrose will be determined by the source and so is equal for the two sinks, as is P_{source}. If A and r are equal for the two sinks, the important variables are l and δP. Other things being equal, a closer sink should receive a greater flux than a distant; and a sink capable of reducing the turgor in phloem below that of its neighbour will receive a greater flux. Clearly the two — distance and turgor — will interact, such that a distant sink with very low P_{sink} may be able to compensate for its remoteness. The data of Cook and Evans (1978), which show an interaction between distance and sink size (the number of grains remaining on an ear of wheat) could be explained qualitatively at least in this way. So could the old observation that reproductive organs make strong sinks, usually explained hormonally, for if a growing vegetative apex unloads symplastically (no apoplastic step between phloem and receiver cells) the phloem will have a turgor similar to that of the expanding cells — above the yield threshold; the transition to a reproductive sink may be accompanied by transition to apoplastic phloem unloading (an apoplastic step between maternal and juvenile tissue), where the turgor of phloem could be kept low by apoplastic solutes, whilst that of expanding cells in a different symplastic domain might be much higher to facilitate their expansion. Direct experimental tests of these suggestions need to be made.

The direction in which sucrose is transported during export from a source leaf may depend on the flux out of it. If flux is lowered, turgor in source leaf phloem will be lower, and so if the turgor in each of two receiving sinks is different and unaltered, the ratio of the turgor gradients into each will change, with a resultant change in relative flux. This may be the origin of the suggestion that extensive chloroplastic storage of starch causes a high S:R (Geiger *et al.*, 1985; Huber, 1983): the flux from leaves at night will be lower and may be disproportionately directed to the closer shoot meristems.

A model of a single source feeding multiple sinks, assuming Munch pressure flow and assimilate utilization in sinks following Michaelis–Menten kinetics, shows that the behaviour of a one source, two sink system can be remarkably complex (Minchin *et al.*, unpublished data). If both sinks are equivalent in all respects, partitioning between them is independent of flux from the source. If the two sinks are not identical, the behaviour of the system depends in part on whether or not the sinks can sustain a higher flux of assimilate than they are receiving at the outset, and on the parameters describing sink function. Such a model is valuable in predicting results of experiments in the highly complex system of a whole plant. That it can predict successfully has been confirmed in our experiments where the shoot system of barley was shaded whilst one half of the split root system was warmed or cooled; only when one half of the root was at a different temperature did the flux change consequent on shading alter the partitioning between the root halves, and in all cases the model could predict the outcome (Minchin *et al.*, unpublished data).

Thus the fine controls that dictate partitioning between organs, and reflect the mechanisms of phloem transport and partitioning, can give rise to a rich behaviour. Currently we can only speculate about how coarse control, operating selectively between sinks, can bring about longer term changes in partitioning. Whilst it is possible that fine controls generate long-term adaptive responses (such as increased S:R on shading or decreased S:R at low nutrient supply) it is more likely to be coarse control acting via gene expression that is the central arbiter of plant form.

8.8 Towards balanced S:R

Brouwer (1962) has shown that, if *Phaseolus* is perturbed by removal of half of the shoot or the root system, subsequent growth favours the mutilated part such that the control S:R is restored. Clearly, achieving balanced S:R is not just a matter of unfolding a developmental sequence, but of constant mutual regulation of shoot and root. The dependence of the allometric constant k on internal changes such as flowering (Troughton, 1955) and its heritability (Troughton and Whittington, 1969) also argue for such regulation.

In outline, such regulation may be achieved as follows. Phloem transport is driven by turgor pressure gradients established by differences in solute potential across the sieve tube bounding membrane in source and sink; in growing tissues with symplastic phloem unloading, phloem may empty directly into expanding cells. Cytosolic sugar pools buffer the consumption of assimilate in receiver cells from sugar arriving in the phloem, and may serve to communicate between them. Internal and external factors affecting phloem water relations or sugar pool sizes may affect partitioning, therefore, but not necessarily to bring about those changes which will become apparent in the longer term. Such changes to the S:R may rather be a consequence of coarse control — where sucrose acts as a messenger between source and sink, taking information on availability of assimilate in one direction and demand for it in the other, and modifying the expression of key genes such that the supply of assimilate from the source and rate of use in sinks are closely matched. Such matching explains why it has always been difficult to tell if growth is source or sink limited.

This model for regulating S:R is both preliminary and in part untested. It is also incomplete: it refers to carbon alone. Since decrease in supply of nutrients results in decreased S:R (Wilson, 1989), any theory of how S:R is regulated must be able to explain the effects of inorganic nutrients. A first attempt at an explanation follows. When nitrogen is scarce, sugars will accumulate in the root but they will not stimulate root growth as nitrogen limits it; the stimulatory effects on respiratory enzymes will not occur (Farrar and Williams, 1990). In the leaf, sugars will accumulate and repress genes of photosynthesis and stimulate those of carbohydrate storage. If the shoot is repressed more than the root, S:R will

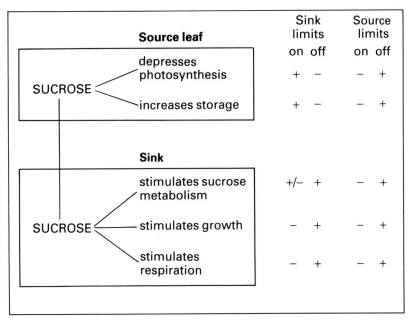

		Sink limits		Source limits	
Source leaf		on	off	on	off
SUCROSE	depresses photosynthesis	+	–	–	+
	increases storage	+	–	–	+
Sink					
SUCROSE	stimulates sucrose metabolism	+/–	+	–	+
	stimulates growth	–	+	–	+
	stimulates respiration	–	+	–	+

Figure 8.3. An hypothesis for the role of sucrose in controlling shoot:root ratio. Within source leaf and sink, sucrose regulates key processes at the level of gene expression. On the right, the consequences of growth being limited by the sink (e.g. low nutrient, low temperature) or source (e.g. low light, low CO_2) are shown for two cases: where the limitation is established (on) or just following its rmeoval (off).

decrease; this may depend on the effects on gene expression of nitrogen status itself (Redinbaugh and Campbell, 1991) or phosphorus status (Fife *et al.*, 1990).

The consequences of source and sink limitations of growth for the model of coarse control are shown in Figure 8.3. Whilst the preliminary nature of this model is all too evident, it does begin to offer a rational basis from which to design work on carbon partitioning in plants, emphasizing as it does the central role of sucrose, a central role which extends from fine to coarse control of partitioning.

Acknowledgements

I would like to thank the AFRC and NERC for financial support; the Wain fund of the AFRC helped with sabbatical leave. To Peter Minchin and Michael Thorpe of DSIR Fruit and Trees, New Zealand, I owe gratitude not only for hospitality during the genesis of this paper but also for prolonged debate and discussion; I have doubtless plagiarized them outrageously.

References

Baker, D.A. and Milburn, J.A. (1989) *Transport of Photoassimilates*. Longman, Harlow.

Barnes, A. (1979) Vegetable plant part relationships. II. A quantitative hypothesis for shoot/storage root development. *Ann. Bot.* **43**, 487–499.

Bingham, I.J. and Farrar, J.F. (1988) Regulation of respiration in roots of barley. *Physiol. Plant.* **73**, 278–285.

Bonnemain, J.-L., Delrot, S., Lucas, W. and Dainty, J. (1991) *Recent Advances in Phloem Transport and Assimilate Compartmentation*. Ouest, Nantes.

Brenner, M.L. (1987) The role of hormones in photosynthate partitioning and seed filling. In: *Plant Hormones and their Role in Plant Growth and Development* (ed. P.J. Davis). Martinus Nijhoff, Dordrecht, pp. 474–493.

Brouwer, R. (1962) Nutritive influences on the distribution of dry matter in the plant. *Neth. J. Agric. Sci.* **10**, 399–408.

Brouwer, R. (1983) Functional equilibrium: sense or nonsense? *Neth. J. Agric. Sci.* **31**, 335–348.

Brown, R. and Broadbent, D. (1950) The development of cells in the growing zones of the root. *J. Exp. Bot.* **1**, 249–263.

Cairns, A.J. and Pollock, C.J. (1988) Fructan biosynthesis in excised leaves of *Lolium temulentum*. *New Phytol.* **109**, 407–413.

Canny, M.J. (1976) Translocation and distance I. The growth of the fruit of the sausage tree, *Kigelia pinnata*. *New Phytol.* **72**, 1269–1280.

Clive, M.G. (1991) Apical dominance. *Bot. Rev.* **57**, 318–350.

Cook, M.G. and Evans, L.T. (1978) Effect of relative size and distance of competing sinks on the distribution of photosynthetic assimilate in wheat. *Aust. J. Plant Physiol.* **5**, 495–509.

Cosgrove, D. (1986) Biophysical control of plant cell growth. *Ann. Rev. Plant Physiol.* **37**, 377–405.

Delrot, S. (1989) Loading of phloem assimilates. In: *Transport of Photoassimilates* (eds D.A. Baker and J.A. Milburn). Longman, Harlow, pp. 167–204.

Esau, K. (1941) Phloem anatomy of tobacco affected with curly top and mosaic. *Hilgardia*, **13**, 437–490.

Evans, P.S. (1972) Root growth of *Lolium perenne* L. III. Investigation of the mechanism of defoliation-induced suppression of elongation. *N. Z. J. Agric. Res.* **15**, 347–355.

Farrar, J.F. (1985) Fluxes of carbon in roots of barley plants. *New Phytol.* **99**, 57–69.

Farrar, J.F. (1989) Temperature and the partitioning and translocation of carbon. In: *Plants and Temperature* (eds S.P. Long and F.I. Woodward) *Symposium of the Society for Experimental Biology* Volume 42, Cambridge University Press, Cambridge, pp. 203–235.

Farrar, J.F. and Farrar, S.C. (1985) Fluxes of carbon in leaves and roots of barley plants. In: *Regulation of Sources and Sinks* (eds B. Jeffcoat and A.D. Stead). British Plant Growth Regulator Group, Bristol, pp. 67–84.

Farrar, J.F. and Williams, J.H.H. (1990) Control of the rate of respiration in roots: compartmentation, demand and the supply of substrate. In: *Compartmentation of Metabolism in Non-Photosynthetic Tissue* (ed. M. Emes). Cambridge University Press, Cambridge, pp. 167–188.

Farrar, J.F. and Minchin, P.E.H. (1991) Carbon partitioning in split root systems of barley: relation to metabolism. *J. Exp. Bot.* **42**, 1261–1269.

Farrar, S.C. and Farrar, J.F. (1985) Carbon fluxes in leaf blades of barley. *New Phytol.* **99**, 57–69.

Farrar, S.C. and Farrar, J.F. (1986) Compartmentation and fluxes of sucrose in intact leaf blades of barley. *New Phytol.* **103**, 645–657.

Farrar, S.C. and Farrar, J.F. (1987) Effects of photon fluence rate on carbon partitioning in barley source leaves. *Plant Physiol. Biochem.* **25**, 541–548.

Fife, C.A., Newcomb, W. and Lefebvre, D.D. (1990) The effect of phosphate deprivation on protein synthesis and fixed carbon storage reserves in *Brassica nigra* suspension cells. *Can. J. Bot.* **68**, 1840–1847.

Francis, D. (1992) The cell cycle in plant development. *New Phytol.* **121**, in press.

Geiger, D.R. and Fondy, B.R. (1985) Responses of export and partitioning to internal and environmental factors. In: *Regulation of Sources and Sinks in Crop Plants* (eds B. Jeffcoat, A.F. Hawkins and A.D. Stead). British Plant Growth Regulator Group, Bristol, pp. 177–194.

Geiger, D.R., Jablonski, L.M. and Ploeger, B.J. (1985) Significance of carbon allocation to starch in growth of *Beta vulgaris*. In: *Regulation of Carbon Partitioning in Photosynthetic Tissue* (eds R.L Heath and J. Preiss). American Society of Plant Physiologists, Maryland, pp. 289–307.

Giaquinta, R.T., Lin, W., Sadler, N.L. and Franceschi, V.R. (1983) Pathway of phloem unloading of sucrose in corn roots. *Plant Physiol.* **72**, 362–367.

Green, P.B., Erickson, R.O. and Buggy, J. (1971) Metabolic and physical control of cell elongation rate. *Plant Physiol.* **47**, 423–430.

Ho, L.C. (1976) The relationship between the rates of carbon transport and of photosynthesis in tomato leaves. *J. Exp. Bot.* **27**, 87–97.

Ho, L.C. (1988) Metabolism and compartmentation of imported sugars in sink organs in relation to sink strength. *Ann. Rev. Plant Physiol. Plant Mol. Biol.* **39**, 355–378.

Ho, L.C. and Thornley, J.H.M. (1978) Energy requirements for assimilate translocation from mature tomato leaves. *Ann. Bot.* **42**, 481–483.

van't Hof, J., Hoppin, D.P. and Yagi, S. (1973) Cell arrest in G1 and G2 of the mitotic cycle of *Vicia faba* root meristems. *Am. J. Bot.* **60**, 889–895.

van't Hof, J. and Kovacs, C.J. (1972) Mitotic cycle regulation in the meristem of cultured roots: the principal control points hypothesis. In: *The Dynamics of Meristem Cell Populations* (eds M.W. Miller and C.C. Kuehnert). Plenum, New York, pp. 15–32.

Huber, S.C. (1983) Relationship between photosynthetic starch formation and dry weight partitioning between shoot and root. *Can. J. Bot.* **61**, 2709–2716.

Jensen, W.A. (1958) A morphological and biochemical analysis of the early phases of cellular growth in the root tip of *Vicia faba*. *Exp. Cell Res.* **8**, 506–522.

Koch, K.E., Nolte, K.D., Duke, E.R., McCarty, D.R. and Avigne, W.T. (1992) Sugar levels modulate differential expression of maize sucrose synthase genes. *Plant Cell*, **4**, 59–69.

Krapp, A., Quick, W.P. and Stitt, M. (1991) There is a dramatic loss of Rubisco, other Calvin cycle enzymes and chlorophyll when glucose is supplied to mature spinach leaves via the transpiration stream. *Planta*, **186**, 58–69.

Kursanov, A.L. (1984) *Transport in Plants*. Elsevier, Amsterdam.

Lambers, H., van der Werf, A., and Konings, H. (1991) Respiratory patterns in roots in relation to their functioning. In: *Plant Roots: the Hidden Half* (eds Y. Waisel, P. Eshel and U. Kafkafi). Marcel Dekker, New York, pp. 229–263.

Mason, T.G. and Maskell, E.J. (1928) Studies on the transport of carbohydrate in the cotton plant. II. The factors determining the rate and the direction of the movement of sugars. *Ann. Bot.* **42**, 571–636.

Milthorpe, F.L. and Moorby, J. (1969) Vascular transport and its significance in plant growth. *Ann. Rev. Plant Physiol.* 20, 117–138.

Minchin, P.E.H. and Thorpe, M.T. (1987) Measurement of unloading and reloading of photoassimilate along the stem of bean. *J. Exp. Bot.* 38, 211–220.

Moorby, J. and Jarman, P.D. (1975) The use of compartment analysis in the study of movement of carbon through leaves. *Planta*, 122, 155–168.

Müller-Röber, B.T., Kossman, J., Hannah, L.C., Willmitzer, L. and Sonnewald, U. (1990) ADP-glucose pyrophosphorylase genes from potato: mode of RNA expression and its relation to starch synthesis. In: *Phloem Transport and Assimilate Compartmentation* (eds J.-L. Bonnemain, S. Delrot, W.J. Lucas and J. Dainty). Ouest, Nantes, pp. 204–208.

Murphy, R. (1989) Water flow across the sieve tube boundary, estimating turgor and some implications for phloem loading and unloading. IV. Root tips and seed coats. *Ann. Bot.* 63, 571–579.

Pamplin, E.J. and Chapman, J.M. (1978) Sucrose suppression of chlorophyll synthesis in tissue culture: changes in the activity of the enzymes of the chlorophyll biosynthetic pathway. *J. Exp. Bot.* 26, 212–218.

Passioura, J.B. and Ashford, A.E. (1974) Rapid translocation in the phloem of wheat roots. *Aust. J. Plant Physiol.* 1, 521–527.

Patrick, J.W. (1987) Are hormones involved in assimilate transport? In: *Hormone Action in Plant Development — A Critical Appraisal* (eds G.V. Hoad, J.R. Lenton, M.B. Jackson and R.K. Atkin). Butterworths, London, pp. 175–188.

Patrick, J.W. (1990) Sieve element unloading: cellular pathway, mechanism and control. *Physiol. Plant.* 78, 298–308.

Patrick, J.W. (1991) Control of phloem transport to and short-distance transfer in sink regions: an overview. In: *Recent Advances in Phloem Transport and Assimilate Compartmentation* (eds J.-L. Bonnemain, S. Delrot, W. Lucas and J. Dainty). Ouest, Nantes, pp. 167–177.

Pearsall, W.H. (1927) Growth studies VI. On the relative sizes of growing plant organs. *Ann. Bot.* 41, 549–556.

Peterson, R.L. (1967) Differentiation and maturation of primary tissues in white mustard root tips. *Can. J. Bot.* 45, 319–331.

Pollock, C.J. and Lloyd, E.J. (1987) The effect of low temperature upon starch, sucrose and fructan synthesis in leaves. *Ann. Bot.* 60, 231–235.

Pritchard, J., Wyn Jones, R.G. and Tomos, A.D. (1991) Turgor, growth and rheological gradients of wheat roots following osmotic stress. *J. Exp. Bot.* 42, 1043–1049.

Redinbaugh, M.G. and Campbell, W.H. (1991) Higher plant responses to environmental nitrate. *Physiol. Plant.* 82, 640–650.

Schafer, C., Simper, H. and Hofman, B. (1992) Glucose feeding results in coordinated changes of chlorophyll content, rubisco activity and photosynthetic capacity in photoautotrophic suspension cultured cells of *Chenopodium rubrum*. *Plant, Cell Environ.* 15, 343–350.

Schmalstig, J.G. and Cosgrove, D.J. (1990) Coupling of solute transport and cell expansion in pea stems. *Plant Physiol.* 94, 1625–1633.

Sheen, J. (1990) Metabolic repression of transcription in higher plants. *Plant Cell*, 2, 1027–1038.

Smith, J.A.C. and Milburn, J.A. (1980) Osmoregulation and the control of phloem-sap composition in *Ricinus communis* L. *Planta*, 148, 28–34.

Stitt, M. (1991) Rising CO_2 levels and their potential significance for carbon flow in photosynthetic cells. *Plant Cell Environ.* 14, 741–762.

Stitt, M., von Schaewen, A. and Willmitzer, L. (1990) 'Sink' regulation of photosynthetic metabolism in transgenic tobacco plants expressing yeast invertase in their cell walls involves a decrease of the Calvin-cycle enzymes and an increase of glycolytic enzymes. *Planta*, 183, 40–50.

Sutcliffe, J.F. and Sexton, R. (1969) Cell differentiation in the root in relation to physiological functioning. In: *Root Growth* (ed. W.J. Whittington). Butterworth, London, pp. 80–102.

Swanson, C.A. and Christy, A.L. (1976) Control of translocation by photosynthesis and carbohydrate concentration of the source leaf. In: *Transport and Transfer Processes in Plants* (eds I.F. Wardlaw and J.B. Passioura). Academic Press, New York, pp. 329–338.

Taiz, L. (1984) Plant cell expansion: regulation of cell wall mechanical properties. *Am. Rev. Plant Physiol.* 35, 585–657.

Tetlow, I.J. and Farrar, J.F. (1992) Apoplastic sugar concentration and pH in barley leaves infected with brown rust. *New Phytol.* in press.

Thomas, T.H. (1986) Hormonal control of assimilate movement and compartmentation. In: *Plant Growth Substances 1985* (ed. M. Bopp). Springer-Verlag, Berlin, pp. 350–359.

Thorne, J.H. (1985) Phloem unloading of C and N assimilates in developing seeds. *Am. Rev. Plant Physiol.* 36, 317–343.

Thornley, J.H.M. (1977) Root:shoot interactions. In: *Integration of Activity in the Higher Plant* (ed. D.H. Jennings). Cambridge University Press, Cambridge, pp. 367–390.

Tomos, A.D. (1988) Cellular water relations of plants. In: *Water Science Reviews*, Volume 3 (ed. F. Franks). Cambridge University Press, Cambridge, pp. 186–277.

Trewavas, A. (1991) How do plant growth substances work? II. *Plant Cell Environ.* 14, 1–12.

Troughton, A. (1955) The application of the allometric formula to the study of the relation between the roots and shoots of young grass plants. *Agric. Progr.* 30, 59–65.

Troughton, A. and Whittington, W.J. (1969) The significance of genetic variation in root systems. In: *Root Growth* (ed. W.J. Whittington). Butterworth, London, pp. 296–314.

Walker, A.J. and Ho, L.C. (1977) Carbon translocation in the tomato: carbon import and fruit growth. *Ann. Bot.* 41, 813–823.

Wardlaw, I.F. (1990) The control of carbon partitioning in plants. *New Phytol.* 116, 341–381.

Wareing, P.F. (1976) Environmental control of development. In: *The Developmental Biology of Plants and Animals* (eds C.F. Graham and P.F. Wareing). Blackwell Scientific Publications, Oxford, pp. 349–352.

Williams, J.H.H. and Farrar, J.F. (1990) Control of barley root respiration. *Physiol. Plant.* 79, 259–266.

Williams, J.H.H., Minchin, P.E.H. and Farrar, J.F. (1991) Carbon partitioning in split root systems of barley: the effect of osmotica. *J. Exp. Bot.* 42, 453–460.

Williams, J.H.H., Winters, A.L., Pollock, C.J. and Farrar, J.F. (1992) Regulation of leaf metabolism by sucrose. *Fiziol. Rast.* in press.

Wilson, J.B. (1989) A review of evidence on the control of shoot:root ratio, in relation to models. *Ann. Bot.* 61, 433–449.

Wyse, R., Briskin, D. and Aloni, B. (1985) Sucrose transport: regulation and mechanism at the tonoplast. In: *Regulation of Carbon Partitioning in Photosynthetic Tissue* (eds R.C. Heath and J. Preiss). American Society of Plant Physiologists Maryland, pp. 231–252.

Carbon metabolism and transport in a biotrophic fungal association

J.L. Hall, J. Aked, A.J. Gregory and T. Storr

9.1 Introduction

Fungal disease in higher plant systems may affect both the carbohydrate metabolism of infected and other tissues and also the transport of carbon to and from the infected areas. Infection usually results in net transfer of reduced carbon from host to pathogen. From the limited evidence available, it appears that microbial infection may enhance solute efflux across the plasma membrane which will presumably require a significant modification of carbon partitioning in the host tissue (Patrick, 1989). Thus, for example, a pathogen on a leaf may be viewed as an additional sink which will compete with host sinks such as roots or shoots (Walters, 1985).

However, there are many uncertainties concerning the carbon nutrition of pathogenic fungi, and some of the more important of these are listed below.

(i) A number of studies have approached the question of altered carbohydrate metabolism by examining the changes occurring in key enzyme activities, particularly invertases which are essential for the mobilization of sucrose. However, no clear pattern in the response to infection or the role of this activity has emerged (see Farrar and Lewis, 1987). Again, when the effects of infection on starch biochemistry are considered, no general response is observed (Storr and Hall, 1992).

(ii) The role of the apoplast and of increased membrane permeability resulting from infection in the nutrition of the pathogen is another area of uncertainty. Some reports show large increases in host membrane permeability as a result of infection, whereas some biotrophic fungal attacks may result in no damage to membranes (Farrar, 1984). The net transfer of solutes to the pathogen could depend on an increase in cytoplasmic levels of the solute resulting from metabolic changes, increased membrane permeability, an attenuation of solute retrieval mechanisms, or any combination of these (see Patrick, 1989).

(iii) Little is known about the detailed mechanism of transport of sugars into pathogenic fungi, either directly or through the host/pathogen interface such as haustoria. The nature of the transport, whether active or passive, the location of the key membranes regulating efflux and uptake, and the identity of the major solutes transported are all areas of considerable uncertainty and may well vary from system to system (see Farrar, 1984; Smith and Smith, 1990).

This paper is largely concerned with the carbon nutrition of the biotrophic powdery mildew (*Erysiphe pisi* D.C.) infection of pea (*Pisum sativum*). The effects of infection on leaf acid and alkaline invertases and on aspects of starch biochemistry have been investigated. The proposal that infection may change host cell membrane permeability has been examined, together with uptake of sugars into the mildew through a leaf disc system. It includes experiments using the Argenteum mutant of pea which has a readily detachable epidermis and so allows the reliable separation of host and mildew activities. Some relevant experiments carried out with isolated haustoria and mycelial suspensions of barley mildew (*E. graminis*) are also discussed.

9.2 The powdery mildew system

The powdery mildew fungi are biotrophic plant pathogens that form a structurally complex interface between host and parasite which presumably regulates nutrient flow from host to mycelium. These interfaces are known as haustoria and, in most powdery mildew infections, they are restricted to the shoot epidermis. Thus a long-term association is established in which the pathogen taps the nutrient sources of the host without leading to the rapid death of the host cells. There is now considerable evidence that the haustoria form the major, perhaps the sole, route for nutrient transfer between host and fungus (Gay, 1984).

The structure of the powdery mildew haustorium has been reviewed on numerous occasions (see Gay, 1984; Manners, 1989 and earlier references therein) and so will only be outlined here. The haustorial interface consists of two membranes, an invaginated and modified host plasma membrane, known as the extrahaustorial membrane (EHM) and the haustorial fungal membrane (Figure 9.1). Between these two membranes is a gel-like extrahaustorial matrix. The haustoria are isolated from the general host apoplast by neckbands to which the epidermal plasma membrane and haustorial plasma membrane are fused. The haustoria are usually distinctly lobed; in barley the lobes are considerably elongated and lie parallel to the epidermal surface.

Clearly this haustorial interface is relatively inaccessible for nutrient transport studies, while studies on intact systems are complicated by the need for exogenously supplied nutrients to pass several membranes before entering the fungal mycelium. Nevertheless, considerable evidence has accumulated, particularly from ultrastructural studies, on the major permeability barriers involved in the transport process (Gay, 1984; Manners, 1989).

The neckbands act to isolate the host apoplast from the extrahaustorial matrix, which means that nutrients must enter the symplasm of the epidermal cell before

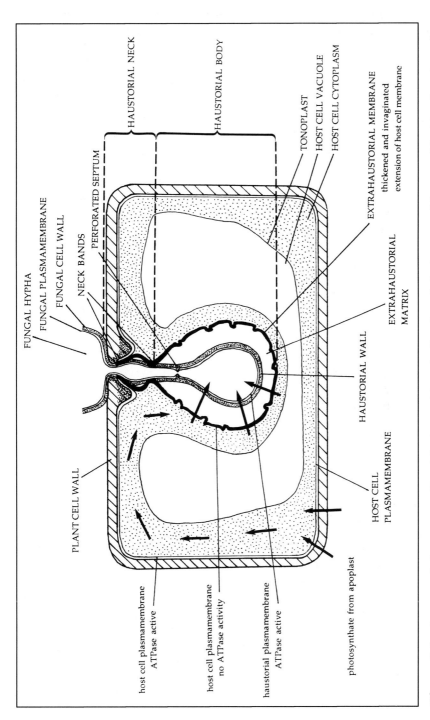

Figure 9.1. Diagram of a haustorium of Erysiphe in an epidermal cell to show the basic structural features of this host–fungal interface and the probable pathways of carbon transport. Reproduced from Isaac (1992) with permission from Chapman and Hall, and from an original drawing by Dr G. N. Greenhalgh.

183

uptake by the haustorium. Under the electron microscope, the EHM is thicker than the normal host plasma membrane, lacks intramembrane particles, and appears to contain considerable amounts of polysaccharide. More critical to the mechanism of nutrient transfer has been the suggestion from cytochemical studies on the *E. pisi* infection that the host plasma membrane surrounding the fungal haustorium lacks adenosine triphosphatase (ATPase) activity whereas the rest of the host plasma membrane has normal staining (Spencer-Phillips and Gay, 1981). In contrast, in other haustorial systems, the fungal plasma membrane also lacks ATPase activity (Gay, 1984). Thus nutrient movement may involve release of solutes from the host plasma membrane into the apoplastic extrahaustorial matrix followed by uptake into the mycelium, either by an active H^+-cotransport mechanism or by facilitated diffusion down a concentration gradient. However, the interpretation of these cytochemical studies should always take account of the uncertainty that has been expressed concerning the validity of the cytochemical stain for ATPase activity (see Chauhan *et al.*, 1991).

A further consideration is the scarcity of plasmodesmatal connections between the epidermis and underlying mesophyll, at least in pea (Bushnell and Gay, 1978). The pathway of nutrient movement from host to fungus presumably involves solute leakage into the leaf apoplast, followed by uptake into the epidermal cytoplasm before uptake by the haustoria. Thus factors which affect the concentrations of nutrients in the apoplast may affect the nutrients available for mildew growth and sporulation.

9.3 The effect of infection by pea powdery mildew on leaf invertase activities and starch content

Changes in the activity of invertase have often been investigated in diseased plants since they may be related to the uptake of host carbohydrate by the fungal pathogen (Farrar and Lewis, 1987). Higher plants may contain both acid and alkaline invertases, but it is the activity of acid invertase that has been investigated most often in plant/pathogen interactions. Both forms of the invertase have been demonstrated in pea leaf tissue, and their characteristics were similar in the two cultivars used, Onward and Argenteum (Storr and Hall, 1992).

The activity of acid and alkaline invertases (assayed at pH 4.5 and 7.3 respectively) and concentrations of reducing sugars and sucrose in whole leaves of Onward were measured, using crude extracts, in control and infected plants 7 days after inoculation. Crude extracts were considered to be free from endogenous inhibitors, as product formation was a function of enzyme concentration and length of incubation period only. Repeated washing and resuspension of the pellet gave no appreciable increase in activity. Increasing the concentration of extraction buffer tenfold did not increase the extractable activity of the invertases. The activities of both acid and alkaline invertases were significantly greater in infected leaves than in control leaves. On average, activity of acid invertase was 1.4 times greater, and activity of alkaline invertase 2.4 times greater, in infected leaves than control leaves (Table 9.1). Activity of alkaline invertase was

Table 9.1. *Effect of infection by* Erysiphe pisi *on activities of acid and alkaline invertases in extracts of mature leaves of* Pisum sativum *cv. Onward or in isolated mesophyll of cv. Argenteum 7 days after inoculation*

	Invertase activity (nkat g^{-1} FW)	
	Acid	Alkaline
Onward whole leaf		
Infected	1.33 ± 0.77*	6.50 ± 4.07***
Control	0.98 ± 0.59	2.73 ± 1.25
Argenteum mesophyll		
Infected	1.40 ± 0.91	5.09 ± 0.98**
Control	1.12 ± 1.06	4.13 ± 1.00

Invertase activity was measured at pH 4.5 (acid) and pH 7.3 (alkaline) in whole leaf homogenates. With infected leaves, mycelia were removed prior to assay by brushing. Results are for eight or nine independent replicate experiments (± SE) and were analysed by the Wilcoxon paired sample test to compare infected with control plants. *$0.05 \geqslant P \geqslant 0.01$; **$0.01 \geqslant P > 0.001$; ***$P \leqslant 0.001$. Data from Storr and Hall (1992).

significantly higher than that of acid invertase in leaves of both infected and control plants. In similar experiments conducted using isolated mesophyll of Argenteum leaves (Table 9.1), only alkaline invertase activity was significantly increased by infection, being on average 1.3 times greater than in similar control plants. Activity of acid invertase and concentrations of reducing sugars and sucrose were not changed by infection (results not shown).

These results describe for the first time increased activity of alkaline invertase in host tissues resulting from a biotrophic fungal infection. Additionally, it has been possible, by the use of the Argenteum mutant, to separate spatially the host and parasite domains and assay invertase activity in the exclusively host domain of the mesophyll. The characteristics of the partially purified acid and alkaline invertases in this study are similar to those in other reports; alkaline invertase activity was, in control tissues, significantly greater than the activity of acid invertase, supporting the suggestion of Masuda *et al.* (1987) that alkaline invertase is predominant in mature tissues. The two previous studies that have measured invertase activities in mildewed peas (Donaldson, 1984; Manners, 1979) have both used the leaf disc method of Bacon *et al.* (1965) for assaying activity of this enzyme. This method has been criticised by Billett *et al.* (1977) because it may only measure the activity of insoluble, cell wall bound acid invertase. In this study, the leaf disc method did not detect activity of alkaline invertase in either control or infected mature leaves (results not shown). This may be because the ethyl acetate wash inhibits activity of alkaline invertase; activity of this enzyme in mature leaf homogenates was completely inhibited by ethyl acetate concentrations of 30% or greater.

The sucrose hydrolase activity of both pea and barley powdery mildew has been shown to be due to the activity of α-glucosidase (Donaldson, 1984; Donaldson and Jørgensen, 1988). By the use of *p*-nitro-α-glucoside (PNPG) as a substrate it is easy to distinguish between the activity of this enzyme and that of invertase, in either plant or fungal tissue. Table 9.2 shows that, at the very most and leaving a large margin for error (α-glucosidase has a greater affinity for PNPG than sucrose), about 0.5 nkat g^{-1} FW of sucrose hydrolase activity may be due to the mildew in infected leaves with the mycelia not removed, and about 0.1 nkat g^{-1} FW sucrose hydrolase activity in infected leaves with the mycelia removed. Thus, in experiments where the mycelia were removed and invertase activity measured in host tissues, it was considered that increases in invertase activity of greater than 0.2–0.3 nkat g^{-1} FW were probably of host origin. Clearly then, the increase in alkaline invertase activity resulting from infection in whole Onward leaves (6.5−2.73 = 3.77 nkat g^{-1} FW) could not be due to activity of α-glucosidase and the *in vivo* increase in activity of this enzyme is probably at least 3 nkat g^{-1} FW.

Amounts of starch, and activities of ADPGlc pyrophosphorylase and α-glucan phosphorylase, in leaves of infected and control plants were measured 2 h into the photoperiod, when amounts of starch are minimal (Stitt *et al.*, 1978). Infection resulted in a significant decrease in the amounts of leaf starch (Table 9.3). On average, leaves of infected plants contained about half the amount of starch found in leaves of healthy plants. Similar results were obtained when the data were expressed on a dry weight basis.

To investigate the enzymic cause of the decrease in leaf starch, the activities of ADPGlc pyrophosphorylase and α-glucan phosphorylase were measured, after characterizing the enzyme assays. The assay of ADPGlc pyrophosphorylase was considered to be free from endogenous inhibitors because product formation was proportional to enzyme concentration; the $K_{mADPGlc}$ of ADPGlc pyro-phosphorylase was 1.3 mol m^{-3} and the pH optimum was 7.5. The assay of α-glucan phosphorylase was also considered to be free from endogenous inhibitors because the formation of product during the assay was proportional to the length of incubation period and enzyme concentration. Infection did not result in a

Table 9.2. Activity of α-glucosidase in mature leaves of control and infected plants of P. sativum cv. Onward, 7 days after inoculation, with and without removal of mycelia of E. pisi

Treatment	α-glucosidase activity (pkat g^{-1} FW)
Control	13
Infected	
Mycelia removed	146
Mycelia retained	542

α-glucosidase activity was measured in leaf homogenates by the release of nitrophenol using PNPG as substrate at pH 6.9. Data are the mean of two independent replicate experiments.

significant change in the activity of α-glucan phosphorylase or in the levels of P_i. However, the activity of ADPGlc pyrophosphorylase in leaves of infected plants was, on average, 11.7 nkat g^{-1} FW, and in control plants 17.8 nkat g^{-1} FW. Of the four experiments conducted, three of these showed less activity of this enzyme in the infected than control treatments. It seems unlikely that the hexoses produced by the increased activity of alkaline invertase are incorporated into starch because, as reported here, mildew infection causes a decrease in leaf starch. Such a decrease in starch has been reported in some other studies, although some tissues show infection-induced increases. The enzymic cause of this decrease in leaf starch is more likely to be due to a reduction in the activity of ADPGlc pyrophosphorylase, the regulatory enzyme of starch synthesis, and not an increase in the activity of α-glucan phosphorylase, the enzyme primarily responsible for starch degradation; that is, the reduction of leaf starch levels is due to a decrease in the rate of starch synthesis, not an increase in the rate of starch degradation. Decreased amounts of starch in whole leaves would tend to suggest that the leaf is experiencing a net increased demand for carbohydrate. The enhanced activity of alkaline invertase on infection tends to suggest increased 'sink' activity (Avigad, 1982). The increased demand for carbohydrate in the diseased leaf may divert metabolism of the host away from starch and towards sucrose.

9.4 The effect of pea powdery mildew infection on leaf apoplastic solute concentrations

As discussed above, the fungus is dependent on nutrients available in the epidermal symplasm. The majority of pea epidermal cells are heterotrophic (except the guard cells) and are therefore dependent on the underlying mesophyll for photosynthates. Plasmodesmatal connections between pea leaf mesophyll and the epidermis are absent or very infrequent (Bushnell and Gay, 1978); thus the movement of photosynthates between these two tissues must be via the apoplast. One technique for studying the contents of the apoplast is that of intercellular

Table 9.3. *Activities of ADPGlc pyrophosphorylase (ADPGPPase) and α-glucan phosporylase (α-GPase), and content of starch in healthy and infected leaves of* P. sativum, *7 days after inoculation with* E. pisi

	Enzyme activity (nkat g^{-1} FW)		Starch (mg g^{-1} Fw; $n=15$)
	ADPGPPase ($n=4$)	α-GPase ($n=5$)	
Infected	11.72 ± 5.59	2.82 ± 0.82 NS	1.38 ± 1.20***
Control	17.80 ± 4.25	2.90 ± 0.72	5.65 ± 6.59

Mycelia were removed and activities in leaf homogenates were assayed as described by Storr and Hall (1992). (±SE; n=number of independent replicate experiments) were analysed by the Wilcoxon paired sample test (***$P \leq 0.001$) comparing infected with control plants. Data from Storr and Hall (1992).

washing. The basic principle is that the whole of the apoplast is filled with water or buffer, usually by vacuum infiltration, followed by collection of the intercellular washing fluid (IWF) by centrifugation at speeds sufficiently low so as not to disrupt the symplast significantly. This technique has been used here to determine metabolite concentrations in the apoplast of pea leaves and to investigate changes in apoplastic sugars that occur when leaves are infected with powdery mildew.

Measurements of malate dehydrogenase activity, a cytoplasmic marker, K^+ concentrations and osmotic potential of IWF, cell sap and leaf homogenate indicated that the degree of cytoplasmic contamination was 0.3% in IWF extracted from healthy leaves (IWF_h) and 0.7% in IWF extracted from leaves with a 4-day-old mildew infection with the mycelium removed prior to vacuum infiltration (IWF_i). These levels of contamination are comparable with those measured in IWF from pea epicotyl sections, that is, less than 1.5% (Terry and Bonner, 1980). The higher contamination in IWF from mildew-infected leaves may be due either to epidermal cell damage during mycelial removal or to leakage from haustorial neck sites. The concentrations of K^+ found in IWF were similar to the values (ranging from 5 to 12 mol m^{-3}) measured in pea leaves by elution analysis (Long and Widders, 1990).

The concentrations of glucose and sucrose in the IWF extracted from healthy and infected leaves over a time period are shown in Table 9.4. The glucose concentration decreased very slightly in IWF_h over 7 days, whereas that in IWF_i showed some increase over the same period. Sucrose concentrations fell fourfold in IWF_h in 7 days but increased by about one-third in IWF_i. Glucose, fructose and sucrose in IWF were also measured in one experiment using GLC. Glucose and sucrose showed similar trends to that described above. Fructose increased fourfold in IWF_i in 7 days but only twofold in IWF_h; however, fructose concentrations were much lower than those of glucose and sucrose. These results may be compared with those of Delrot *et al.* (1983) for nycthemeral changes in

Table 9.4. *Changes in the concentration of glucose and sucrose in intercellular washing fluid from healthy and infected leaves of* P. sativum *over 7 days*

Days after inoculation	Sugar concentration (mol m^{-3})			
	Glucose		Sucrose	
	Healthy	Infected	Healthy	Infected
0	0.52	0.52	0.38	0.37
3	0.3	0.56	0.23	0.16
5	0.17	0.21	0.27	0.46
7	0.39	0.87	0.08	0.49

IWF was extracted by centrifugation at 700 *g* for 1 h. Glucose and sucrose contents were measured before and after invertase hydrolysis by the glucose oxidase assay. Results are the means of two independent replicated experiments.

the apoplastic sugars of *Vicia faba* leaves. In this tissue, glucose concentrations were nearly constant despite variations of intracellular glucose. On the other hand, apoplastic sucrose varied fivefold during the day/night cycle, and there was also virtually no apoplastic fructose. These workers pointed out that the estimation of mean sucrose concentration in the apoplast (1.0–4.5 mol m^{-3}) is only tentative as nothing is known about apoplastic microlocalization of sucrose in the leaf. The concentrations presented in this paper are likely to be considerably lower than actual apoplastic or water-filled space concentrations. Dot matrix analysis of transverse sections of pea leaves shows that intercellular space occupies about 40% of the leaf volume (results not shown). The process of vacuum infiltration of whole leaves for the preparation of IWF caused an average increase in leaf weight of 30%. Therefore a rough estimate of water-filled space might be 10%, that is 0.1 cm^3 g^{-1} FW. This falls within the range of 4–20% determined for other tissues (e.g. Minchin and Thorpe, 1984); therefore the apoplastic sugar concentrations for pea leaves might be fourfold higher than the levels measured in IWF.

To investigate further the effect of infection on membrane permeability, the efflux of sucrose from preloaded leaf discs was studied. These experiments suggested that the loss of cytoplasmic label may be more rapid from infected tissue, although the loss of vacuolar label may occur more slowly (results not shown). Previous studies do not allow a clear conclusion to be reached regarding the effect of infection on membrane permeability (Farrar, 1984). For example, Manners (1979) found that healthy pea leaf discs lost more photosynthate in efflux experiments than those infected with mildew, whereas Ayres (1977) reported an increased loss of electrolytes from leaves as a result of mildew infection. These effects may be due in part to the phytoalexin, pisatin, which can accumulate to concentrations of approximately 1 mmol m^{-3} in mildewed peas (Oku *et al.*, 1975). This concentration of pisatin alters the semipermeability of isolated pea epidermis and induces rupturing of isolated pea protoplasts at similar concentrations *in vitro* (Shiraishi *et al.*, 1975). If the permeability of the leaf cell membranes is increased, it is possible that the mildew can benefit from the greater availability of nutrients in the apoplast since the increased metabolic demand of the infected cells could make them a strong competitive sink. Thus the efficiency of the host–parasite interface at removing nutrients from the epidermal cytoplasm, combined with increases in host membrane permeability due to pisatin, could maintain a flow of carbohydrate from host to parasite that sustains this biotrophic relationship.

9.5 The uptake of glucose, fructose and sucrose into pea powdery mildew from the leaf apoplast

Two basic approaches have been adopted to study the steps involved in the uptake of nutrients into powdery mildew mycelia. One is to attempt to isolate viable and active haustorial complexes and to measure uptake of exogenously supplied solutes directly (see Section 9.6). The alternative approach is to use a relatively intact system, such as leaf discs, and to study the kinetics of solute transport to the mycelium. In this section, results are described using a method involving uptake into pea leaf discs floating on radiolabelled sugars and a Sello-tape method to collect the mycelium for scintillation counting at the end of the uptake period (Wyness and Ayres, 1987). However, the interpretation of such results are clearly complicated by the need for solutes first to enter the cytoplasm of the epidermal cells; thus three membranes must be crossed before exogenous solutes enter the fungal mycelium.

9.5.1 *Sugar uptake into leaf epidermal tissue*

To provide background information to an investigation of mycelial uptake, the mutant cv. Argenteum was used to investigate the sugar uptake characteristics of leaf epidermal tissue. The uptake rates from either 1 mol m^{-3} glucose, fructose or sucrose into whole leaf discs of Argenteum and of *P. sativum* cv. Onward were shown to be very similar; uptake was linear over a 3 h period with glucose taken up faster than the other two sugars in both cultivars. There was a similar uptake into the lower epidermis of Argenteum. Furthermore, uptake from 1 mol m^{-3} O-methylglucose (3-OMG) was markedly slower than from 1 mol m^{-3} glucose. Kinetic analysis of sugar uptake revealed the presence of two transport components. At low exogenous concentrations (<1 mol m^{-3}), a saturable component dominated the uptake. At concentrations greater than 1 mol m^{-3} a linear component predominated, with fructose entering the epidermis faster than glucose and sucrose. An increase in the pH of the exogenous medium from 5 to 8 caused a slight but steady drop in the rate of uptake of glucose, fructose and sucrose; glucose uptake had a pH optimum of about 5. Glucose and fructose competed with one another and also with sucrose for uptake into the epidermis. Sucrose was not, however, an effective competitor against glucose and fructose.

9.5.2 *Sugar uptake into the fungal mycelium*

Patterns of sugar uptake into the mycelium of *E. pisi* were investigated by float-ing discs of mildewed pea leaves on radiolabelled sugar solutions, followed by removal of the mycelium with Sellotape for radiolabel counting. The time course of sugar uptake from 1 mol m^{-3} glucose, fructose, sucrose or 3-OMG into the mycelium showed that the uptake of fructose was approximately linear over a 7 hour period whereas that of glucose, and possibly sucrose, showed an increase in the rate of uptake with time (Figure 9.2). The uptake rates of fructose and sucrose were very similar over the first 5 h of uptake, whereas glucose was taken up almost twofold faster than the other two sugars. In contrast, 3-OMG showed linear uptake over the same time interval, and the rate of uptake was very low by comparison with that of glucose. After an uptake period of 30 min, glucose uptake was over threefold greater than 3-OMG uptake and by 3 h this difference had risen to eightfold. The ability of the fungus to compete with other host sinks will depend on the rate of uptake by the haustorium and thus, at least in part, on

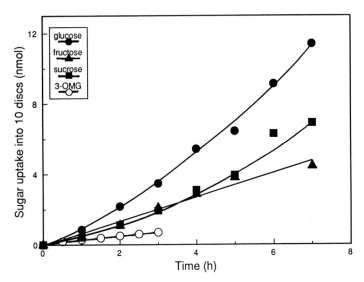

Figure 9.2. *The uptake of glucose, fructose, sucrose, and 3-O-methylglucose (3-OMG) into the mycelium of* E. pisi. *Infected leaf discs were floated on 1 mol m^{-3} ^{14}C-sugar. At set intervals, discs were rinsed, blotted dry and the mycelium removed with Sellotape. The Sellotape was placed directly into scintillation fluid for counting. The results are the average of four replicated experiments, except for 3-OMG (two experiments).*

the rate of metabolism within the fungal haustorium and mycelium. The fungus could maintain a flow of photosynthate from the epidermal cytoplasm by creating a strong carbohydrate gradient by rapid conversion of incoming host sugars into other compounds. The ratio of uptake of glucose to 3-OMG at 1 mol m^{-3} suggests this. 3-OMG is a non-metabolized glucose derivative thought to be transported on the same carrier as glucose (Reinhold and Eshhar, 1968). There is a gradual increase in the rate of glucose uptake with time that is not seen with 3-OMG. This suggests that glucose-metabolizing processes are being switched on in the cytoplasm of the haustorium or mycelium.

The kinetics of glucose and fructose uptake were biphasic and indicated saturable uptake at lower sugar concentrations and a linear diffusive phase at higher sugar concentrations. This could merely reflect what is happening at the epidermal membrane. Sucrose uptake shows rather different kinetics to that at the epidermal membrane (see Section 9.5.1). Although there may be two phases, as is the case with the hexoses, the increase in the rate of sugar uptake at the higher sugar concentration range is extremely small. A sucrose carrier may be present on the haustorial membrane which is active at lower sugar concentrations but there is very little facilitated diffusion or no second carrier operative at a higher sugar concentration. Wyness and Ayres (1987), using strips of mildew-infected pea leaves to investigate sucrose uptake into the mycelium, found that uptake was biphasic over a range of external sugar concentrations and suggested that uptake was predominantly active in the range 0.05–10 mol m^{-3}. They estimated a K_m of

7.65 mol m^{-3} for sucrose uptake into the mycelium which is in good agreement with the value of 7.83 mol m^{-3} estimated by a similar method in this study. The effect of pH on the uptake of sugars into the mildew mycelium was very similar to that measured for the pH response of uptake at the epidermal plasma membrane (see Section 9.5.1). If pH had a markedly different effect on mycelial uptake it would suggest that the haustorial membranes were coming into direct contact with the host apoplast and that sugars could also be leaking to the mycelium or haustorium directly from the apoplast. The cytoplasm, however, acts as a buffer between the apoplast and the haustorium.

The effect of a range of inhibitor treatments on uptake into the mycelium is shown in Table 9.5. The H$^+$-ATPase inhibitor erythrosin B (ERB) (Ball *et al.*, 1987) applied at 0.05 mol m^{-3} affected the uptake from 1 mol m^{-3} or 200 mol m^{-3} glucose less than that of fructose and sucrose. However, when the concentration of ERB was increased to 0.4 mol m^{-3}, the inhibition of uptake from 1 mol m^{-3} sugar increased with all three sugars, showing the same degree of inhibition at around 64%. The sulphydryl binding inhibitor *p*-chloromercuribenzene sulphonic acid (PCMBS) showed different degrees of inhibition of uptake of the three sugars (Table 9.5). Fructose was inhibited least, with sucrose being the most heavily inhibited. Phloridzin, an inhibitor of glucose transport in animal cells, caused about a 25% inhibition of all three sugars. The protonophore CCCP had a more marked effect on sugar uptake, particularly on that of glucose. However, low temperature (4°C) had the most marked effect on the uptake of all three sugars. Overall these results indicate that much of the sugar uptake into the mycelium could be by facilitated diffusion driven by a concentration gradient which is maintained by the rapid metabolism of glucose in the fungal cytoplasm.

Table 9.5. *The effect of a range of inhibitors on the uptake of sugars into the mycelium of* E. pisi

Treatment	Inhibition of uptake into mycelium (%)		
	Glucose	Fructose	Sucrose
PCMBS (1 mol m^{-3})	20.9	10.5	46.9
CCCP (0.05 mol m^{-3})	73.4	55.3	45.7
Phloridzin (5 mol m^{-3})	21.5	28.5	24.0
ERB (0.05 mol m^{-3})	15.1	32.0	32.9
ERB (0.4 mol m^{-3})	63.0	64.0	64.0
Temperature (4°C)	80.6	85.5	75.9

Leaf discs were incubated for 2 h on 1 mol m^{-3} glucose, fructose or sucrose with or without inhibitors. In the ERB experiments, leaf discs were pre-incubated for 30 min with or without ERB. For the treatment at 4°C, the incubation was carried out on ice and was compared with a control at 25°C. All results are the mean of at least three independent experiments except the low temperature treatment which was performed once. Data from Aked and Hall (manuscript submitted).

9.6 Solute uptake by isolated haustorial complexes

Haustoria have been isolated from a number of plant/pathogen combinations
and have been used to study their structure, and their cytochemical and permeability properties (see Gay and Manners, 1987; Stumpf and Gay, 1989). However,
doubts have been raised over their value in transport studies since damage may
occur during isolation (Manners, 1989). In our studies, the problem of the contamination of isolated fractions by chloroplasts and other organelles can also
present considerable difficulties. Earlier attempts to measure the uptake of sugars
by isolated haustoria reported only low rates of glucose and sucrose incorporation (e.g. Manners and Gay, 1982) although it should be noted that the range of
protectants included in the initial homogenization medium in these studies was
limited. Here, we report on some preliminary uptake studies using haustoria
isolated from barley leaves infected with *E. graminis*. The homogenization
medium contained a range of protectants and haustoria were isolated from epidermal strips by a combination of differential centrifugation and phase partitioning. After isolation, about 50% of the haustoria appeared to be viable as
measured by the fluorescent dye, fluorescein diacetate, although viability fell
steadily over a 6 h period.

Uptake into isolated haustoria was normally measured from 0.1 or 1 mol m^{-3}
sugar solutions, at pH 5 or 7, and at 22°C. To test for possible uptake by con-
taminating organelles, controls consisted of fractions isolated in the same way
from uninfected leaves. When glucose uptake was measured into isolated
haustorial fractions, uptake showed a clear increase with time and was sensitive
to PCMBS (Table 9.6). In contrast, uptake from 0.1 mol m^{-3} sucrose was very
low or undetectable and showed no sensitivity to PCMBS. Uptake of glucose by
the control fractions from healthy leaves was also extremely low. When
measured at pH 5, uptake over 60 min by the haustorial fractions was at least 30
times greater than that of the control fractions.

Table 9.6. *Uptake of glucose by isolated haustoria from barley
leaves infected with* E. graminis

Uptake period	Uptake (nmol glucose μg^{-1} protein)		
(min)	pH 5.0	pH 7.0	pH 5.0 plus 2 mol m^{-3} PCMBS
15	1.21	1.18	0.33
60	5.30	6.36	0.53

Haustoria were isolated from infected barley leaves by homogenization,
filtration and centrifugation. Haustorial fractions were incubated in radiolabelled glucose (0.1 mol m^{-3}) for the times indicated and separated by
filtration before scintillation counting. Results are the mean of two replicated experiments.

9.7 Solute uptake by powdery mildew mycelia

In order to gain a complete understanding of the uptake processes involved in this biotrophic system, some knowledge of the transport systems present in the fungal mycelium would be very helpful. However, surprisingly little is known about the sugar transport mechanisms of pathogenic fungi, in general, and of biotrophs in particular (Farrar, 1985; Farrar and Lewis, 1987). In general, fungi take up a wide range of solutes by coupling transport to the simultaneous uptake of protons (Sanders, 1988). However, there is little evidence that fungi possess high affinity systems for sucrose transport, although this is usually the most directly available carbon source for pathogenic fungi (Farrar and Lewis, 1987; Komor, 1982). Thus, in this paper, we describe some preliminary experiments on the transport properties of mycelia of *E. graminis*.

Mycelial suspensions took up glucose from a 0.1 mol m^{-3} glucose solution steadily over a 2 h period. Uptake was sensitive to pH, being much more rapid at pH 5 than at pH 7, and was inhibited markedly by 0.1 mol m^{-3} PCMBS and CCCP, but less so by vanadate (Table 9.7). These results were consistent with proton-coupled transport; the relative insensitivity to vanadate is presumed to reflect a penetration problem. Uptake from 0.1 mol m^{-3} glucose was not affected by the presence of 10 mol m^{-3} mannitol, fructose or sucrose. When the uptake of several solutes was compared, glucose transport was very much greater than that of sucrose, fructose or glutamine (Table 9.8).

9.8 Conclusions

The results obtained from these powdery mildew systems allow some answers to be suggested to the uncertainties raised in the Introduction, and a general scheme proposed for the transfer of carbon from host to mycelium.

Infection of leaves by pathogens is frequently associated with a significant increase in invertase activity (Lewis, 1991), although the source of this increase (whether host or fungus) and its role are far from clear (Farrar and Lewis, 1987). In the biotrophic infection described here, the major increase is associated with alkaline invertase, which is generally considered to be cytosolic. In addition, the use of the Argenteum mutant, which permits a clear separation of host mesophyll from epidermis, demonstrates that at least a proportion of the increase in invertase is associated with host cells that are physically separated from the site of infection. The hexoses produced from the increased invertase activity do not appear to be incorporated into starch since, overall, the amounts of leaf starch fell with infection. However, infection does appear to cause an increase in sugar concentrations in the leaf apoplast. This could result from increased cytoplasmic hexose levels, due to invertase activity, and also to an increase in host cell membrane permeability. This study supports a number of other reports which suggest that infection may result in an increased loss of solutes from host cells (see Section 9.4). It is also of interest to note that apoplastic fructose concentrations appear to be much lower than those of glucose and sucrose.

Table 9.7. *Effect of inhibitors on the uptake of glucose by mycelial fraction from* E. graminis

Inhibitor	Uptake (nmol glucose μg^{-1} protein h^{-1})
Control	9.3
+ PCMBS	1.5 (16%)
+ SW26	3.9 (42%)
+ Vanadate	6.6 (71%)
+ CCCP	0.1 (1%)
+ DMSO	2.9 (31%)

Mycelial preparations were incubated in radiolabelled glucose (0.1 mol m^{-3}) at pH 5 for 60 min and separated by filtration before scintillation counting. Inhibitors were applied at a concentration of 0.1 mol m^{-3}. CCCP was applied in DMSO and the effect of this solvent alone is also shown. Results are the mean of two replicated experiments.

Table 9.8. *Uptake of solutes by mycelial fractions from* E. graminis

Solute	Uptake (nmol μg^{-1} protein h^{-1})	
	pH 5	pH 7
Glucose	8.0	0.6
Fructose	0.2	0.9
Sucrose	0.3	0.0
Glutamine	0.3	0.0

Mycelial preparations were incubated in radiolabelled solutions (0.1 mol m^{-3}) at pH 5 or pH 7 for 60 min and separated by filtration before scintillation counting.

Leaf cells commonly contain a mechanism for sugar retrieval from the apoplast (Maynard and Lucas, 1982). The epidermal cells, in particular, do not contain functioning chloroplasts (except the guard cells) and are normally symplastically isolated from the mesophyll. Therefore uptake of sugars at the plasma membrane must be a major source of nutrients for the epidermis. Thus the mesophyll and epidermal cells are unlikely to show a net efflux of solutes unless their permeability properties are altered in some way. This could occur in two ways: increased leakage as described above, and the development of haustoria. The EHM appears to lack both normal transmembrane particles (Bushnell and Gay, 1978) and staining for ATPase activity (Spencer-Phillips and Gay, 1981) (see Section 9.2). These observations indicate that the normal transport functions linked to the ATP-driven proton pump are not maintained at the EHM

(Manners, 1989), and uptake into the fungal mycelium may proceed by facilitated diffusion or even simple diffusion down a concentration gradient maintained by the release of sugars from the host cells and their rapid metabolism in the fungal cytoplasm. This is supported by the relatively limited effect reported for a range of inhibitors on uptake into the mycelium and by the marked difference in uptake rates between glucose and the non-metabolized 3-OMG.

Glucose is preferentially taken up by the mycelium when compared to sucrose and fructose. It is of interest to note that, in mycorrhizal associations, sucrose is hydrolysed prior to uptake and glucose is then preferentially selected from the resultant mixture (Lewis, 1991). Although mildews possess an α-glucosidase capable of hydrolysing sucrose (Donaldson and Jørgensen, 1988), the evidence for the uptake of sucrose intact by the fungus is quite limited (Farrar and Lewis, 1987). This work shows that glucose is taken up more rapidly than the other sugars and no evidence for direct sucrose uptake was obtained by the use of isolated haustorial complexes or suspensions of mildew mycelia.

The demand for nutrients in the cytoplasm of infected epidermal cells will be far greater than in the uninfected tissue. The increase will be due to the demands from the fungus and to the increase in host metabolism which occurs as a response to infection (Kosuge and Kimpel, 1981). Thus there will be a requirement for an increased flow of solutes from the apoplast into the epidermis. The following steps could be involved in meeting this demand and are supported by the observations described in this paper. First, increased alkaline invertase activity and decreased starch synthesis lead to the increased availability of hexoses in the cytoplasm of host cells. Other factors, such as an increase in phloem unloading, may also be involved. Apoplastic sugar concentrations are raised, perhaps by a change in mesophyll cell permeability which could be caused by the phytoalexin, pisatin. Uptake of sugars, mainly in the form of glucose, into the mycelium is driven by simple or facilitated diffusion across the modified EHM membrane. The concentration gradient for this flow of nutrients is maintained by the increased invertase activity in the host cells and the rapid conversion of sugars into fungal-specific metabolites in the haustorial and mycelial cytoplasms.

Acknowledgements

We would like to thank the SERC for studentships awarded to J.A. and T.S. and the SERC and ICI for a case award to A.J.G.

References

Avigad, G. (1982) Sucrose and other disaccharides. In: *Encyclopedia of Plant Physiology*, Volume 13A, *Plant carbohydrates* (eds F.A. Loewus and W. Tanner). Springer-Verlag, Berlin, pp. 234–235.

Ayres, P.G. (1977) Effects of powdery mildew *Erysiphe pisi* and water stress upon the water relations of pea. *Physiol. Plant Pathol.* 10, 139–149.

Bacon, J.S.D., Macdonald, I.R. and Knight, A.H. (1965) The development of invertase activity in slices of the root of *Betula vulgaris* L. washed under aseptic conditions. *Biochem. J.* 94, 175–182.

Ball, J.H., Williams, L. and Hall, J.L. (1987) Effect of SW26 and erythrosin B on ATPase activity and related processes in *Ricinus* cotyledons and cucumber hypocotyls. *Plant Sci.* 52, 1–5.

Billett, E.E., Billett, M.A. and Burnett, J.H. (1977) Stimulation of maize invertase activity following infection by *Ustilago maydis*. *Phytochemistry* 16, 1163–1166.

Bushnell, W.R. and Gay, J.L. (1978) Accumulation of solutes in relation to structure and function of haustoria in powdery mildews. In: *The Powdery Mildews* (ed. D.M. Spencer). Academic Press, London, pp. 183–235.

Chauhan, E., Cowan, D.S. and Hall, J.L. (1991) Cytochemical localization of plasma membrane ATPase activity in plant cells. *Protoplasma*, 165, 27–36.

Delrot, S., Faucher, M., Bonnemain, J.-L. and Bonmort, J. (1983) Nycthemeral changes in intracellular and apoplastic sugars in *Vicia faba* leaves. *Physiol. Végét.* 21, 459–467.

Donaldson, I.A. (1984) Erysiphe pisi, *a model for fungal/higher plant biotrophic parasitism*. Ph.D. Thesis, University of Oxford.

Donaldson, I.A. and Jørgensen, J.H. (1988) Barley powdery mildew 'invertase' is an alpha-glucosidase. *Carlsberg Res. Comm.* 53, 421–430.

Farrar, J.F. (1984) Effects of pathogens on plant transport systems. In: *Plant Diseases Infection, Damage and Loss* (eds R.K.S. Wood and G.J. Jellis). Blackwell Scientific Publlications, Oxford, pp. 87–104.

Farrar, J.F. (1985) Carbohydrate metabolism in biotrophic plant pathogens. *Microbiol. Sci.* 2, 314–317.

Farrar, J.F. and Lewis, D.H. (1987) Nutrient relations in biotrophic infections. In: *Fungal Infections of Plants* (eds G.F. Pegg and P.G. Ayres). Cambridge University Press, Cambridge, pp. 92–132.

Gay, J.L. (1984) Mechanisms of biotrophy in fungal pathogens. In: *Plant Diseases: Infection, Damage and Loss* (eds R.K.S. Wood and G.J. Jellis). Blackwell Scientific Publications, Oxford, pp. 49–59.

Gay, J.L. and Manners, J.M. (1987) Permeability of the parasitic interface in powdery mildews. *Physiol. Molec. Plant Pathol.* 30, 389–399.

Isaac, S. (1992) *Fungal-Plant Interactions*. Chapman and Hall, London.

Komor, E. (1982) Transport of sugar. In: *Encyclopedia of Plant Physiology*, Volume 13A, *Plant Carbohydrates I* (eds F.A. Loewus and W. Tanner). Springer-Verlag, Berlin, pp. 635–675.

Kosuge, T. and Kimpel, J.A. (1981) Energy use and metabolic regulation in plant-pathogen interactions. In: *Effects of Disease on the Physiology of the Growing Plant* (ed. P.G. Ayres). Cambridge University Press, Cambridge, pp. 29–46.

Lewis, D.H. (1991) Fungi and sugars — a suite of interactions. *Mycol. Res.* 95, 897–904.

Long, J.M. and Widders, I.E. (1990) Quantification of apoplastic potassium content by elution analysis of leaf lamina tissue from pea (*Pisum sativum* L. cv Argenteum). *Plant Physiol.* 94, 1040–1047.

Manners, J.M. (1979) *Physiology of fungal haustoria,* (Erysiphales). Ph.D. Thesis, University of London.

Manners, J.M. (1989) The host-haustorium interface in powdery mildew. *Aust. J. Plant Physiol.* 16, 45–52.

Manners, J.M. and Gay, J.L. (1982) Transport, translocation and metabolism of ^{14}C

photosynthates at the host-parasite interface of *Pisum sativum* and *Erysiphe pisi*. *New Phytol.* 91, 221–244.

Masuda, T., Takahashi, T. and Sugawara, S. (1987) The occurrence and properties of alkaline invertase in mature roots of sugar beets. *Agric. Biol. Chem.* 51, 2309–2314.

Maynard, J.W. and Lucas, W.J. (1982) Sucrose and glucose uptake into *Beta vulgaris* leaf tissues. *Plant Physiol.* 70, 1436–1443.

Minchin, P.E.H. and Thorpe, M.R. (1984) Apoplastic phloem unloading in the stem of bean. *J. Exp. Bot.* 35, 538–550.

Oku, H., Ouchi, S., Shiraishi, T. and Baba, T. (1975) Pisatin production in powdery mildewed pea seedlings. *Phytopathology*, 65, 1263–1267.

Patrick, J.W. (1989) Solute efflux from the host at plant-microorganism interfaces. *Aust. J. Plant Physiol.* 16, 53–67.

Reinhold, L. and Eshhar, Z. (1968) Transport of 3-O-methylglucose into and out of storage cells of *Daucus carota*. *Plant Physiol.* 43, 1023–1030.

Sanders, D. (1988) Fungi. In: *Solute Transport in Plant Cells and Tissues* (eds D.A. Baker and J.L. Hall). Longman, Harlow, pp. 106–165.

Shiraishi, T., Oku, H., Isono, M. and Ouchi, S. (1975) The injurious effect of pisatin on the plasma membrane of pea. *Plant Cell Physiol.* 16, 939–942.

Smith, S.E. and Smith, F.A. (1990) Structure and function of the interfaces in biotrophic symbioses as they relate to nutrient transport. *New Phytol.* 114, 1–38.

Spencer-Phillips, P.T.N. and Gay, J.L. (1981) Domains of ATPase in plasma membranes and transport through infected plant cells. *New Phytol.* 89, 393–400.

Stitt, M., Bulpin, P.V. and ap Rees, T. (1978) Pathway of starch breakdown in photosynthetic tissues of *Pisum sativum*. *Biochim. Biophys. Acta*, 544, 200–214.

Storr, T. and Hall, J.L. (1992) The effect of infection by *Erysiphe pisi* D.C. on acid and alkaline invertases and aspects of starch biochemistry in leaves of *Pisum sativum* L. *New Phytol.* in press.

Stumpf, M.A. and Gay, J.L. (1989) The haustorial interface in a resistant interaction of *Erysiphe pisi* with *Pisum sativum*. *Physiol. Molec. Plant Pathol.* 35, 519–533.

Terry, M.E. and Bonner, A.B. (1980) An examination of centrifugation as a method of extracting an extracellular solution from peas, and its use for the study of indoleacetic acid-induced growth. *Plant Physiol.* 66, 321–325.

Walters, D.R. (1985) Shoot:root interrelationship: the effects of obligately biotrophic fungal pathogens. *Biol. Rev.* 60, 47–79.

Wyness, L.E. and Ayres, P.G. (1987) Plant-fungus water relations affect carbohydrate transport from pea leaf to powdery mildew (*Erysiphe pisi*) mycelium. *Trans. Br. Mycol. Soc.* 88, 97–104.

Carbon partitioning and transport in parasitic angiosperms and their hosts

W.E. Seel, I. Cechin, C.A. Vincent and M.C. Press

10.1 Introduction

There are more than 3000 known species of parasitic angiosperm, with a wide taxonomic and geographical distribution. Our understanding of carbon partitioning in these associations is still at a rudimentary stage and has lagged behind that of other symbiotic associations involving angiosperms (*sensu* de Bary, see Smith and Douglas, 1987). Many of the more commonly studied parasitic angiosperms are those with agriculturally important hosts. Some parasitic angiosperms are important weeds, especially in semi-arid tropical and Mediterranean areas. Parasitic witchweeds (*Striga* spp.), for example, can reduce the yield of their C_4 cereal hosts to zero, and also cause marked changes in allometry (see, for example, Press *et al.*, 1990).

Host dependency varies enormously between parasitic taxa. Some chlorophyllous (hemiparasitic) angiosperms are facultative parasites and can grow, flower and set seed in the absence of a host, albeit less vigorously (e.g. *Rhinanthus*, *Melampyrum* and *Euphrasia*). In nature, unattached plants are rarely found beyond the seedling stage and, following attachment, these species supplement autotrophic carbon with heterotrophic carbon from the host (Hodgson, 1973). Other genera, including *Striga*, are obligate parasites, and again, despite the presence of chlorophyll receive a proportion of their carbon from the host. The balance between autotrophic and heterotrophic carbon in *Striga* and probably other hemiparasites, is host-dependent (Press *et al.*, 1991). Achlorophyllous (holoparasitic) angiosperms depend absolutely on a heterotrophic supply of carbon. An understanding of carbon partitioning and transport in parasitic angiosperm associations is essential in order to understand further the range of host dependencies encountered within this group of organisms, and to account for the response of host plants to infection.

10.2 Parasite sink strength

10.2.1 *Dependency on host carbon supply*

The extent to which a parasitic angiosperm competes with existing host sinks for carbon will depend on the sink strength of the parasite in relation to the sink strength of the host tissue, and hence on its 'competitive' ability. The regulation of sink competition determines assimilate partitioning and sink competition between host and parasite is thus a major determinant of the performance (e.g. productivity and reproductive output) of both partners.

Defining the sink strength of hemiparasites is complicated by their varying degrees of dependency on host-derived carbon. In the chlorophyllous genus *Striga*, dependency on heterotrophic carbon is reflected in the leaf at the morphological, ultrastructural and biochemical level (Press *et al.*, 1991). In comparison with non-parasitic plants, *S. hermonthica* has poorly developed palisade mesophyll cells, fewer air spaces between the spongy mesophyll cells, lower chloroplast density and lower ribulose-1,5-bisphosphate carboxylase/oxygenase (Rubisco) activity (Press *et al.*, 1986; Tuohy *et al.*, 1986). Achlorophyllous holoparasites represent a greater sink because almost all their net carbon must be host derived. No net carbon assimilation is detectable in the Orobanchaceae, although phosphoenolpyruvate (PEP) carboxylase activity in this family (and also in hemiparasitic Loranthaceae, Viscaceae, Santalaceae and Scrophulariaceae) is not dissimilar from that in other C_3 plants (see Stewart and Press, 1990). However, *Cytinus hypocistus* (Rafflesiaceae), which shows extreme reductions in morphology, has very low levels of PEP carboxylase activity (Renaudin *et al.*, 1982).

In order to make quantitative estimations of parasite sink strength it is necessary to know the balance between autotrophic and heterotrophic carbon supply. Unfortunately host carbon dependency has been quantified in relatively few species. ^{14}C labelling has been used to illustrate the movement of host-derived photosynthate to parasites (see, for example, Okonkwo, 1966), although principally in a qualitative rather than quantitative manner. However, this technique has allowed the identification of transported substances. More recently use has been made of measurements of the natural abundance of stable carbon isotopes to estimate the proportion of host-derived carbon in *Striga* (C_3 plants) parasitic on C_4 hosts (Press *et al.*, 1987a). C_4 plants are enriched in ^{13}C compared to C_3 plants, and the two groups have $\delta^{13}C$ values in the range -10 to $-18‰$ and -22 to $-35‰$, respectively (O'Leary, 1988). The expected $\delta^{13}C$ value of *Striga* leaves not in receipt of host carbon can be estimated from measurements of the quotient of the intercellular/ambient CO_2 concentration (C_i/C_a), using the models of Farquhar *et al.* (1982). Thus C_i/C_a quotients and $\delta^{13}C$ measurements can be used in conjunction to estimate the proportion of heterotrophic carbon in *Striga* leaves. Using this approach Press *et al.* (1987a) have shown that approximately 28% and 35% of carbon is derived from the host (sorghum) in mature leaves of *S. hermonthica* and *S. asiatica*, respectively. Although *Striga* lacks a direct phloem link in the haustorium, this study clearly demonstrates that it can derive significant amounts of carbon from the host. In a study following the development of *Striga* from attachment to emergence above ground and maturity, Press *et al.* (1987a) were able to demonstrate that until the parasite developed chlorophyll and began to photosynthesize it had a carbon isotope signature close to that of the host, suggesting that all the carbon in the immature parasite came from the host. Carbon budget models of three associations (*S. hermonthica*-sorghum, *S. hermonthica*–maize and *S. gesnerioides*–cowpea, Graves *et al.*, 1989, 1990, 1992) suggest that the estimates from stable carbon isotope measurements represent the minimum amount of carbon transferred from the host (see also Press *et al.*, 1991). This work prompted a re-examination of the carbon nutrition of hemiparasites mistletoes, which were previously thought to obtain only water and inorganic solutes from their hosts (see, for example, Hull and Leonard, 1964). The C_i/C_a quotient of the photosynthetic stems of the mistletoe *Phoradendron juniperinum* is significantly higher than that of the leaves of its coniferous host, *Juniperus ostersperma*, and $\delta^{13}C$ and gas exchange measurements suggest that in the mistletoe approximately 60% of parasite carbon is heterotrophic (Marshall and Ehleringer, 1990). This estimate is in close agreement with calculations made using transpiration rates and the concentration of carbon in host xylem fluid (Marshall and Ehleringer, 1990).

It is clear from these studies that, like holoparasites, photosynthetic hemiparasites also represent a sink for host carbon. It is also clear that there is no simple distinction between holoparasites and hemiparasites which can easily be used to predict parasite sink strength. Further, the relative demand of the parasite on the host appears to change with development (Press *et al.*, 1987a), and host species (Press and Graves, 1991), and is also influenced by environmental factors (see Section 10.7).

10.2.2 *Determinants of parasite sink strength*

Wardlaw (1990) describes the following characteristics which can determine sink strength in non-parasitic plants: (i) size of surface (membrane) area across which metabolites are transferred from the vascular system to the zone of utilization (unloading area); (ii) efficiency of transfer of carbon (rate per unit area) from the vascular system to the sink; and (iii) spatial or biochemical isolation of assimilates in the growing (or storage) organs once they leave the vascular system.

In parasitic angiosperms the surface area across which photo-assimilates are transferred is determined by the number and structure of haustoria present. There is a considerable range in the size of haustoria, which can vary by more than an order of magnitude even within a single genus (from approximately 1 mm in *S. hermonthica* to approximately 2 cm in *S. gesneriodies* (Visser and Dörr, 1987). Anatomical and ultrastructural studies of haustoria (see, for example, Kuijt, 1977) suggest that they are composed of a number of different cell types, and the cross-sectional area of the organ may not necessarily correspond to the surface area available for transport. There is also variation in the number of haustoria formed. Pate *et al.* (1990) report only eight per meter of root for *Olax phyllanthi* while *Cuscuta* spp. and *Cassytha* spp. can form many (10s–100s) haustoria as they coil around the host stems and branches. Since a single host plant may be infected by more than one individual, the potential surface area for transfer of host photosynthate is large.

Strong sinks are characterized by an efficient transfer of carbon from the vascular system. Wolswinkel (1974) has demonstrated that *Cuscuta* (a plant with a low chlorophyll content which is often regarded as intermediate between holo- and hemiparasites), withdraws almost 100% of the assimilates which normally move from a photosynthesizing host leaf (*Vicia faba*) to growing pods and seeds. Quantitative data of this sort are lacking for other host–parasite associations, and there is a need to study the carbon transfer dynamics across haustoria. Despite this, the large and rapid reductions in host productivity which can result upon infection, and the growing body of data quantifying carbon fluxes from hosts (see, for example, Schulze *et al.*, 1991), make it tempting to suggest that a consistent feature of parasitic angiosperms is an ability to conduct an efficient transfer of host carbon.

Spatial or biochemical isolation of assimilates is important since it prevents 'backflow' from the sink. In terms of spatial isolation, both holo- and hemiparasites appear to be near perfect sinks. Of the many reports of transfer of assimilates between host and parasite there are none which record carbon transfer, under any circumstances, from parasite to host. Once assimilates are transferred across the haustorium into the parasite body they are effectively lost to the host. This situation contrasts with that of developing leaves. In clover, for example, developing leaves act as sinks until they are 35–50% fully expanded, then they become sources (Chapman *et al.*, 1990). Some leaves of autotrophic plants (e.g. cucumber) can simultaneously import and export carbon as they mature (Turgeon and Webb, 1973). It is possible that unlike the petiole, the haustorium may act as an effective barrier, either physically or biochemically, to carbon export from the parasite, whilst at the same time conducting an efficient import system.

The mechanics of carbon transfer across the haustoria, and any physical systems which may exist to prevent export from the parasite, are not well understood. It is known, however, that a chemical conversion of sucrose (the form of carbon most commonly transported in the phloem) to hexoses can prevent reloading into the (host) phloem (Eschrich, 1989). Similarly, conversion to starch can reduce the solute concentration of the sink (thereby maintaining a favourable osmotic gradient for the transfer of assimilates from host to parasite). The conversion of sucrose to either hexoses or starch can thus aid biochemical isolation of the sink. In the holoparasite *Orobanche crenata*, Whitney (1972) reported that ^{14}C-labelled sucrose, traced from the host plant (*V. faba*), was hydrolysed to glucose and fructose, which accumulated in the parasite. Whitney (1972) noted that mature *Orobanche* tubers, which consist largely of thin-walled parenchyma cells, have very high starch contents. A similar observation was made for the haustoria of *S. hermonthica* (Mallaburn and Stewart, 1987). A characteristic feature of symbiotic associations is the accumulation of carbohydrate 'reserves' in the symbiont which differ from those in the host (Smith and Douglas, 1987). High concentrations of alditols (sugar alcohols, acyclic polyols) are present in many parasitic angiosperms (Lewis, 1984). In hemiparasites these are principally photosynthetic products (Hodgson, 1973), although in holoparasites they may be synthesized from heterotrophic carbon in both haustorial and non-haustorial tissue (Wegmann, 1986).

From the limited data available it seems that both holo- and hemiparasites satisfy criteria for strong sinks, with potentially large surface areas for transport of host assimilates, efficient uptake of host assimilates and effective spatial and biochemical isolation from the host vascular system. They can also grow very quickly once attached to the host plant, and have a rapid dry weight gain.

Ho et al. (1989) have argued that dry weight gain, which is sometimes seen as a measure of sink strength, is a result rather than a determinant of sink strength. These authors put forward the view that in an organ such as the wheat grain, sink strength is closely related to the number of plastids or endosperm cells present at an early stage of development. Starch accumulation, dry weight gain and sink strength are subsequently limited by the number of plastids. No similar developmental studies have been carried out with parasitic angiosperms. Whilst it is likely that the potential sink size of the parasite is determined genetically, there are several lines of evidence which suggest that actual dry weight gain, and therefore sink strength, is influenced by host solute supply. First, total parasite dry weight per host is inversely correlated with the number of individuals supported, hence sink strength of any one parasite must be dampened by assimilate availability. Secondly, the growth of parasites with a wide host range is influenced by the species to which they attach (Atsatt and Strong, 1970; Gibson and Watkinson, 1991; Govier et al., 1967; Hodgson, 1973). Hence sink strength may be influenced by host-derived factors in addition to carbon supply, for example, nitrogen (see Section 10.7.2).

10.3 Carbon transfer

10.3.1 Hemiparasitic associations

Early studies by Okonkwo (1966) show that despite the presence of photosynthetic tissue, hemiparasites receive carbon from the host. It had previously been assumed that they were largely autotrophic for carbon, and parasitic only for water and inorganic solutes. Raven (1983) has since pointed out that a potentially large transfer of carbon from host to parasite can occur even through the xylem stream in the form of amino and organic acids.

In many associations apoplastic continuity exists between host and parasite, although not necessarily through direct xylem-to-xylem contact (Fineran, 1991; see also Press et al., 1990). The haustorium contains a large number of parenchyma cells in relation to those which are differentiated into vascular tissue. Pate et al. (1990) made a study of the numbers and types of parasite cells at haustorial interfaces between the hemiparasite Olax phyllanthi and a number of host species, including estimates of the extent of direct contact between conducting elements of host and parasite. Terminating xylem elements of the parasite directly contacting xylem elements of the host at the haustorial interface accounted for only 8–10% of the total parasite cell contacts (which ranged from 12 880 on *Lysinema ciliatum* to 44 230 on *Bossiaea linophylla*). Parasite haustorial parenchyma was rarely in contact with host tracheids and vessels, and was mostly in association with either host schlerenchyma or xylem/ray parenchyma.

Ultrastructural studies of haustorial parenchyma cells have shown the presence of micro-convolutions and tubules in the plasma membrane, and it has been suggested that these might reflect active transport between the parenchyma and the apoplastic space of the cell walls (Fineran, 1991). Despite this, the possible importance of solute movement from host to parasite via parenchyma, which makes up the largest part of the haustorium in both hemiparasite and holoparasite associations, has yet to be explored.

High rates of transpiration are a characteristic feature of many hemiparasites, and transpiration rates can exceed those of the host by more than an order of magnitude (see Press *et al.*, 1990). A consequence of high transpiration rates is a large flux of solutes through the xylem stream from host to parasite. Despite lack of a correlation between transpiration rates and solute acquisition in some non-parasitic plants (Schulze and Bloom, 1984; Tanner and Beevers, 1990), and the relatively few direct xylem-to-xylem connections in the haustorium, it been proposed (Schulze *et al.*, 1984) that high transpiration rates play a key role in solute acquisition from the host.

Movement of solutes in the xylem may be important in other instances where there is no direct phloem link between source and sink. It has been demonstrated, for example, that phloem-mobile solutes may bypass a phloem blockage (such as that caused by heat treatment) by transferring to the xylem (Martin, 1982). Transport in the xylem is thought to be important during the early development of organs: Sauter and Ambrosius (1986) have suggested that in *Betula pendula*, movement of sugars in the transpiration stream through the xylem to the canopy has an important role in supplying sugars to developing catkins in early spring (i.e. before new phloem has become functional). The efficiency with which carbon gained through the xylem stream subsequently moves to the phloem by lateral movement is unknown, as is the importance of retrieval of apoplastic solutes by phloem parenchyma at leaf hydathodes (Wilson *et al.*, 1991), which are often a prominent feature in parasitic angiosperms.

10.3.2 *Holoparasitic associations*

Holoparasites do not possess the high transpiration rates observed in hemiparasites, and hence 'passive' transfer of solutes from the host is unlikely to account for heterotrophic carbon acquisition in these species. Whitney (1972) reports indications of efficient phloem-type transport of host assimilates across the haustorium of *Orobanche crenata*, despite the fact that he was unable to find a direct phloem connection. Similarly Aber *et al.* (1983) noted that very young haustoria of *O. crenata* were capable of abstracting carbohydrates from *V. faba* despite the fact that they were mostly undifferentiated.

It now seems that some holoparasitic genera such as *Orobanche*, and the 'intermediate' *Cuscuta* have transfer cells in the haustorium, similar to those found in the haustoria at junctions between gametophyte and sporophyte in mosses (Browning and Gunning, 1979), lycopods (Peterson and Whitter, 1991) and parts of higher plants (Pate and Gunning, 1972). These transfer cells are formed from contact parenchyma cells adjacent to the host vascular system. Wolswinkel (1978a, b) has suggested that the transfer cells of *Cuscuta* operate at high efficiency in the absorption of solutes from the apoplast of the host. It seems that in *Orobanche*, transfer cells form only in association with the host xylem, whilst in *Cuscuta* they form only in association with host phloem (Lamont, 1982 and references therein; Dörr, 1975). The remaining parenchyma cells in the haustorium are largely undifferentiated. In *Cuscuta*, development of a transfer cell from a contact cell is accompanied by a 20-fold increase in surface area of the plasma membrane, which may be important in increasing the area available for transport of solutes from host to parasite, as in the transfer cells of pea minor veins (Wimmers and Turgeon, 1991).

Solute transfer from the host plant to *Cuscuta* may occur in two stages (Wolswinkel, 1974): solutes move out of the sieve tube lumina of the host into the free space (of the cell walls and intercellular spaces), and are then absorbed from the free space by the transfer cells of the parasite. Since the parasite is a foreign sink, this phenomenon immediately prompts two questions. First, by what means does the parasite acquire solutes from the host which are not normally released from the phloem *en route* to a host sink? Secondly, by what means does the parasite reload solutes in the absence of a direct phloem connection?

Wolswinkel (1978a) used a 'washing-out' technique (in which the solute content of the apoplast could be analysed) to show that, at and near (within 1 cm) the site of infection, *V. faba* plants parasitized with *Cuscuta* have a larger proportion of ^{14}C-labelled assimilates in the free space than do similarly labelled but uninfected plants. Moreover, Wolswinkel (1978b) also demonstrated that this accumulation of solutes near the site of infection contained only those inorganic solutes which are phloem mobile (K, P, Mg, but not the phloem immobile Ca), suggesting a specific effect of the parasite on phloem rather than xylem transport. Jacob and Neumann (1968) suggest that in the presence of *Cuscuta*, phloem of the host plant becomes 'leaky', allowing solutes which would otherwise be retained to enter the apoplast. This seems unlikely for at least two reasons. First, it appears that infection strongly stimulates the release of sucrose into the host free space, but has little effect on the efflux of either glucose or fructose (Wolswinkel and Ammerlaan, 1983). As the effect of *Cuscuta* appears to be sucrose specific, it might be that the stimulating influence of the parasite on host sugar release is restricted to the sieve tube mechanism. Secondly, it appears that the enhanced unloading of photo-assimilates is under metabolic control. It has been demonstrated that the enhanced phloem unloading which occurs in *V.*

faba infected with *Cuscuta* does not occur at 0°C, or in the presence of dinitrophenol or sodium azide (Wolswinkel, 1978a). Similarly, the unloading of labelled L-valine and L-asparagine from infected stem segments of *V. faba* does not occur at 0°C, or after addition of *p*-chloromercuribenzene sulphonic acid (PCMBS; Wolswinkel *et al.*, 1984). This, combined with the apparent specificity of solutes unloaded, argues against a general degradation of host phloem integrity by the parasite.

As phloem loading requires a supply of energy, unloading should be able to proceed without further energy input, so active unloading of photo-assimilates is probably not usual in non-parasitized plants. Farrar and Minchin (1991) demonstrated that the import of ^{11}C-labelled assimilates into barley roots was independent of respiration (at least in the short term). The addition of metabolic inhibitors in this case had no effect on phloem unloading. This is in contrast to the situation described above for parasitic angiosperms. However, the apparent requirement for energy during unloading in parasitised plants is not unique. Phloem unloading into soybean fruits was shown to be temperature- and O_2-dependent (Thorne, 1982), and Williams *et al.* (1991) noted an increase in root respiration associated with artificial increases in the solute concentration of the apoplast (which should stimulate sink activity), but it was not clear if this was a consequence of osmoticum-induced phloem unloading, or represented a cost of osmoregulation.

Gifford and Evans (1981) proposed that sinks may permit phloem unloading, not by active processes *per se*, but by local inhibition of phloem reloading. In path tissues, phloem usually retains solutes with minimal radial exchange. This is because the sieve tubes have a high affinity for sucrose. Sucrose molecules released into the apoplast will be reloaded unless they are taken out of the pathway by hydrolysis (Eschrich, 1989). It appears that the attachment of the parasite is sufficient to stimulate sink activity. High free space acid invertase activity, similar to that in young stem segments of *V. faba*, has been reported in *Cuscuta* (Wolswinkel and Ammerlaan, 1983). This will help create a gradient across the apoplast which favours phloem unloading. Hydrolysis of unloaded sucrose into glucose and fructose will lead to the generation of sink activity (Eschrich, 1989) since accumulation of hexoses in the apoplast lowers the water potential of the free space, leading to a movement of water out of the phloem, so lowering phloem turgor pressure. There is then a lowering of sieve tube osmotic potential, leading to a further movement of sucrose along the phloem towards the sink. It has been suggested that high solute concentration in the sink apoplast (or low sieve tube solute concentration) is possibly one of the most important factors controlling the rate of assimilate transport from source to sink (Wolswinkel, 1985). If this is correct then the enhanced phloem unloading initiated by parasites such as *Cuscuta* will be significant in maintaining a flux of solutes from the host.

Phloem loading of solutes from the apoplast by the parasite, like phloem loading in the host, is dependent on a supply of energy. Wolswinkel (1978b) found enhanced concentration of Mg and P near the site of attachment of *Cuscuta* on

V. faba and suggested that this was related to a more intensive metabolism in these areas, and it seems likely that the high rates of respiration of the haustoria (see, for example, Press and Graves, 1991) are associated at least partly with an active phloem loading mechanism. It is difficult to determine the origin of the haustorial tissue (i.e. whether it is host or parasite derived). Hence it is not possible to show high respiration rates in association with the 'uptake side' of the haustoria. Moreover, it appears that infection leads to considerably enhanced respiration rates in the host root, for a distance of up to 1–2 cm from the site of attachment (Singh and Krishnan, 1971; Singh and Singh, 1971; Vincent and Press, unpublished data). High root respiration rates will maintain a flux of solutes towards the site of infection, and they will also utilize some of the assimilates before they reach the parasite. This response could also be associated with repair mechanisms induced in the host following infection by the parasite.

10.4 Effects of parasite demand on host carbon acquisition

Parasitic angiosperms are a potentially large additional sink for host assimilates. The question of whether plant growth is sink (or demand) limited is contentious. If there is a mechanistic relationship between assimilate demand and supply (Herold, 1980; Neales and Incoll, 1968), then a higher rate of photosynthesis in host plants compared with uninfected plants might be expected. There is some preliminary evidence to support this view. There is often an initial (although not always significant) increase in the light-saturated rate of photosynthesis in C_4 cereals infected with *S. hermonthica*, particularly at low levels of infection and nitrogen supply (Section 10.7.2; Cechin and Press, unpublished data). ter Borg (1986) and ter Borg and van Ast (1991) report stimulation of growth in *V. faba* plants infected with *O. crenata* (photosynthetic measurements are not reported). However, stimulation of growth was observed in a cultivar of *V. faba* which is relatively tolerant to *O. crenata* infection, and only during the early stages of infection when inorganic solutes were not limiting growth. A similar phenomenon has been recorded for some cultivars of sugar cane parasitized by *S. asiatica* (Visser, 1987).

Both positive and, more commonly, negative photosynthetic responses in host leaf tissue to fungal infection have been reported. Although they have been better described than responses to parasitic angiosperm infection, the mechanisms by which fungi affect photosynthesis are uncertain (Isaac, 1992). Stimulation of photosynthesis has also been observed in plants parasitized by animals: for example, a positive effect of the presence of aphids on the photosynthetic activity of wheat (Rabbinge *et al.*, 1981). The promotion of the assimilation rate was attributed to the removal of assimilates by the aphids.

Any initial stimulation of photosynthesis in *Striga*-infected plants is not sustained and photosynthesis decreases in host plants below rates in uninfected plants as the parasite matures (Press and Stewart, 1987; Press *et al.*, 1987b). Host plants can show dramatic reductions in growth, and parasite-induced effects on photosynthesis have been estimated to account for up to 80% of the difference in dry matter accumulation between infected and uninfected plants (Press and Graves, 1991).

Mansfield *et al.* (1990) used $^{14}CO_2$ pulse-chase techniques to investigate further the effect of *Striga* infection on the photosynthetic metabolism of maize and concluded that the impaired photosynthesis of the host may result from a reduced rate of malate decarboxylation or a lowered rate of pyruvate export from the bundle sheath chloroplasts. Measurements of quantum yield, CO_2 fixation at different intercellular concentrations of CO_2 and host stable isotope signatures also suggest that photosynthetic impairment may result from changes in metabolic shuttling between mesophyll and bundle sheath chloroplasts and/or effects on the rate of photosynthetic electron transport (Press *et al.*, 1990).

The absence of any detectable movement of solutes from parasite to host suggests that the photosynthetic dysfunction associated with *Striga* infection may be associated with parasite-induced changes in resource partitioning within the host, rather than the production of a toxin by the parasite (cf. Musselman, 1980; Parker, 1984). No reductions in photosynthesis were observed in tomato plants infected with the holoparasites from the genus *Orobanche*, despite large reductions in host growth (Vincent and Press, unpublished data; see also Table 10.1). It remains to be seen whether the effects of *Striga* on host carbon assimilation are the rule or the exception for plants supporting parasitic angiosperms.

10.5 Carbon accumulation and partitioning in host plants

Carbon partitioning is ultimately determined by the competitive ability of the various sinks within the plant to attract assimilates. Since parasitic angiosperms remove carbon from the host plant, they must be a stronger sink than one, or a number, of host sinks. It follows from this that carbon partitioning within the host plant may be affected by the presence of parasitic angiosperms, and this has

Table 10.1. *Light-saturated rates of photosynthesis ($\mu mol\ m^{-2}\ sec^{-1}$) for tomato grown and measured at either approximately 350 or 550 p.p.m. CO_2, in the presence or absence of O. aegyptiaca.*

Age (days)	CO_2 (p.p.m.)	Photosynthesis ($\mu mol\ m^{-2}\ sec^{-1}$)	
		Uninfected	Orobanche-infected
31	350	3.9 ± 0.3	4.8 ± 0.4
31	550	13.4 ± 1.1	8.5 ± 0.8
48	350	10.4 ± 1.0	10.5 ± 0.8
48	550	9.8 ± 0.7	11.2 ± 1.0
78	350	4.8 ± 0.5	4.9 ± 0.6
78	550	5.2 ± 0.8	6.6 ± 0.7

Measurements are for the youngest fully developed leaves. Means ± standard errors are reported, $n = 25$ for 31-day-old plants, and 15 for 48- and 78-day-old plants (Vincent and Press, unpublished data).

been demonstrated for both *Striga-* and *Orobanche*-infected plants (Graves *et al.*, 1989, 1990, 1992; Vincent and Press, unpublished data).

S. hermonthica markedly affects partitioning between above-ground (shoot) and below-ground tissues in cereal hosts, as can be seen from changes in the allometric coefficient (Ledig *et al.*, 1970) of sorghum plants grown in sand culture with a low nitrogen supply (Figure 10.1). Such differences between infected and uninfected plants disappear at higher nitrogen concentrations (Figure 10.1, see Section 10.7.2). In infected plants the most striking effects of the parasite are lower stem weights and a lower seed (grain) yield, sometimes result-ing in zero reproductive effort. Root biomass appears to remain unaltered in some hosts (see below). These allometric responses may also characterize infec-tion by other parasitic angiosperms. Figure 10.2 illustrates the response of tomato to infection by three species of *Orobanche*.

In wheat, stem reserves usually supply one-third of the dry matter accumulated by grain, but this proportion is known to increase greatly when photosynthesis is depressed (Apel and Natr, 1976). The dramatic effects of parasitic angiosperms on carbon allocation to stems in infected plants (which coincide with reductions in photosynthesis in *Striga*-infected plants), suggest that mobilization of stem reserves may be an important source of carbohydrates for the parasite.

Partitioning of dry weight to the roots increases in both maize and sorghum infected with *S. hermonthica*, but decreases significantly in cowpea infected by *S. gesnerioides* (Graves *et al.*, 1992). Unlike *S. hermonthica*, *S. gesnerioides* has very reduced, almost scale-like leaves, and in this respect resembles the holoparasite *Orobanche* (a major functional difference being that the latter is achloro-phyllous). *Orobanche*-infected tomato plants also have significantly less dry

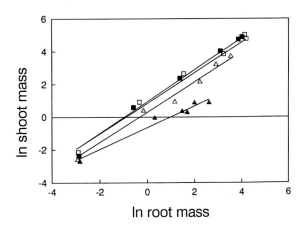

Figure 10.1. *The relationship between shoot and root mass in uninfected (open symbols) and* S. hermonthica-*infected (closed symbols) sorghum plants, grown in sand culture and supplied with 40% full strength nitrogen free Long Ashton solution containing either 0.5 mM (triangles) or 4 mM (squares) N as NH$_4$NO$_3$ (see Figure 10.6) (Cechin and Press, unpublished data). The regression equations are as follows: y = 0.30 + 0.92x (R^2 = 0.98) (△); y = − 0.64 + 0.66x (R^2 = 0.92) (▲); y = 0.94 + 0.97x (R^2 = 0.98) (□); y = 0.83 + 0.93x (R^2 = 0.98) (■).*

Figure 10.2. *Partitioning of dry matter in tomato (L. esculentum cv. Moneymaker) either uninfected (−) or infected with O. cernua (+ c), O. aegyptiaca (+ a), and O. ramosa subsp mutelli (+ r). Total dry matter accumulation of the parasites is also shown ((c), (a) and (r)). Plants were grown in a glasshouse from April 1991 at 26 ± 2° C (day), 16 ± 2° C (night) and were harvested 98 days after host germination. The dry weight (g) (means of five replicates) is shown for root (⊠), stem (▭), photosynthetic leaves (▦), fruit (▧) and parasite (▭). (Vincent and Press, unpublished data.)*

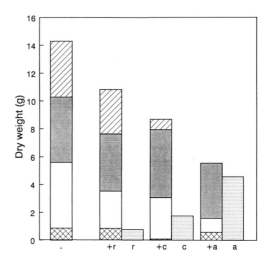

matter allocated to roots, in comparison to uninfected plants (Figure 10.2). The reason why some parasites should apparently stimulate host root growth, whilst others apparently inhibit host root growth is not yet clear. It may be of some significance that *S. hermonthica* has high rates of transpiration, while *S. gesnerioides*, like *Orobanche*, with scale-like leaves, has much lower rates of transpiration. *S. hermonthica* (with only a vestigial root system) will therefore need access to larger volumes of water than *S. gesnerioides* and *Orobanche*, possibly accounting for the different carbon allocation patterns in hosts supporting these parasites.

Lower root biomass in plants infected with *Orobanche* and *S. gesnerioides* may be a product of the unusually high rates of respiration, especially close to the sites of infection. Whitney (1972) noted that parasitism of *V. faba* by *O. crenata* caused sudden wilting, yellowing and leaf-fall. This was attributed to reduced carbohydrate supply to host roots, rendering them less efficient at water uptake and reducing their growth. Vincent and Press (unpublished data) observed differences in the mechanical properties of the roots of tomato plants infected with *Orobanche* spp., with the roots becoming progressively more stunted and friable as the parasite matured.

Orobanche aegyptiaca has a more severe effect on tomato than either *O. cernua* or *O. ramosa* subsp. *mutelii* (Figure 10.2). The reasons for this are as yet uncertain, but may be because this parasite attached to the host earlier (and therefore emerged above ground earlier and accumulated dry matter more quickly) than the other two species. Other experiments with *S. hermonthica* parasitizing sorghum have shown that there is a differential effect on host dry weight accumulation and partitioning which is dependent on the age of the host plant at the time of parasite attachment. Sorghum plants which are less than 1 week old at infection are more severely affected than those which are at least 3 weeks old at infection (Cechin and Press, unpublished data). It is possible that older plants, with their greater carbon reserves, may have some 'buffering capacity', which lessens the effect of the parasite. The greater effect of O. *aegyptiaca* may thus be a function of the lesser degree of host development before attachment, and so could also be related to the stages of vascular development connecting path tissues with host sinks. If the vascular connection between host source and sink is established before infection there could be a greater probability of host sinks out-competing the parasite. This has yet to be investigated in greater detail, but it is of interest since tomato plants which have begun to fruit before infection will continue to develop those fruits, whilst those which have not begun to fruit will not do so. A study of the relationship between timing of host fruit phloem development and parasite attachment would be of interest.

A further point which merits attention is the effect of distance of point of attachment of the parasite from the host fruit on competition for assimilates. Whilst it is generally accepted that the distance for transfer of assimilates is not a major limiting factor in sink development (Wardlaw, 1990), there is some evidence that in uninfected tomatoes the position of a fruit in a truss affects the import rate of assimilates (Ho *et al.*, 1989). It may be of some significance that *Cuscuta*, which can apparently halt development of fruits already initiated, is a stem parasite which can often attach close to or even on the host fruit (Wolswinkel, 1985), whilst *Orobanche* is a root parasite, and cannot stop development of fruits which are already growing. The question of whether or not this, like the differential effects of parasitism on partitioning in hosts of different ages, is mainly a product of vascular development has yet to be addressed conclusively.

10.6 Mechanistic basis of host response

10.6.1 *Sink competition*

The mechanisms which account for the responses described in Section 10.5 are only poorly understood. One potential mechanism is direct sink competition where the parasite diverts host resources away from host sinks. However, sink competition is not an explanation for altered carbon partitioning. It is itself a result of perhaps several factors. The ability of the parasite to generate and maintain sink activity, possibly by high acid invertase activity in the apoplast (converting sucrose to hexoses) and by the sequestration of carbon in a relatively osmotically inert form have been discussed in Section 10.2.2.

Parasitic angiosperms have K^+ concentrations which can exceed those of their host by more than an order of magnitude (see Stewart and Press, 1990). The accumulation of K^+ in the parasite is presumably a reflection of the unidirectional flow of solutes from the host, thus preventing circulation and redistribution. K^+ is significant as an osmoticum, and is more osmotically important than sucrose (Wardlaw, 1990). It seems that the turgor driven (water relations) phloem transport model could work equally well with any osmotically active solute (Lang, 1983) and phloem transport could be influenced by K^+ status. High K^+ concentrations at the sink will enhance the turgor gradient, and so increase the import of assimilates. Lang (1983) has proposed that K^+ gradients may be an additional mechanism by which plants can control partitioning. It has also been suggested that increased K^+ accumulation by sinks and mature leaves may cause a greater rate of synthetic metabolism which will in turn lead to enhanced import of solutes (Conti and Geiger, 1982; Geiger and Conti, 1983). Furthermore, K^+ has been implicated in the loading process at the SE (Baker and Chaudry, 1986).

For hemiparasites, traditionally thought of as being independent of phloem-derived host solutes, the high K^+ concentrations have been explained in association with maintenance of an osmotic gradient to ensure water flux through the xylem. However, as noted by Raven (1983) (see also Section 10.3.1) the xylem stream contains a significant amount of host carbon (as well as water and inorganic solutes) in the form of amino and organic acids. The uninfected host plant would presumably transport this carbon to a sink, either passively in the xylem stream, or in the phloem following lateral movement between xylem and phloem. Removal of this carbon by the parasite will prevent its utilization by the host, and will consequently affect host carbon partitioning. It is clear from this that parasitic plants are capable of influencing the growth and sink development of their hosts in ways which are not normally addressed in discussions of source–sink interactions.

10.6.2 *Plant growth regulators*

One way in which parasitic angiosperms may have a direct effect on host sinks could be through altering metabolism of plant growth regulators (PGRs). Drennan and El Hiweris (1979) have suggested that diversion of host solutes by parasitic plants could occur through the reduction of host growth rates by the induction of growth-inhibiting substances, or through a similar effect resulting from a shortage of growth-promoting substances. Their study of *S. hermonthica* on sorghum showed that infected plants had lower xylem sap cytokinin and gibberellin concentrations, by 90–95% and 30–80%, respectively, from those measured in uninfected sorghum. Higher abscisic acid (ABA) and farnesol concentration were also reported. These changes led Drennan and El Hiweris (1979) to suggest that the parasite induces host tissues to respond to infection by modifying production of its own growth regulating products. There was no evidence to suggest a production and movement of growth inhibitors from parasite to host. However it is not clear whether the changes in PGRs reported in sorghum are a primary response to *S. hermonthica*, or represent a secondary response to parasite-induced water stress, as a consequence of the high transpiration rate of the parasite.

A reduction in host growth, mediated by PGRs would reduce potential host sink size, thus allowing a diversion of solutes towards the parasite, in the same way that removal of apical and auxiliary vegetative shoots at flowering can increase seed set (Wardlaw, 1990). It is possible that the marked suppression of lateral branching in *Orobanche*-infected tomato plants (Vincent and Press, unpublished data) and in *Cuscuta*-infested pea plants (Tsivion, 1981), could be caused by parasite-induced decreases in cytokinins.

Parasitic angiosperms are able to synthesize PGRs in addition to those which might be received from the host (Hall *et al.*, 1987). The precise role of cytokinins in competitive solute acquisition remains to be determined, but evidence suggests that it is important at least in inducing the formation of haustoria in *Cuscuta* (Paliyath *et al.*, 1978). Gupta and Singh (1985) showed that the cytokinin concentration of *Cuscuta* apices was only 1% of that in the haustorial region, and moreover, that in the haustorium itself the concentration of cytokinin in the concave half (in contact with the host) was 40-fold greater than in the corresponding convex half. In mistletoe associations, cytokinins may be important in the formation of 'witches brooms' and assimilate sinks in infected trees (Paquet, 1978).

The higher ABA and farnesol concentrations reported by Drennan and El Hiweris (1979) in *Striga*-infected sorghum are consistent with lower rates of transpiration and stomatal conductance in infected plants (Press *et al.*, 1987c). Root hemiparasites, in particular, may benefit from lower fluxes of water and dissolved carbon in the host xylem. These hemiparasites are able to maintain high transpiration rates, even in the presence of ABA [possibly as a result of high leaf K^+ concentrations (Smith and Stewart, 1990)], and can thus divert water flow, and carbon, from the host. Pickard *et al.* (1979) report decreased export from the source leaf of moonflower when transpiration was increased. If the converse is true, and reduced transpiration favours export from source leaves, as implied by Milburn and Kallarackal (1989), then parasite-induced reductions in host transpiration could increase carbon export and further favour carbon flux to the parasite. There is also the possibility that increased ABA production in infected roots could stimulate sucrose unloading (Tanner, 1980), increasing the possibility of carbon allocation to the parasite by transfer through the haustoria.

The potential benefits to the parasite which may result from induced changes in host PGRs are, however, questionable. Too great a reduction in growth stimulators, or too great an induction of growth inhibitors will ultimately deprive the parasite of its solute and water supply by halting or reducing generation at the source. This is an area which would benefit from further study, and could prove useful in helping to determine the possible roles of PGRs in the control of carbon allocation suggested by Thomas (1986) and Brenner (1987).

10.7 Interactions with environment

Carbon partitioning in parasitic angiosperms and their hosts may be influenced by both the degree of autotrophy of the parasite and by the photosynthetic capacity and carbon balance of the host. Carbon partitioning cannot therefore be considered in isolation from environmental factors which might influence carbon acquisition by either partner. In some other parasitic associations, the symbiont has been shown to have beneficial effects on the host in one environment while being detrimental in another (see, for example, Michalakis *et al.*, 1992). Although there is no unequivocal evidence to support this view with respect to parasitic angiosperm associations, environment can play a major role in determining the outcome of the interaction. The influence of atmospheric and edaphic factors is illustrated here by examining the response of *Orobanche* associations to elevated concentrations of CO_2, and *Striga* associations to nitrogen supply, respectively.

10.7.1 *Atmospheric factors: responses to elevated CO_2 concentrations*

Although exposure of plants to higher CO_2 concentrations usually results in an initial stimulation of photosynthesis, rates may decline in some species as plants acclimate (Bazzaz, 1990). The reasons for this decline are not fully understood, but sink capacity for photosynthate storage and/or utilization may strongly influence its extent (Bazzaz, 1990; Sharkey, 1985). Studies involving manipulation of sink strength under elevated CO_2 atmospheres support this hypothesis (see, for example, Clough *et al.*, 1981; Thomas and Strain, 1991), showing sink limitation to be more marked at elevated CO_2 concentrations. Many of these studies involve artificial disturbance of the plant (e.g. pod removal and root restriction). *Orobanche* associations potentially provide a model system for the study of these responses, since sink strength can readily be manipulated by varying the parasite load supported by the host, and this system has the advantage that the host plant can be left intact.

Tomato was grown in ambient (approximately 350 p.p.m.) and elevated (550 p.p.m.) CO_2 atmospheres, either in the presence or absence of *O. aegyptiaca*, and light-saturated rates of photosynthesis were measured in 31-, 48- and 78-day-old plants (Table 10.1). Significant differences were seen only in 31-day-old plants. Photosynthesis in *O. aegyptiaca*-infected plants grown and measured at approximately 350 p.p.m. CO_2 exceeded those in uninfected plants by 23%. Photosynthetic rates in plants grown and measured at 550 p.p.m. CO_2 exceeded those at approximately 350 p.p.m. CO_2, although at the higher CO_2 concentration photosynthetic activity in uninfected plants exceeded that in infected plants by 58%. No significant differences were observed between treatments in either 48- or 78-day-old plants.

It seems unlikely that the higher photosynthetic rates measured in 31-day-old infected tomato at approximately 350 p.p.m. CO_2 result from additional sink demand imposed by *O. aegyptiaca*, because there is no such response at 550 p.p.m. CO_2, where it would be expected to be more pronounced. It is perhaps more likely that the different photosynthetic rates result from the effects of both the parasite and CO_2 on the phenology of the host. There are significant differences in photosynthetic activity between plants of different ages, and measurements at more frequent intervals would enable photosynthetic and phenological responses to CO_2 and *O. aegyptiaca* to be determined.

Elevated atmospheric CO_2-stimulated growth in uninfected plants, and dry matter accumulation at 550 p.p.m. CO_2 exceeded that at approximately 350 p.p.m. up to at least 78 days from sowing (Figure 10.3). This response to CO_2 was not seen in infected plants. However, there was a difference at the higher CO_2 concentration in the allocation of dry matter between infected and un-infected plants. The most significant difference was the formation of fruit only in elevated CO_2-grown plants. No significant differences in parasite productivity were observed between CO_2 treatments.

The reasons for the differences between host responses at ambient and elevated CO_2 concentrations are by no means clear, but the data suggest that elevated CO_2 may reduce the influence of the parasite on the host sink hierarchy. Greater substrate availability at elevated CO_2 seems to allow the development of both host and parasite sinks (fruit). Growth in elevated CO_2 also delayed the onset of senescence in tomato (see also Table 10.1), which is usually a striking feature in the apical regions of *Orobanche*-infected plants. Together, these data

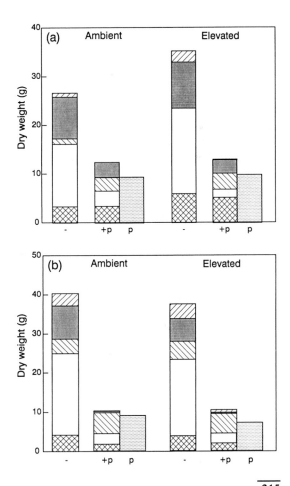

Figure 10.3. *Effect of elevated atmospheric CO_2 on partitioning of dry matter in tomato (L. esculentum cv. Moneymaker) either uninfected (−), or infected with O. aegyptiaca (+p). Plants were grown at 26° C (day), 16° C (night), photosynthetic photon flux density 1000 μmol m⁻² sec⁻¹ and 16h photoperiod in either ambient (approximately 350 p.p.m.) or elevated (550 p.p.m.) CO_2. Dry weight (g) (means of five replicates) is shown for root (▨), stem (☐), dead leaves (▨), photosynthetic leaves (■), fruit (▨) and parasite (☐) for plants harvested (a) 78 days and (b) 112 days after host germination. (Vincent and Press, unpublished data.)*

suggest that elevated CO_2 concentrations may influence any role of PGRs in control of carbon partitioning in these associations.

Foliar carbohydrate contents did not differ significantly between infected and uninfected tomato plants (Figure 10.4), in agreement with the response of sorghum to *S. hermonthica* (Press *et al.*, 1991). However, significant differences were observed in roots, with lower contents of starch, sucrose and glucose in infected roots at both ambient and elevated CO_2 concentrations (Figure 10.4). The most notable change in infected roots was a decrease in sucrose and glucose content, supporting the observations of Wolswinkel and Ammerlaan (1983) for *Cuscuta*. Elevated atmospheric CO_2 concentrations markedly affected carbon partitioning between carbohydrates in the parasite (Figure 10.5).

The mechanisms which control the differential response of infected and uninfected plants to elevated CO_2 remain to be determined. It appears that the extra sink activity generated by *Orobanche* stimulates neither photosynthesis nor growth in the host plant. Elevated CO_2 does not significantly reduce the effect of the parasite on host dry weight accumulation, and does not result in stimulation of parasite growth.

Figure 10.4. *Effect of elevated atmospheric CO_2 on carbohydrate content in tomato (*L. esculentum cv. Moneymaker*) either uninfected (control), or infected with* O. aegyptiaca. *Details of growth conditions are given in the legend to Figure 10.2. Starch (▨), glucose (▦) and sucrose (▢) are expressed on a dry weight basis for (a) photosynthetic leaves and (b) roots, collected during the middle of the photoperiod. Oven dried (100° C for 1 h followed by 60° C for 48 h) material was extracted into 80% ethanol (sucrose and glucose) or 0.2 M sodium hydroxide (starch), and assayed using commercial methods (Boeringer Mannheim, FRG). Means of separate extractions from three plants are shown. (Vincent and Press, unpublished data.)*

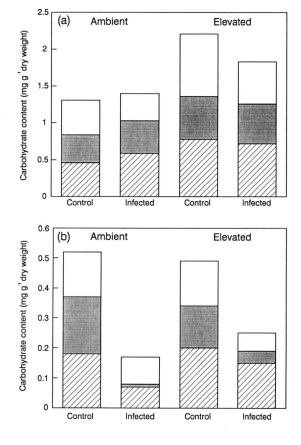

10.7.2. *Soil factors: responses to nitrogen supply*

Crop losses associated with agriculturally important parasitic angiosperms, such as the genera *Striga* and *Orobanche*, are negatively correlated with soil fertility in general and nitrogen in particular. In natural and semi-natural communities, root hemiparasites (at least) tend to be more abundant in ecosystems where soil nitrogen availability is low. These observations suggest that nitrogen supply plays an important role in determining the outcome of parasitic angiosperm interactions.

In the *S. hermonthica*–sorghum association, the partitioning of dry matter between host and parasite is dependent on the concentration of nitrogen supplied to the plants (Figure 10.6), as is the difference in dry matter accumulated between infected and uninfected sorghum plants. Higher nitrogen concentrations favour the performance of the host relative to the parasite for a combination of at least two reasons (Cechin and Press, unpublished data). First, higher concentrations of nitrogen have a negative effect on germination of *Striga* seed and subsequent attachment to host roots, resulting in fewer parasites per host plant. This only results in very large differences in parasite dry weight per host plant when *Striga* seed density in the soil is low. Secondly, impairment of host photosynthesis by *Striga* is much more marked at lower nitrogen concentrations, the effect being almost negligible when nitrogen supply is non-limiting (Cechin and Press, unpublished data). More than 50% of foliar nitrogen in C_3 plants is invested in photosynthetic apparatus, and photosynthetic capacity is strongly correlated with foliar nitrogen concentration (Evans, 1989). This correlation is also seen in *S. hermonthica*, and when nitrogen is supplied at higher concentrations it seems that the parasite can make a greater contribution to its carbon budget, possibly reducing its reliance on the host. At low nitrogen supply, competition for nitrogen between host and parasite may affect partitioning of nitrogen between photosynthetic apparatus in the host leaf, resulting in lower net CO_2 fixation rates. It is clear that the carbon and nitrogen economies of the symbiosis are inextricably linked, and should be considered together

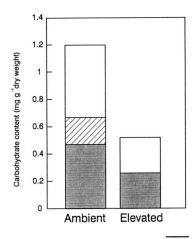

Figure 10.5. *Effect of elevated atmospheric CO_2 on the carbohydrate content of O.* aegyptiaca *grown on tomato (*L. esculentum *cv. Moneymaker). See Figure 10.2 for experimental details. Means of separate extractions from three plants are shown. (Vincent and Press, unpublished data.)*

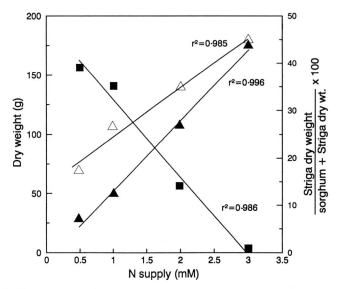

Figure 10.6. *Effect of nitrogen concentration on total dry weight (g) (means of five plants) of uninfected sorghum (△), S.* hermonthica-*infected sorghum plus* S. hermonthica *(▲), and the percentage of total dry weight in the* S. hermonthica-*sorghum association which is contained within* S. hermonthica *plants (■). Plants were grown in sand and watered with 0.4-strength Long Ashton solution containing either 0.5, 1, 2, or 3 mM nitrogen as ammonium nitrate, supplied three times per week. The experiment was conducted in a glasshouse in the summer with supplementary light and heating (36–39° C day, 24–27° C night). Lines were fitted using linear regression. The slope and the intercept of lines corresponding to uninfected sorghum and sorghum plus* Striga *are significantly different (Student's* t-test, $P \leqslant 0.05$).*

rather than in isolation. The carbon budget models of *Striga* associations described by Graves *et al.* (1989, 1990, 1992) therefore need to be expanded to include nitrogen and possibly other environmental parameters.

10.8 Summary

Parasitic angiosperms are an important sink for host carbon. In addition to depressing carbon acquisition and accumulation they can also influence carbon partitioning, with greatest effects on stem dry weight and reproductive output. The effectiveness of the parasite as a sink depends on a number of factors, including degree of autotrophy, developmental stage of host and parasite, and the supply of resources from the atmosphere and soil.

Acknowledgements

We thank Dr A.N. Parsons for commenting on the paper and Ms R. Cooper for technical assistance.

References

Aber, M., Fer, A. and Sallé, G. (1983) Transfer of organic substances from the host plant *Vicia faba* L. to the parasite *Orobanche crenata* Forsk. *Z. Pflanzenphysiol.* 112, 297–308.

Apel, P. and Natr, L. (1976) Carbohydrate content and grain growth in wheat and barley. *Biochem. Physiol. Pflanzen.* 169, 437–446.

Atsatt, P.R. and Strong, D.R. (1970) The population biology of annual grassland hemi-parasites I. The host environment. *Evolution* 24, 278–291.

Baker, D.A. and Chaudry, S.B. (1986) Active sucrose transport in the castor bean: effect of monovalent cations. In: *Plant Biology* Volume, 1, *Phloem Transport* (eds J. Cronshaw, W.J. Lucas and B.T. Giaquinta). Alan R. Liss, New York, pp. 77–88.

Bazzaz, F.A. (1990) The response of natural ecosystems to the rising global CO_2 levels. *Ann. Rev. Ecol. Syst.* 21, 167–196.

ter Borg, S.J. (1986) Effects of environmental factors on *Orobanche*-host relationships, a review and some recent results. In: *Proceedings of a Workshop on the Biology and Control of Orobanche* (ed. S.J. ter Borg). LH\VPO, Wageningen, pp. 57–69.

ter Borg, S.J. and van Ast, A. (1991) Parasitic plants as stimulants of host growth. In: *5th International Symposium on Parasitic Weeds* (eds J.K. Ransom and L.J. Musselman). CIMMYT, Nairobi, pp. 442–446.

Brenner, M.L. (1987) The role of hormones in photosynthetic partitioning and seed filling. In: *Plant Hormones and their Role in Plant Growth and Development* (ed. P.J. Davies). Kluwer, Dordrecht, pp. 474–493.

Browning, A.J. and Gunning, B.E.S. (1979) Structure and function of transfer cells in the sporophyte haustorium of *Funaria hygrometrica* Hedw. I. The development and ultrastructure of the haustorium. *J. Exp. Bot.* 30, 1233–1246.

Chapman, D.F., Robson, M.J. and Snaydon, R.W. (1990) The carbon economy of developing leaves of white clover (*Trifolium repens* L.). *Ann. Bot.* 66, 623–628.

Clough, J.M., Peet, M.M. and Kramer, P.J. (1981) Effects of high atmospheric CO_2 and sink size on rates of photosynthesis of a soybean cultivar. *Plant Physiol.* 67, 1007–1010.

Conti, T.R. and Geiger, D.R. (1982) Potassium nutrition and translocation in sugar beet. *Plant Physiol.* 70, 168–172.

Dörr, I. (1975) Development of transfer cells in higher parasitic plants. In: *Phloem transport* (eds J.J. Dainty, P.R. Gorham, L.M. Srivastava and C.A. Swanson). Plenum Press, New York, pp. 177–186.

Drennan, D.S.H. and El Hiweris, S.O. (1979) Changes in growth regulating substances in *Sorghum vulgare* infected by *Striga hermonthica*. In: *Proceedings of The Second Symposium of Parasitic Weeds* (ed. R.C. Raleigh). North Carolina State University, pp. 144–155.

Eschrich, W. (1989) Phloem unloading of photo assimilates. In: *Transport of Photoassimilates* (eds D.A. Baker and J.A. Milburn). John Wiley and Sons, New York, pp. 206–263.

Evans, J.R. (1989) Photosynthesis and nitrogen relations in leaves of C_3 plants. *Oecologia* 78, 9–19.

Farrar, J.F. and Minchin, P.E.H. (1991) Carbon partitioning in split root systems of barley: relation to metabolism. *J. Exp. Bot.* 42, 1261–1269.

Farquhar, G.D., O'Leary, M.H. and Berry, J.A. (1982) On the relationship between carbon isotope discrimination and the intercellular carbon dioxide concentration in leaves. *Aust. J. Plant Physiol.* 9, 121–137.

Fineran, B.A. (1991) A structural approach towards investigating transport systems between host and parasite, as exempified by some mistletoes and root parasites. In: *Parasitic Flowering Plants* (eds H. Chr. Weber and W. Forstreuter). Philipps University, Marburg, pp. 201–220.

Geiger, D.R. and Conti, T.R. (1983) Relation of increased potassium nutrition to photosynthesis and translocation of carbon. *Plant Physiol.* 71, 141–144.

Gibson, C.C. and Watkinson, A.R. (1991) Host selectivity and the mediation of competition by the root hemiparasite *Rhinanthus minor. Oecologia,* 86, 81–87.

Gifford, R.M. and Evans, L.T. (1981) Photosynthesis, carbon partitioning and yield. *Ann. Rev. Plant Physiol.* 32, 485–509.

Govier, R.N., Nelson, M.D. and Pate, J.S. (1967) Hemiparasitic nutrition in Angiosperms I. The transfer of organic compounds from host to *Odontites verna* (Bell.) Dum. (Scrophulariaceae). *New Phytol.* 66, 285–297.

Graves, J.D., Press, M.C. and Stewart, G.R. (1989) A carbon balance model of the sorghum-*Striga hermonthica* host-parasite association. *Plant Cell Environ.* 12, 101–107.

Graves, J.D., Wylde, A., Press, M.C. and Stewart, G.R. (1990) Growth and carbon allocation in *Pennisetum typhoides* infected with the parasite angiosperm *Striga hermonthica. Plant Cell Environ.* 13, 367–373.

Graves, J.D., Press, M.C., Smith, S. and Stewart, G.R. (1992) The carbon canopy economy of the association between cowpea and the parasitic angiosperm *Striga gesnerioides. Plant Cell Environ.* 15, 283–288.

Gupta, A. and Singh, M. (1985) Mechanism of parasitism by *Cuscuta reflexa*: distribution of cytokinins in different regions of the parasitic vine. *Physiol. Plant.* 63, 76–78.

Hall, P.J., Badenoch-Jones, J., Parker, C.W., Letham, D.J. and Barlow B.A. (1987) Identification and quantification of cytokinins in the xylem sap of mistletoes and their hosts in relation to leaf mimicry. *Aust. J. Plant Physiol.* 14, 429–438.

Herold, A. (1980) Regulation of photosynthesis by sink activity — the missing link. *New Phytol.* 86, 131–144.

Ho, L.C., Grange, R.I. and Shaw, A.F. (1989) Source/sink regulation. In: *Transport of Photoassimilates* (eds D.A. Baker and J.A. Milburn). John Wiley and Sons, New York, pp. 306–343.

Hodgson, J.F. (1973) *Aspects of the carbon nutrition of angiospermous parasites.* Ph.D Thesis. University of Sheffield.

Hull, R.J. and Leonard, O.A. (1964) Physiological aspects of parasitism in mistletoes (*Arceuthobium* and *Phoradendron*). I. The carbohydrate nutrition of mistletoe. *Plant Physiol.* 39, 996–1007.

Isaac, S. (1992) *Fungal-Plant Interactions.* Chapman and Hall, London.

Jacob, F. and Neumann, S. (1968) Studien an *Cuscuta reflexa* Roxb.I.Zur Funktion der Haustorian bei der Aufnahme von Saccharose. *Flora abt.A.* 159, 191–203.

Kuijt, J. (1977) Haustoria of phanerogamic parasites. *Ann. Rev. Phytopathol.* 17, 91–118.

Lamont, B. (1982) Mechanisms for enhancing nutrient uptake in plants with particular reference to Mediterranean South Africa and Western Australia. *Bot. Rev.* 48, 597–689.

Lang, A. (1983) Turgor-regulated translocation. *Plant Cell Environ.* 6, 683–689.

Ledig, F.T., Bormann, F.H. and Wenger, K.F. (1970) The distribution of of dry matter growth between shoot and roots in loblolly pine. *Bot. Gaz.* 131, 349–359.

Lewis, D.H. (1984) Physiology and metabolism of alditols. In: *Storage Carbohydrates in Vascular Plants* (ed. D.H. Lewis). Cambridge University Press, Cambridge, pp. 157–179.

Mallaburn, P.S. and Stewart, G.R. (1987) Haustorial function in *Striga*: comparative anatomy of *S. asistica* (L) Kuntz and *S. hermonthica* (Del.) Benth. (Scrophulariaceae). In: *Parasitic Flowering Plants* (eds H. Chr. Weber and W. Forstreuter). Philipps University, Marburg, pp. 523–536.

Mansfield, F., Stewart, G. and Keys A. (1990) Investigation of the photosynthetic carbon metabolism of maize infected with *Striga hermonthica*. *J. Exp. Bot.* 41 (Suppl.), 420.

Marshall, J.D. and Ehleringer, J.R. (1990) Are xylem-tapping mistletoes partially heterotrophic? *Oecologia* 84, 244–248.

Martin, P. (1982) Stem xylem as a possible pathway for mineral retranslocation from senescing leaves to the ear in wheat. *Aust. J. Plant Physiol.* 9, 197–207.

Michalakis, Y., Olivieri, I., Renaud, F. and Raymond, M. (1992) Pleiotropic action of parasites: how to be good for the host. *Trends Ecol. Evolut.* 7, 59–62.

Milburn, J.A. and Kallarackal, J. (1989) Physiological aspects of phloem translocation. In: *Transport of Photoassimilates* (eds D.A. Baker and J.A. Milburn). John Wiley and Sons, New York, pp. 264–305.

Musselman, L. J. (1980) The biology of *Striga*, *Orobanche*, and other root-parasitic weeds. *Ann. Rev. Phytopathol.* 18, 463–489.

Neales, T.F. and Incoll, L.D. (1968) The control of leaf photosynthesis rate by the level of assimilate concentration in the leaf: a review of the hypothesis. *Bot. Rev.* 34, 107–124.

O'Leary, M.H. (1988) Carbon isotopes in photosynthesis. *BioSci.* 38, 328–336.

Okonkwo, S.N.C. (1966) Studies on *Striga senagalensis* II. Translocation of ^{14}C-labelled photosynthate, urea-^{14}C and sulphur-35 between host and parasite. *Am. J. Bot.* 53, 142–148.

Paliyath, G., Maheshwari, R. and Mahadecan, S. (1978) Initiation of haustoria in *Cuscuta* by cytokinin application. *Curr. Sci.* 47, 427–429.

Paquet, P. J. (1978) Cytokinins in Douglas-Fir dwarf mistletoe and its host. *Plant Physiol.* 61 (Suppl.), 49.

Parker, C. (1984) The physiology of *Striga* species: present state of knowledge and priorities for future research. In: *Striga, Biology and Control* (eds E.S. Ayensu, H. Doggett, R.D. Keynes, J. Marton-Lefevre, L.J. Musselman, C. Parker and A. Pickering). ICSU/IRL Press, Oxford, pp. 179–193.

Pate, J.S. and Gunning, B.E.S. (1972) Transfer cells. *Ann. Rev. Plant Physiol.* 23, 173–196.

Pate, J.S., Kuo, J. and Davidson, D.J. (1990) Morphology and anatomy of the haustorium of the root hemiparasite *Olax phyllanthi* (Olacaceae), with special reference to the haustorial interface. *Ann. Bot.* 65, 425–436.

Peterson, R.L. and Whitter, D.P. (1991) Transfer cells in the sporophyte-gametophyte junction of *Lycopodium appressum*. *Can. J. Bot.* 69, 222–226.

Pickard, W.F., Minchin, P.E.H. and Troughton, J.H. (1979) Real time studies of carbon-11 translocation in moonflower. II. Further experiments on the effects of a nitrogen atmosphere, water stress and chilling, and a qualitative theory of stem translocation. *J. Exp. Bot.* 30, 307–318.

Press, M.C. and Stewart, G.R. (1987) Growth and photosynthesis in *Sorghum bicolor* infected with *Striga hermonthica*. *Ann. Bot.* 60, 657–662.

Press, M.C. and Graves, J.D. (1991) Carbon relations of angiosperm parasites and their hosts. In: *Progress in* Orobanche *Research* (eds K. Wegmann and L.J. Musselman).

Eberhard-Karls-Universitat, Tubingen, pp. 55–65.

Press, M.C., Shah, N. and Stewart, G.R. (1986) The parasitic habit: trends in metabolic reductionism. In: *Proceedings of a Workshop on the Biology and Control of Orobanche* (ed. S.J. ter Borg). LH\VPO, Wageningen, pp. 96–106.

Press, M.C., Shah, N., Tuohy, J.M. and Stewart, G.R. (1987a) Carbon isotope ratios demonstrate carbon flux from C_4 host to C_3 parasite. *Plant Physiol.* 85, 1143–1145.

Press, M.C., Tuohy, J.M. and Stewart, G.R. (1987b) Gas exchange characteristics of the sorghum-*Striga* host-parasite association. *Plant Physiol.* 84, 814–819.

Press, M.C., Tuohy, J.M. and Stewart, G.R. (1987c) Leaf conductance, transpiration and relative water content of *Striga hermonthica* (Del.) Benth and *S. asiatica* (L.) Kuntze (Scrophulariaceae) and their host *Sorghum bicolor* (L.) Moench (Gramineae). In: *Parasitic Flowering Plants* (eds H. Chr. Weber and W. Forstreuter). Philipps University, Marburg, pp. 631–636.

Press, M.C., Graves, J.D. and Stewart, G.R. (1990) Physiology of the interaction of angiosperm parasites and their higher plant hosts. *Plant Cell Environ.* 13, 91–104.

Press, M.C., Smith, S. and Stewart, G.R. (1991) Carbon acquisition and assimilation in parasitic plants. *Funct. Ecol.* 5, 278–283.

Rabbinge, R., Drees, E.M., van de Graaf, M., Verberne, F.C.M. and Wesselo, A. (1981) Damage effects of cereal aphids in wheat. *Neth. J. Plant Pathol.* 87, 217–232.

Raven, J.A. (1983) Phytophages of xylem and phloem: a comparison of animal and plant sap-feeders. *Adv. Ecol. Res.* 13, 135–234.

Renaudin, S., Vidal, J. and Larher, F. (1983) Characterization of phosophenolpyruvate carboxylase in a range of parasitic phanerogames. *Z. Pflanzenphysiol.* 106, 229–237.

Sauter, J.J. and Ambrosius, T. (1986) Changes in the partitioning of carbohydrate in the wood during bud break in *Betula pendula* Roth. *J. Plant Physiol.* 124, 31–43.

Schulze, E.-D. and Bloom, A.J. (1984) Relationship between mineral nitrogen influx and transpiration in radish and tomato. *Plant Physiol.* 76, 827–828.

Schulze, E.-D., Turner, N.C. and Glatzel, G. (1984) Carbon, water and nutrient relations of two mistletoes and their hosts: a hypothesis. *Plant Cell Environ.* 7, 293–299.

Schulze, E.-D., Lange, O.L., Ziegler, H. and Gebauer, G. (1991) Carbon and nitrogen isotope ratios of mistletoes growing on nitrogen and non-nitrogen fixing hosts and on CAM plants in the Namib desert confirm partial heterotrophy. *Oecologia*, 88, 457–462.

Sharkey, T.D. (1985) Photosynthesis in intact leaves of C_3 plants: physics, physiology and rate limitations. *Bot. Rev.* 51, 53–106.

Singh, J.N. and Singh, J.N. (1971) Studies on the physiology of host-parasite relationship in *Orobanche* I. Respiratory metabolism of host and parasite. *Physiol. Plant.* 24, 380–386.

Singh, P. and Krishnan, P.S. (1971) Effect of root parasitism by *Orobanche* on the respiration and chlorophyll content of *Petunia*. *Phytochemistry*, 10, 315–318.

Smith, D.C. and Douglas, A.E. (1987) *The Biology of Symbiosis*. Edward Arnold, London.

Smith, S. and Stewart, G.R. (1990) Effect of potassium levels on the stomatal behaviour of the hemi-parasite *Striga hermonthica*. *Plant Physiol.* 94, 1472–1476.

Stewart, G.R. and Press, M.C. (1990) The physiology and biochemistry of parasitic angiosperms. *Ann. Rev. Plant Physiol. Plant Mol. Biol.* 41, 127–151.

Tanner, W. (1980) On the possible role of ABA in phloem unloading. In: *Phloem Loading and Related Processes* (eds W. Eschrich and H. Lorenzen). Gustar Fischer Verlag, New York, pp. 349–351.

Tanner, W. and Beevers, H. (1990) Does transpiration have an essential function in long-distance ion transport in plants? *Plant Cell Environ.* 13, 745–750.

Thomas, R.B. and Strain, B.R. (1991) Root restriction as a factor in photosynthetic acclimation of cotton seedlings grown in elevated CO_2. *Plant Physiol.* 96, 627–634.

Thomas, T.H. (1986) Hormonal control of assimilate movement and compartmentation. In: *Plant Growth Substances* (ed. M. Bopp). Springer-Verlag, Berlin, pp. 350–359.

Thorne, J.H. (1982) Characterization of the active sucrose transport system of immature soybean embryos. *Plant Physiol.* 70, 953–958.

Tsivion, Y. (1981) Suppression of auxilliary buds of its host by parasitising *Cuscuta* I. Competition amoung sinks and indirect inhibition. *New Phytol.* 87, 91–99.

Tuohy, J., Smith, E.A. and Stewart, G.R. (1986) The parasitic habit: trends in morphological and ultrastructural reductionism. In: *Proceedings of a Workshop on the Biology and Control of Orobanche* (ed. S.J. ter Borg). LH\VPO, Wageningen, pp. 86–95.

Turgeon, R. and Webb, J.A. (1973) Leaf development and phloem transport in *Curcurbita pepo*: transition from import to export. *Planta,* 113, 179–191.

Visser, J.H. (1987) The susceptibility of some sugar cane cultivars to witchweed (*Striga asiatica* (L.) Kuntze). In: *Parasitic Flowering Plants* (eds H. Chr. Weber and W. Forstreuter). Philipps University, Marburg, pp. 789–795.

Visser, J.H. and Dorr, I. (1987) The haustorium. In: *Parasitic Weeds in Agriculture I.* Striga (ed. L.J. Musselman). CRC Press, Florida, pp. 91–106.

Wardlaw, I.F. (1990) The control of carbon partitioning in plants. *New Phytol.* 116, 341–381.

Wegmann, K. (1986) Biochemistry of osmoregulation and possible biochemical reasons of resistance against *Orobanche*. In: *Proceedings of a Workshop on the Biology and Control of Orobanche* (ed. S.J. ter Borg). LH\VPO,Wageningen, pp. 107–113.

Whitney, P.J. (1972) The carbohydrate and water balance of beans (*Vicia faba*) attacked by broomrape (*Orobanche crenata*). *Ann. Appl. Biol.* 70, 59–66.

Williams, J.H.H., Minchin, P.E.H. and Farrar, J.F. (1991) Carbon partitioning in split root systems of barley: the effect of osmotica approximately *J. Exp. Bot.* 42, 453–460.

Wilson, T.P., Canny, M.J. and McCully, M.E. (1991) Leaf teeth, transpiration and the retrieval of apoplastic solutes in balsam poplar. *Physiol. Plant.* 83, 225–232.

Wimmers, L.E. and Turgeon, R. (1991) Transfer cells and solute uptake in minor veins of *Pisum sativum* leaves. *Planta,* 186, 2–12.

Wolswinkel, P. (1974) Complete inhibition of setting and growth of fruits of *Vicia faba* L. resulting from the draining of the phloem system by *Cuscuta* species. *Acta Bot. Neerl,* 23, 48–60.

Wolswinkel, P. (1978a) Phloem unloading in stem parts parasitised by *Cuscuta*: the release of ^{14}C and K^+ to the free space at 0°C and 25°C. *Physiol. Plant.* 42, 167–172.

Wolswinkel, P. (1978b) Accumulation of phloem mobile mineral elements at the site of attachment of *Cuscuta europaea* L. *Z. Pflanzenphysiol.* 86, 77–84.

Wolswinkel, P. (1985) Phloem unloading and turgor-sensitive transport: factors involved in sink control of assimilate partitioning. *Physiol. Plant.* 65, 331–339.

Wolswinkel, P. and Ammerlaan, A. (1983) Sucrose and hexose release by excised stem segments of *Vicia faba* L. The sucrose specific stimulating influence of *Cuscuta* on sugar release and the activity of acid invertase. *J. Exp. Bot.* 34, 1516–1527.

Wolswinkel, P., Ammerlaan, A. and Peters, H.F.C. (1984) Phloem unloading of amino acids at the site of attachment of *Cuscuta europaea*. *Physiol. Plant.* 75, 13–20.

11

Carbon-11 in the study of phloem translocation

P.E.H. Minchin and M.R. Thorpe

11.1 Introduction

Few laboratories have access to the radioactive isotope ^{11}C because its half-life is only 20.4 min. Therefore, its unique advantages in phloem studies are not well known, and so in this review we hope to demonstrate the advantages of ^{11}C to a wider audience. ^{11}C was first used as a tracer in biology in a study of photosynthesis by Ruben et al. (1939) but with the development of ^{14}C production in the mid-1940s interest in ^{11}C vanished. It was next used by Moorby et al. (1963) for phloem studies in soybean, and after a further delay of 10 years by More and Troughton (1973). Since then there has been a growing interest by groups in Canada (Fensom et al., 1977), the USA (Fares et al., 1978), Scotland (Williams et al., 1979), Germany (Jahnke et al., 1981; Roeb and Britz, 1991) and New Zealand.

Two properties of ^{11}C make it ideal for certain types of phloem study. These are: (i) the high energy of its decay radiation enables in vivo measurement in any part of a plant, and (ii) its short half-life allows multiple experiments on the same plant as there is no accumulation of label from previous measurements. Consequently a time series of tracer levels within various parts of a whole plant can be obtained before, during and after an experimental treatment. This enables detailed data on the movement of labelled photosynthate to be collected from a single plant.

With biological systems there is variability from sample to sample which raises questions about the relevance of such a detailed examination of only a few plants. When one's interest is in function, we believe that this is not an important issue, as it is accepted that the mechanisms are the same in each sample; it is the degree of quantitative expression which varies. Hence detailed studies on a small number of samples are appropriate in determining function. When the interest is in expression of this function within the population, larger samples are clearly necessary.

Experimentally, one applies $^{11}CO_2$ to a leaf and then observes the time series of the amount of tracer at various positions along the phloem transport pathway. These profiles contain information about the processes transporting this tracer and in the first part of this chapter we discuss how to use the profiles to gain understanding about these processes. The shape of a tracer profile is determined by all the processes through which the tracer passes up to the site of measurement, and so explanations of this shape must account for all these processes. The change in shape of a profile between consecutive detectors is a consequence of the transport processes occurring between detectors and reflects the dynamics of transport. Analysis of the dynamics of tracer movement gives information on these flows. Mechanistic understanding of the flow processes is gained by postulating processes which would give rise to similar tracer dynamics. For example, if a tracer pulse is seen to broaden while moving along a stem, this is consistent with axial flow at a single speed together with radial unloading and reloading along the pathway. Further, by applying experimental treatments while monitoring the flows, changes in these flows can be followed.

Flows are determined by the mechanisms involved, though in a given steady state they could arise from a wide variety of mechanisms. More diagnostic of the mechanisms are the flow transients induced during the change from one steady state to another. A powerful method of determining the mechanisms driving and controlling the flows is therefore to observe the flow changes while an experimental treatment is applied. Because plants are able to adapt quickly to changed conditions, there is a need to ensure that during an experiment one is observing the process of interest and not new processes brought about by adaptation to the experimental treatment. On perturbing a plant experimentally, measurements will initially contain information on the mechanisms currently at play (e.g. flows, rate coefficients), medium term measurements will give information on the characteristics which change more slowly (e.g. pool sizes), while measurements made after a long time contain information about adaptive processes (e.g. changes in transfer coefficients, changed pathways). Hence after applying a treatment, we can expect changes in different aspects of the system to occur at different times, with the short-term changes being most useful in identifying the mechanisms functioning at the time of treatment.

We believe that a major role for ^{11}C is in following the transient changes in flow after an experimental treatment, which requires *in vivo* measurement over a time span long enough for the transients to be characterized. When the interest is in the flow patterns some time after the treatment, then simpler destructive measurements using ^{14}C may be more appropriate. Phloem studies based upon following ^{13}C or ^{14}C transport usually require tissue to be destroyed to allow measurement of the tracer, preventing the observation of a time series of tracer movement needed for a dynamic study. An important exception to this is the experimental system for ^{14}C developed by Swanson and Geiger (1967) using a thin sink leaf which allows *in vivo* measurement of ^{14}C accumulation. More recently this method has been extended to follow export from a thin source leaf (Geiger and Fondy, 1979). This experimental system has led to a major increase in understanding of phloem flow but is restricted to thin sources and sinks, and pathway detection is not possible. With ^{11}C these limitations do not exist; for example, we have recently monitored phloem flow into a developing apple fruit.

11.2 Experimental techniques

The application of short-lived isotopes in biological research has been fully reviewed at a specialized conference (Minchin, 1986) so the experimental techniques will be mentioned only briefly here.

^{11}C production. Because of its short half-life, ^{11}C must be produced near the site of its application, which is the major difficulty in its use as a low energy nuclear accelerator is needed. All current methods produce label as $^{11}CO_2$.

^{11}C detection. ^{11}C decay is by emission of 0.92 MeV positrons which have a maximum path length of about 4 mm in water and 4 m in air. On coming to rest each positron annihilates creating two 0.511 MeV γ-rays which are attenuated by about 9% cm^{-1} in water. The positrons themselves can be detected using Geiger–Müller tubes, or the annihilation radiation detected with scintillation detectors.

Two methods of measuring temporal tracer profiles are currently in use, involving slit or sink collimation (Figure 11.1). A slit-collimated detector monitors tracer within a small segment of leaf or stem, while sink collimation measures the total amount of tracer downstream from the boundary of the collimator. Slit collimation is achieved using a slit milled through a block of lead (e.g. Troughton *et al.*, 1977) or by using a pair of detectors to observe the pair of coincident γ-rays produced by positron annihilation (e.g. Fares *et al.*, 1978). Slit collimation monitors the amount of tracer within a segment of the plant, which includes tracer within the transport system (sieve tubes) and tracer within the surrounding tissues. Hence a change in observed tracer level can be interpreted as a change either in input or output from this segment, leading to ambiguities. Sink collimation measures the tracer within an entire sink region, be it the whole plant, the whole plant minus the labelled leaf, or a distal terminal sink. Sink collimation requires that all tracer within that sink is monitored with uniform sensitivity such that movement of tracer within the field of view does not result

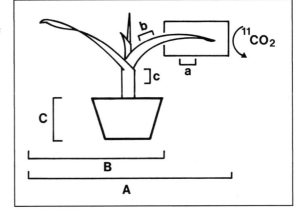

Figure 11.1. *Collimation of radiation detectors to obtain tracer profiles. Alongside the plant are shown possible fields of view for radiation detectors monitoring tracer movement. Each detector monitors tracer within its field of view as indicated. Lower case letters label slit collimation; a = source, b = pathways, c = sink. Uppercase letters label sink collimation; A = whole plant, B = mobilized, C = sink.*

in a change in count rate, and also there must be only one route in the plant for tracer movement into the field of view. Apart from decay, any change in count rate can then be due only to flow through the one available pathway into the field of view. Used this way sink collimation gives an unambiguous measurement of tracer movement.

^{11}C *application to plants*. Both pulse and continuous labelling have been used. Pulse labelling allows a large quantity of label to be applied at a defined time, allowing ^{11}C to be used to study processes which occur over a time span of up to 25 tracer half-lives (500 min). For example, seed coat unloading within pea ovules has been followed for 200 min after pulse labelling (Minchin and McNaughton, 1986), and using a number of lead attenuators which could be removed sequentially as the high level of applied ^{11}C decayed we have followed leaf efflux for up to 500 min (unpublished data).

Continuous loading has been used with both constant and variable amounts of $^{11}CO_2$. It is possible for tracer activity in all parts of the plant to become constant when the net arrival of tracer is exactly balanced by decay (Fares *et al.*, 1978). Unlike the long-lived isotope ^{14}C, with ^{11}C different parts of the plant will always be at different specific activities due to decay of ^{11}C during transport. Recently we have begun to use continuous ^{11}C labelling but with controlled variation in activity (Minchin and Thorpe, 1989). With continuous stimulation of the processes by the varying input of tracer, we can follow carbon flow throughout a plant for an indefinite period by analysis of the tracer dynamics.

11.3 Data analysis

The aim is to gain understanding of the phloem transport system so we must be quite clear what the tracer tells us concerning the movement of photosynthate. Because ^{11}C decays so quickly, the specific activity will change with time following fixation, so knowledge of the applied specific activity cannot be used to link tracer quantities with photosynthate, unless a decay correction is possible. With pulse labelling, simple decay correction of the data is possible, but with continuous labelling this is not the case, since, because of dispersion (Christy and Fisher, 1978; Minchin and Troughton, 1980), label within the plant at some specific time will have been fixed by the source leaf over an unknown range of times. Thain (1984) has shown that tracing flows on the basis of specific activity is often invalid, but that rate constants obtained from tracer movement also describe the movement of photosynthate. Thus we need a method of estimating these rate constants from tracer profiles.

To obtain flow parameters from the tracer profiles, some form of data analysis is needed. Not all analysis requires a detailed quantitative approach. A qualitative approach based on observing major changes in the shape of a tracer profile has been extensively used with various pathway, sink and source treatments (e.g. Fensom *et al.*, 1990; Jahnke *et al.*, 1989; Pickard *et al.*, 1978a,b; Troughton *et al.*, 1977). Quantitative analysis is necessary with slow or delayed responses, when the changes occurring in profile shape are difficult to identify with changes in transport processes.

As ^{11}C is a short-lived isotope, a further difficulty arises. Transport processes with characteristic times greater than several half-lives are not observable, as any tracer following such a route will decay to below the measurement threshold before traversing this route. This is why the efflux curve for a ^{11}C labelled leaf reveals only the short term pool, while with ^{14}C up to three pools are seen (Bauermeister et al., 1980; Rocher and Prioul, 1987). Hence flow parameters obtained using ^{11}C refer only to recently fixed photosynthate, even if continuous loading is used.

We will now consider two distinct methods which have been used for quantitative analysis of ^{11}C data, one based upon efflux analysis and the other on dynamic analysis.

11.3.1 Efflux analysis

Compartmental models have played a major role in describing tracer flow through biological systems. A frequently used technique is that of efflux analysis, which is a limited form of compartmental analysis. This involves labelling a system with tracer, and following the tracer loss when labelling stops. Since it is common to find that efflux curves can be expressed as the sum of exponentials, and compartmental models predict efflux curves which are a sum of exponentials, then compartmental interpretation appears natural. Over the time scale readily followed with ^{11}C, the efflux curves from all species examined has been representable, after decay correction, by the sum of a single exponential and a constant (Figure 11.2). This has been interpreted by a two-compartmental model, one of these being a storage compartment (Dyer et al., 1991; Fares et al., 1978; Kays et al., 1987). Parameter estimates for this model have been based on constant loading which is stopped when there is equilibrium of ^{11}C throughout

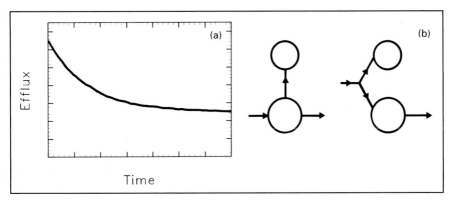

Figure 11.2. (a) Typical ^{11}C-efflux curve, corrected for decay, which can be represented as a single exponential plus a constant. (b) Compartmental models which give rise to an efflux curve of the form shown in (a).

the plant. The source leaf efflux is then followed, the entire measurement taking 4–5.5 h. From the data, parameters of the compartmental model are obtained (e.g. rates of storage and export, size of export pool). But analysis of efflux curves by compartmental analysis has been shown to lack the power to provide reliable information about multicompartmental systems (Cheesemann, 1986) and the basic assumptions underlying compartmental analysis are rarely checked (Zierler, 1981). Therefore, not all properties of the model can confidently be attributed to the leaf, particularly pool sizes which are very sensitive to model structure. One can have more confidence that the rate of export is a property of the leaf. As discussed above, properties derived from ^{11}C movement describe movement of recently fixed photosynthate. Total carbon flow may be different.

Efflux analysis implicity assumes that the system being modelled is not changing during data collection. When diel changes in the transport properties of leaves have been reported (Bell and Incoll, 1982; Farrar and Farrar, 1985; Kays *et al.*, 1987) on the basis of compartmental analysis, there must be immediate doubt about the validity of analysis. Hence we cannot agree with the claim of Dyer *et al.* (1991) that they were able to 'measure carbon turnover times of the photo-synthate pool and velocity of internal photosynthate transport minute by minute'. Also, because efflux analysis is concerned only with tracer loss when there is no inflow, it gives us no information on the flow of tracer through the system when there is inflow. Efflux analysis on intact plants is restricted to the source leaf, as nowhere else can one terminate tracer import experimentally.

11.3.2 *Dynamic analysis*

Dynamic analysis is based on the idea that a varying amount of tracer entering a system (e.g. leaf, length of phloem pathway) will appear at the output with a modified time course. The change in shape between input and output is described mathematically by an 'input–output' equation, describing the dynamics of the system. The input–output equation is nothing more than a terse and precise description of the observed change in profile shape without any assumptions as to the mechanisms involved. Methods for determining the best input–output equation for a pair of profiles, and the techniques used with both pulse and continuous ^{11}C labelling, including the method to account for decay, have been fully described (Minchin and Troughton, 1980; Minchin and Thorpe, 1989) so will not be repeated here. An equivalent description to the input–output equation is the transfer function of the system, which is simply the system output calculated for a theoretical input which is unity at time zero and zero at all other times (Figure 11.3a). The transfer function for the two-compartment model found by efflux analysis (Figure 11.2; Fares *et al.*, 1978) is a single decay-ing exponential as shown in Figure 11.3a. The transfer function h_k is defined at times k = 0, T, 2T, 3T...(T being the sampling interval) and has a number of interesting properties.

(i) The individual values h_k of the transfer function are the fraction of the input at time zero which appears at the output at time k. For bulk flow the transfer

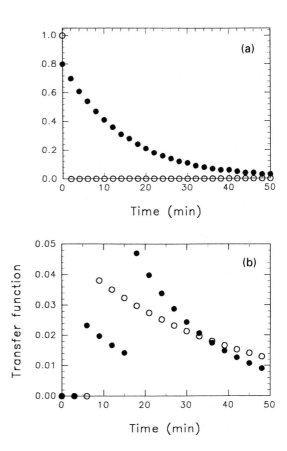

Figure 11.3. *(a) The transfer function of a system is the output (●) for an input (○) which is zero at all times except for time zero when it takes the value of 1 (the unit pulse). The transfer function is usually found by observing the response of the system to a much more complex input, because the ideal unit pulse can rarely be achieved. (b) Transfer functions calculated from tracer profiles (whole plant, mobilized) appropriate for study of photosynthate flow through a leaf. The complex shaped transfer function (●) was derived from the best input–output description of the data while the simpler shaped transfer function (○) was calculated from the best input–output equation of the form describing flow through the compartmental models shown in Figure 11.2b.*

function would only be non-zero at one time corresponding to the time delay in traversing the system. The transfer function can be interpreted as a direct statement of temporal dispersion caused by the system.

(ii) The average time taken to traverse the system is $\Sigma kh_k / \Sigma h_k$.

(iii) The fraction of the input which eventually appears at the output, the system gain, is the sum the fractions which appear at each time; i.e. Σh_k.

Interpretation of gain depends upon the system under study. For a leaf where the input is the total photosynthetic uptake and output is the total tracer mobilized through the petiole, the gain is the fraction of tracer eventually exported, which we call the export fraction. For transport to a sink, where the input is the total tracer supplied to the plant from the load leaf, and the output is tracer within the sink, the gain is the fraction of mobilized tracer eventually transported into the sink, that is the partitioning fraction to that sink.

Implicit in the above is the assumption that the system dynamics, and hence the dynamic relationship, is constant. Such a system is referred to as stationary. If a system is non-stationary, then it must be described by a time-varying dynamic relationship. In practice, one does not assume a stationary system, but tests for this. There are well developed techniques for determination of the most appropriate variation in the dynamic relationship (Minchin and Troughton, 1980).

Dynamic analysis can be used on all parts of the phloem system, source, pathway and sinks, so combined with ^{11}C we can continuously monitor the entire phloem system *in vivo*. The precision in the quantitative analysis of tracer movement, and therefore the flow parameters, depends entirely upon the shape of the tracer profiles. The reason is that to observe a process requires that it be adequately excited by the incoming tracer. With poor excitation the process will have little effect upon the observations and remain undiscovered.

11.4 Applications

Having outlined various ways of analysing profiles of ^{11}C, we now give some examples of its use. These are organized into source, path and sink. Finally, their interdependence is illustrated.

11.4.1 *Source*

Efflux analysis has been used in many species to investigate carbon flow in source leaves (e.g. Bauermeister *et al.*, 1980; Bell and Incoll, 1982; Farrar and Farrar, 1985; Hofstra and Nelson, 1969; Moorby and Jarman, 1975). Seasonal and treatment effects have been found, with C_4 plants tending to have much faster dynamics than C_3. But several of these reports showed that the leaf processes were not stationary, even when the plant was in a constant environment for many hours, with the parameters of the efflux models varying through the day (Bell and Incoll, 1982; Farrar and Farrar, 1985). With ^{11}C and efflux analysis Kays *et al.* (1987) also showed strong diurnal variation in model coefficients. We pointed out earlier that in these non-stationary situations efflux analysis is invalid (Section 11.3.1). However, dynamic analysis is well suited to dealing with systems that are not stationary. An illustrative example follows, similar to that in Thorpe and Minchin (1991).

Flow through a barley leaf was followed using ^{11}C, the tracer observed being both that in the whole plant and that mobilized from the labelled leaf (Figure 11.4a). The best dynamic description of the relationship between the two time series gave a reasonably good prediction of the output profile (Figure 11.4a). The transfer function commonly found for flow through leaves (Figure 11.3b) has more structure than the simple exponential decay that is suggested by efflux analysis (as also reported in Thorpe and Minchin, 1991; and unpublished results). The interpretation is that there is a dual pathway, with some of the flow taking a pathway with a time delay much longer than the other. A detailed interpretation in terms of leaf anatomy and processes would be speculative, but this illustration shows how the dynamics of the flow gives detail that must be inherent in a mechanistic model.

Data analysis assumed that the system was stationary (i.e. not changing with time), but the fit to the data was systematically low in the first part of the day,

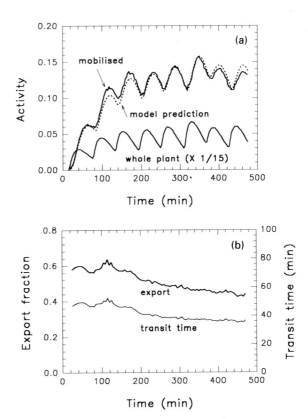

Figure 11.4. *Experiment to monitor carbon transport through a barley leaf. (a) tracer activity in the whole plant (the input), and activity mobilized through the petiole (the output) being activity in the whole plant less the labelled leaf not corrected for decay. The detectors were collimated as A and B in Figure 11.1. The predicted profile of mobilized activity was calculated from the whole plant profile using the dynamic relationship between the two observed profiles. The calculations assumed that the leaf dynamics were stationary. (b) Time courses of export fraction and transit time for flows of photoassimilate through the leaf, calculated assuming the dynamics were non-stationary.*

and high towards the end (Figure 11.4a). This suggests a non-stationary system and illustrates the value of dynamic analysis in obtaining estimates of changing flows. The export fraction was higher in the earlier part of the day (Figure 11.4b), suggested by the systematic disparity between observations and predicted values of mobilized tracer. The time of transit of recent assimilate through the leaf was also changing. We believe that the variation in export fraction and transit time are real phenomena, not experimental artifacts, because the plant was not touched after it was set up for measurement the evening before, and the measurements started 5 h into the photoperiod. The conclusion is that as well as the dual pathway for carbon flow in the barley leaf that was suggested by the dynamic model, there were diel changes in leaf export and transit time (Thorpe and Minchin, 1991).

Further information on flow dynamics can be obtained from analysis of transients in the flow after imposing a treatment. For example, export from leaves of C_3 plants reduced dramatically within 60 sec when the surrounding atmosphere was made anoxic, showing that the process of phloem loading is oxygen-dependent (Figure 11.5; Thorpe and Minchin, 1987). In C_4 species there was usually no effect of anoxia as the Kranz anatomy may protect the loading process from the anoxic treatments (Thorpe and Minchin, 1987). Phloem loading

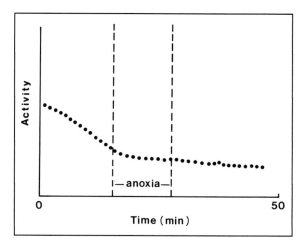

Figure 11.5. *Effect of anoxia on time course of decay-corrected ^{11}C activity in the labelled region of a leaf of* Panicum bisulcatum. *Reproduced from Thorpe and Minchin (1987) by permission of Oxford University Press.*

in C_4 leaves seems to be protected from the effects of SO_2 in the same way (Gould *et al.*, 1988). Similar experiments suggested that phloem loading did not respond to a shading treatment, but the data were not analysed quantitatively, and gave only immediate responses. With quantitative analysis, shading the shoot of a barley plant was found to induce a delayed response. We found the best dynamic description for the data up to the time of shading, assuming a stationary system, and then used this description to predict what the mobilized profile would have been with constant source leaf dynamics (Figure 11.6a). Clearly, shading induced an increase in the mobilized tracer over that which was predicted, that is shading increased the export fraction. Allowing the estimates of transport properties to vary quantified the increase in export fraction after shading the whole shoot (Figure 11.6b). Transit time through the leaf also increased on shading. Shading immediately reduced the photosynthetic rate of the source leaf by about 70%, but there was a slow increase in the export fraction of recently fixed photosynthate from about 0.5 to 0.68 which helped to compensate for the reduction in photosynthetic rate. The increase in export fraction represents a decrease of storage of recent photosynthate in the leaf.

It is well known that light level has a profound effect upon carbon allocation in barley leaves (Farrar and Farrar, 1987) with starch and sucrose production rates, translocation rate, rates of sucrose turnover and sucrose fluxes all changing. The above results demonstrate that changes in the flow coefficients begin to occur within about 20 min of changes while the flows themselves changed more rapidly (see Section 11.4.4). The very different responses to shade and to anoxia show that different processes in the leaf were affected.

11.4.2 *Pathway*

Dispersion has been found to be a feature of flow in the phloem, although a great deal less than in the processes of transport in leaves between mesophyll-

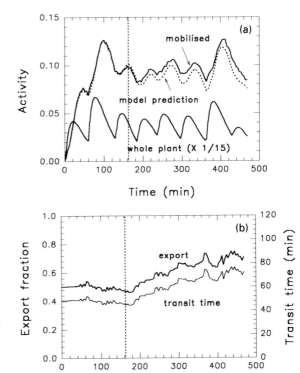

Figure 11.6. *Export of ^{11}C-labelled photosynthate from a barley leaf before and after the entire shoot was shaded. (a) Observed and calculated tracer profiles, the predicted profile being calculated assuming that the leaf dynamics remained the same after shading as they were before shading. (b) Time course of export and transit time for the source leaf calculated assuming non-stationary dynamics. Further details in Figure 11.4.*

chloroplast and the site of phloem loading (Christy and Fisher, 1978; Moorby *et al.*, 1963; Troughton *et al.*, 1977). Consequently, the shape of the temporal profile at a point on the pathway is largely determined in the leaf. As well as axial flow of solution in the pathway, there is radial leakage of tracer into the surrounding tissue, giving rise to dispersion (Evans *et al.*, 1963). We have found that the leakage of carbon can be a combination of both unloading and reloading. Using PCMBS to inhibit carrier systems, these component flows were quantified in young bean stems (Minchin and Thorpe, 1987a). Unloading was 6% per cm and reloading was 3% per cm, a net leakage of 3% per cm, compatible with the measurement of 0.8% per cm leakage in soybean by Evans *et al.* (1963).

Unloading and reloading into the phloem pathway means that the effects of an experimental treatment will be damped when observed downstream from the affected process. A good example of this is the effect of temperature. Accumulation of ^{14}C tracer at a sink leaf did not stop for at least 100 min after application of a cold block to a bean petiole (Geiger and Sovonick, 1970). However, Minchin *et al.* (1983), also using bean, demonstrated that phloem flow stopped immediately at the position where a cold block was applied (Figure 11.7), while flow into the sink, 150 mm further downstream, took over an hour to stop. This delay in response at the sink has been interpreted as the result of reloading from

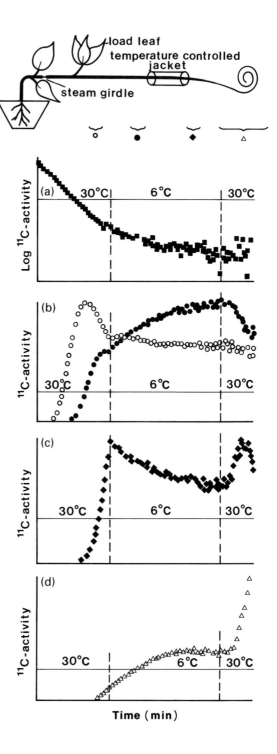

Figure 11.7. *Dynamics of cold-induced inhibition of phloem transport observed in profiles of ^{11}C activity at various places on the plant. The data are corrected for decay from time zero. The location of each detector is shown on the schematic diagram. (a) Source leaf efflux on a logarithmic scale. (b) Slit-collimated detectors upstream of the cold block. Note how the temporal profiles increased as soon as the cold block was applied. (c) Slit-collimated detector just downstream of the cold block. Note the immediate response to the cold block, both on application and removal. (d) Sink-collimated detector on the shoot apex. On applying the cold block there was an immediate reduction in tracer import into the shoot apex, but this did not stop for about 100 min. On removing the cold block, tracer flow quickly recommenced with vigour. Reproduced from Minchin et al. (1983) by permission of Oxford University Press.*

buffering pools in tissue just downstream of the blockage which allowed axial flow to continue until the buffer was exhausted. Flows also continued in the upsteam region because of phloem unloading into the buffering pool (Figure 11.7). The buffering pool has been identified with the apoplast of the stem (Minchin *et al.*, 1984). The observation of immediate cessation of flow implies that there is a blockage mechanism, yet to be identified, which can function within about a minute. Giaquinta and Geiger (1973) suggested displacement of sieve tube contents causing blockage of the sieve plates. Unfortunately, they only did the necessary microscopy on tissue which had been cooled for 30 min, and in this tissue found the sieve plate pores to be plugged. The time-scale associated with this mechanism of blockage has never been investigated.

Further detail about the temperature effect itself has been revealed by the use of ^{11}C in studies of the effect of rate of cooling on phloem transport. Transport stops for a period that depends on both the rate of cooling (Minchin and Thorpe, 1983; Pickard *et al.*, 1978c) and the recovery temperature. Other stimuli such as osmotic, mechanical or electric shocks also stop transport (Pickard and Minchin, 1990). Goeschl *et al.* (1984) reported similar transient blockages of phloem transport in cotton, with some stoppages that were apparently spontaneous. The physiological significance of these phenomena remains unknown, but one possible consequence is the reduction in plant growth caused by handling (Beardsell, 1977). The stimuli may cause local disruption of flow, possibly via calcium channel signalling, as lanthanum ions prevented the electrical response (Pickard and Minchin, 1990).

The importance of pressure gradients to phloem transport has been directly demonstrated by increasing the hydrostatic pressure around the shoot apex of a young bean plant using a pressure chamber, thereby reducing the pressure gradient and the flow of carbon into the apical sink (Minchin and Thorpe, 1987b). When the sieve tube pressure in bean roots was reduced by means of osmotica in the rooting medium, thereby increasing the pressure gradient, the phloem flow increased (Lang and Thorpe, 1986). In each case, the effect was observed within a few minutes of the treatment. Using continuous labelling and dynamic analysis, which allowed the effects of an osmotic treatment to be followed quantitatively and for much longer, Williams *et al.* (1991) have shown that there are pathway and sink processes which quickly started to regulate the flow, emphasizing the strong interaction of processes in source, path and sink which regulate carbon partitioning.

Most ^{11}C work reviewed in this chapter used sampling intervals of 1–5 min, and spatial scales of 10 mm or more. With ^{11}C it is possible to work with a time resolution of 10 sec or less, because of the high activities available. At this resolution and with the fine spatial resolution possible using Geiger–Müller tubes further details of tracer movement are revealed (Figure 11.8). Small fluctuations suggest that transport may involve a random succession of small parcels of sugar, perhaps in different sieve tubes, which are not resolved at the time scale usually used. The speed of these packets is much higher than the speed of bulk flow.

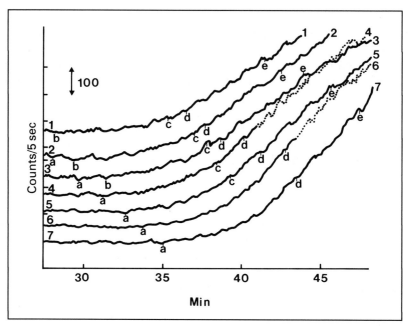

Figure 11.8. *Fine structure of ^{11}C temporal tracer profiles observed on a nasturtium petiole using Geiger–Müller detectors positioned in the sequence 1..7 from the labelled leaf and 10 cm apart. The data have not been corrected for isotopic decay. The profiles are displaced vertically for clarity. Note that pulse 'a' moves slower than pulse 'c' and that pulse 'a' is seen well before there is any bulk flow of label indicated by the sequential general rise in profiles commencing at about 33 min. Reproduced from Thompson et al. (1979) by permission of the National Research Council of Canada.*

This behaviour may account for the frequent observation (e.g. Bishop *et al.*, 1986) that the probability distribution of tracer activity detected in the stem is broader than expected from Poisson statistics: this suggests that the amount of tracer within small regions of pathway is fluctuating, about a mean level, as a result of fine scale transport processes. At coarser time resolutions, the fine detail is lost and flow appears to be convective.

11.4.3 *Sink*

Thorne and Rainbird (1983) showed that careful removal of the embryo and cotyledons from a soybean ovule, still attached to the plant, provided a good system for the study of unloading within developing seeds The empty seed coat is filled with a buffered solution and accumulation of labelled photosynthate into this solution is followed to study the pathway and control of unloading within seeds. Wolswinkel and Ammerlaan (1984) and Minchin and McNaughton (1986) reported increases in seed coat unloading when the solute level of the bathing solution was increased; on the other hand, Patrick *et al.* (1986) using seed coats which had been labelled with ¹⁴C-photosynthate and then detached, observed the opposite effect of solute concentration. These reports were all based upon observing tracer within the bathing solution. With ¹¹C labelling, Grusak and Minchin (1988) were also able to observe tracer within the ovule, and thereby to resolve this apparent conflict. They showed that an increase in bathing solution osmolarity had two effects. The increase in import of carbon to the seed coat from the plant was accompanied by an increase in the amount of unloading into the bathing solution, but the fraction of that imported which reached the bathing solution decreased. Thus there was no conflict between the apparently opposite responses, but with detached and attached ovules different phenomena were being observed. This example demonstrates the advantage of *in vivo* measurement of carbon flow, and also shows the need to be clear as to what is being measured. Minchin and Thorpe (1989) extended the method of dynamic analysis for use with continuous loading and first used this in the study of seed coat unloading in pea ovules. This enabled the long term effects of the surgery on the ovules to be seen, demonstrating a major decline of import into a surgically modified ovule.

Barley seedlings grown with a split-root system have been very useful as a system for studying import into sinks (Farrar and Minchin, 1991; Williams *et al.*, 1991). For example, FCCP affected root respiration within 1 min of application, while ¹¹C import was unchanged for about 1 h suggesting that phloem unloading within the barley root does not require metabolic energy (Farrar and Minchin, 1991).

A regular oscillation with a period of about an hour has been observed in ¹¹C flow into the ear and lower parts of wheat plants, despite steady flow from the labelled flag leaf (Figure 11.9; Roeb and Britz, 1991). Fluctuations with a similar period also occurred in the partitioning between ovules in a pea pod (Minchin and Thorpe, 1989). The physiological significance of these oscillations is unclear; in neither case was there any suggestion of a sinusoidal driving force which could generate a propagating wave of sugar (Ferrier *et al.*, 1975).

Figure 11.9. Waves of ¹¹C activity (not decay corrected) appearing in the ear of wheat. Tracer was applied to the flag leaf at a continuous and constant level and so label within the leaf became constant after about 2 h. Reproduced from Roeb and Britz (1991) by permission of Oxford University Press.

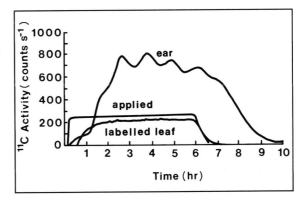

11.4.4 *Whole plant*

So far we have followed a reductionist approach by considering source, pathway and sink processes separately. But ^{11}C makes it is possible to monitor tracer simultaneously within source leaf, pathways and a number of sinks. In the experiment on the effect of shade on source function (Figure 11.6) we also monitored tracer within the root, and hence partitioning of recently fixed carbon to the root. It is not possible to monitor the shoot sink in a grass, because the sheath of the labelled leaf surrounds the shoot meristem, but we can monitor the root and infer shoot partitioning by difference. It can be seen from the observed and predicted profiles of activity in the root that there was a change in the root partitioning fraction after the entire shoot was shaded (Figure 11.10). Analysis of this pair of tracer profiles showed that the partitioning fraction to the root was constant up to the time of shading and then fell, so partitioning to the shoot must have increased when it was shaded. Interestingly, the time of transit was unaffected. The response seen in Figure 11.10 occurred more rapidly than was expected, and this may assist our understanding of the mechanism of the response. A simple model of phloem transport with one source and two sinks

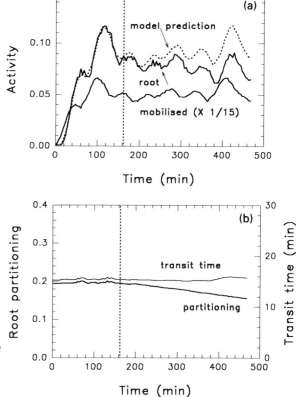

Figure 11.10.

Partitioning of mobilized photosynthate to the root of barley before and after the entire shoot was shaded. Detectors were collimated as B and C in Figure 11.1. For details of the analysis see Figure 11.6.

based upon osmotically driven bulk flow with saturable unloading predicts this observed response in partitioning between root and shoot (Minchin *et al.*, unpublished data). The short-term response reported here is also seen in the long term, that is, an increased growth rate of the shoot relative to that of the root (Brouwer, 1962; Davidson, 1969).

To bring together these separate studies of response to shade in source and sink, we can calculate fluxes of recent assimilate from the labelled leaf, using the dynamic descriptions of transport in the various parts of the plant. The calculations are simple if the plant is in the steady state, when transients need not be considered. In this case the net photosynthetic rate multiplied by the leaf export fraction gives the rate of flow of recent assimilate from the labelled leaf; then multiplying by the appropriate partitioning coefficient gives the rate of flow of recent assimilate into a sink. If the photosynthetic rate or plant dynamics are not constant then similar calculations can be carried out including the full dynamic behaviour as contained in the transfer function. For example, after shading the entire shoot of a barley plant and measuring export fraction and partitioning to the root, we calculated the fluxes of recent assimilate from the labelled leaf (Figure 11.11). The calculation employed the full dynamics, therefore showing

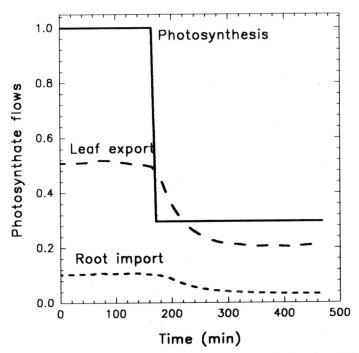

Figure 11.11. *The fluxes of recently fixed photosynthate in a barley plant, before and after shading the entire shoot, expressed as a fraction of the initial rate of carbon fixation. The fluxes of leaf export and root import were calculated from the inferred dynamics of tracer movement through the plant (Figures 11.6 and 11.10).*

transients when the photosynthetic rate changed. The major feature in the response to shading, when viewed as carbon flows, is that the flows in the plant follow the photosynthetic inflow, just as shown by Servaites and Geiger (1974). We must emphasize that ^{11}C measurements can only tell us about the movement of recently fixed photosynthate so cannot include any flows due to remobilization. Unless long-term labelling is used, this is also true for ^{14}C measurements.

For mechanistic studies there is more information in the parameters describing the dynamics of the system which represents the processes, rather than in the flows of carbon which result from them. So even though it is possible to construct a carbon balance, the descriptions of data in terms of the dynamics are of more interest.

Electroshocking a short segment of a bean stem caused an immediate inhibition of phloem transport through the shocked segment, probably due to local disruption of flow (Pickard and Minchin, 1990). However, it has recently been observed that electroshocking a short segment of a pea peduncle induced an immediate change in the relative partitioning to each pea fruit and simultaneously a reduction in export from the labelled leaf (Figure 11.12). This is an example of a pathway treatment having an immediate and sustained effect at both source and sink. To investigate the causal chain of events, we used anoxia at the source leaf as an alternative way to induce changes in phloem loading (Thorpe and Minchin, 1987), and found that the change in export flux caused a change in the relative

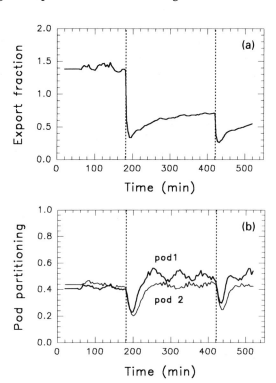

Figure 11.12. *Effect of electric shocks to a peduncle supplying two pea pods on export of recent assimilate from the leaf at that node, and on partitioning between the two pods. Electric shocks were administered twice at the times indicated.*

partitioning between sinks supplied from that source. This effect is predicted by a simple model of phloem transport from a single source to two sinks based upon osmotically driven bulk flow with saturable unloading (Minchin *et al.*, unpublished data). The implication is that the effect of the electrical shock on partitioning between the pods was remote, giving rise first to a reduction in export from the labelled leaf, which in turn affected partitioning between the pods. Remote effects of stimuli have been reported, both on electrophysiology (Malone and Stankovic, 1991) and on carbon transport (Fromm, 1991; Grusak and Lucas, 1986; Pickard *et al.*, unpublished data), suggesting that electrical signalling in plants may have a significant role.

Detailed analysis of the Munch theory of phloem transport predicts that with increased phloem loading, the speed of phloem transport will first increase, as will amounts of photosynthate along the pathway, until unloading becomes saturated when the transport speed begins to fall (Goeschl and Magnuson, 1986). In all species tested, increased loading gave rise to increased levels of ^{11}C-labelled photosynthate along the pathway and to an increased transport speed calculated from the first time of tracer arrival (Magnuson *et al.*, 1986). Further increase in loading rate caused a fall in transport speed as predicted when unloading becomes saturated. Their interpretation of ^{11}C activity observed along the pathway as a direct measurement of photosynthate concentration within the sieve tubes is in conflict with the currently accepted idea that photosynthate is continuously unloaded and reloaded along the pathway (Delrot and Bonnemain, 1985; Gifford and Evans, 1981; Patrick, 1988).

Observations of partitioning throughout a plant using ^{11}C have been made for long-term responses (days or more). For example, Wang *et al.* (1989) observed that mycorrhizal infection increased photosynthetic rate by 20%, doubled the partitioning of recently fixed carbon to stems, and increased the rate of storage of recent photosynthate within the leaves by 45%. Also, Spence *et al.* (1990) have shown that exposure of pine seedlings to ozone for 12 weeks reduced photosynthesis by about 16%, speed of phloem transport by 11%, photosynthate concentration within the phloem by 40%, and transport of recently fixed photosynthate to the roots by 45%. Photosynthate also accumulated within stems so that ozone-treated plants had heavier primary stems than control plants. Not all of these observations required non-destructive, *in vivo*, measurement, and could have been equally well made using simpler and cheaper techniques. Destructive measurement of export from the labelled leaf and partitioning between sinks using ^{14}C are routine procedures, as are gas exchange measurements of photosynthetic rates. *In vivo* measurement of speed and photosynthate concentrations within the phloem required ^{11}C, though reservations on the latter are expressed above.

11.5 Conclusion

The state of knowledge concerning the control of carbon partitioning throughout plants is at a very early stage. Most research on source–sink interactions has used pruning treatments followed by observation of the induced changes in source–sink relations made hours or even days later. Over this time scale adaptive changes occur, altering the system observed. The sequence of changes that occur whenever there is a perturbation from the current growth pattern includes effects with a range of time scales, but we believe that the immediate responses are those which will reveal the mechanisms currently responsible for effecting and controlling carbon distribution, longer term responses being the cumulative effect of changes which lead to adaptation. Further, the entire carbon transport system and associated processes in the plant are highly interactive, so it is essential that loading, pathway and unloading processes are simultaneously monitored.

Because of its unique suitability for such studies we believe that ^{11}C has a bright future as a major tool in elucidating the mechanisms and control of phloem transport and carbon partitioning. Only a few facilities are available for this work, so it is essential that these continue to be funded and are available for collaborative work.

References

Bauermeister, A., Dale, J.E., Williams, E.J. and Scobie, J. (1980) Movement of ^{14}C and ^{11}C labelled assimilate in wheat leaves: the effect of IAA. *J. Exp. Bot.* 31, 1199-1209.

Beardsell, M.F. (1977) Effects of routine handling on maize growth. *Aust. J. Plant Physiol.* 4, 857-861.

Bell, C.J. and Incoll, L.D. (1982) Translocation from the flag leaf of winter wheat in the field. *J. Exp. Bot.* 33, 896-909.

Bishop, H.T., Thompson, R.G., Aikman, D.P. and Fensom, D.S. (1986) Fine structure aberrations in the movement of ^{11}C and ^{13}N in the stems of plants. *J. Exp. Bot.* 37, 1780-1794.

Brouwer, R. (1962) Distribution of dry matter in the plant. *Neth. J. Agric. Sci.* 10, 361-376.

Cheeseman, J.M. (1986) Compartmental efflux analysis: an evaluation of the technique and its limitations. *Plant Physiol.* 80, 1006-1011.

Christy, A.L. and Fisher, D.B. (1978) Kinetics of ^{14}C photosynthate translocation in morning glory vines. *Plant Physiol.* 61, 283-290.

Davidson, R.L. (1969) Effect of root/leaf temperature differentials on root/shoot ratios in some pasture grasses and clover. *Ann. Bot.* 33, 561-569.

Delrot, S. and Bonnemain, J. (1985) Mechanism and control of phloem transport. *Physiol. Veg.* 23, 199-220.

Dyer, M.I., Acra, M.A., Wang, G.M., Coleman, D.C., Freckman, D.W., McNaughton, S.J. and Strain, B.R. (1991) Source-sink carbon relations in two *Panicum coloratum* ecotypes in response to herbivory. *Ecology,* 72, 1472-1483.

Evans, N.T.S., Ebert, M. and Moorby, J. (1963) A model for the translocation of photosynthate in the soybean. *J. Exp. Bot.* 14, 221-231.

Fares, Y., DeMichele, D.W., Goeschl, J.D. and Baltuskonis, D.A. (1978) Continuously produced high specific activity ^{11}C for studies of photosynthesis, transport and metabolism. *Int. J. Appl. Rad. Isotopes* 29, 431-441.

Farrar, S.C. and Farrar, J.F. (1985) Carbon fluxes in leaf blades of barley leaves. *New Phytol.* 271–283.

Farrar, S.C. and Farrar, J.F. (1987) Effects of photon fluence rate on carbon partitioning in barley source leaves. *Plant Physiol. Biochem.* 25, 541–548.

Farrar, J.F. and Minchin, P.E.H. (1991). Carbon partitioning in split root systems of barley: relation to metabolism. *J. Exp. Bot.* 42, 1261–1269.

Fensom, D.S., Williams, E.J., Aikman, D., Dale, J.E., Scobie, J., Ledingham, W.O., Drinkwater, A. and Moorby, J. (1977) Translocation of ¹¹C from leaves of *Helianthus*: preliminary results. *Can.J. Bot.* 55, 1787–1793.

Fensom, D.S., Thompson, R.G. and Caldwell, C.D. (1990) Ammonia gas temporarily interrupts translocation of ¹¹C photosynthate in sunflower. *J. Exp. Bot.* 41, 11–14.

Ferrier, J.M., Tyree, M.T. and Christy, A.L. (1975) A theoretical time-dependent behavior of a munch pressure-flow system: the effect of sinusoidal time variation in sucrose loading and water potential. *Can. J. Bot.* 53, 1120–1127.

Fromm, J. (1991) Control of phloem unloading by action potentials in *Mimosa*. *Physiol. Plant.* 83, 529–533.

Geiger, D.R. and Fondy, B.R. (1979) A method for continuous measurement of export from a leaf. *Plant Physiol.* 64, 361–365.

Giaquinta, R.T. and Geiger, D.R. (1973) Mechanism of inhibition of translocation by localized chilling. *Plant Physiol.* 51, 372–377.

Gifford, R.M. and Evans, L.T. (1981) Photosynthesis, carbon partitioning and yield. *Ann. Rev. Plant Physiol.* 32, 485–509.

Goeschl, J.D. and Magnuson, C.E. (1986) Physiological implications of the Münch–Horowitz theory of phloem transport: effects of loading rates. *Plant Cell Environ.* 9, 95–102.

Goeschl, J.D., Magnuson, C.E., Fares, Y., Jaeger, C.H., Nelson, C.E. and Strain, B.R. (1984) Spontaneous and induced blocking and unblocking of phloem transport. *Plant Cell Environ.* 7, 607–613.

Gould, R.P., Minchin, P.E.H. and Young, P.C. (1988) The effects of sulphur dioxide on phloem loading and transport — evidence for inhibition of these processes. *J. Exp. Bot.*, 39, 997–1007.

Grusak, M.A. and Lucas, W.J. (1986) Cold-inhibited phloem translocation in sugar beet. III The involvement of the phloem pathway in source-sink partitioning. *J. Exp. Bot.* 37, 277–288.

Grusak, M.A. and Minchin, P.E.H. (1988) Seed coat unloading in *Pisum sativum* — osmotic effects in attached versus excised empty ovules. *J. Exp. Bot.* 39, 543–559.

Hofstra, G. and Nelson, C.D. (1969) A comparative study of translocation of assimilated ¹⁴C from leaves of different species. *Planta*, 88, 103–112.

Jahnke, S., Stocklin, G. and Willenbrick, J. (1981) Translocation profiles of ¹¹C-assimilates in the petiole of *Marsilea quadrifolia* L. *Planta*, 153, 56–63.

Jahnke, S., Bier, D., Estruch, J.J. and Beltran, J.P. (1989) Distribution of photoassimilates in the pea plant: chronology of events in non-fertilized ovaries and effects of gibberellic acid. *Planta*, 180, 53–60.

Kays, S.J., Goeschl, J.D., Magnuson, C.E. and Fares, Y. (1987) Diurnal changes in fixation, transport, and allocation of carbon in the sweet potato using ¹¹C tracer. *J. Am. Soc. Hort. Sci.* 112, 545–554.

Lang, A. and Thorpe, M.R. (1986) Water potential, translocation and assimilate partitioning. *J. Exp. Bot.* 37, 495–503.

Magnuson, C.E., Goeschl, J.D. and Fares, Y. (1986) Experimental tests of the Munch-Horowitz theory of phloem transport: effects of loading rates. *Plant Cell Environ.* **9**, 103–109.

Malone, M. and Stankovic, B. (1991) Surface potentials and hydraulic signals in wheat leaves following localized wounding by heat. *Plant Cell Environ.* **14**, 431–436.

Minchin, P.E.H. (1986). *Short-Lived Isotopes in Biology.* Proceedings of an international workshop on biological research with short-lived isotopes. DSIR, Wellington.

Minchin, P.E.H. and Troughton, J.H. (1980) Quantitative interpretation of phloem translocation data. *Ann. Rev. Plant Physiol.* **31**, 191–215.

Minchin, P.E.H. and Thorpe, M.R. (1983) A rate of cooling response in phloem translocation. *J. Exp. Bot.* **34**, 529–536.

Minchin, P.E.H. and McNaughton, G.S. (1986) Phloem unloading within the seed coat of *Pisum sativum* observed using surgically modified seeds *J. Exp. Bot.* **37**, 1151–1163.

Minchin, P.E.H. and Thorpe, M.R. (1987a) Measurement of unloading and reloading of photo-assimilate within the stem of bean. *J. Exp. Bot.* **38**, 211–220.

Minchin, P.E.H. and Thorpe, M.R., (1987b). Is phloem transport due to a hydrostatic pressure gradient? Supporting evidence from pressure chamber experiments. *Aust. J. Plant Physiol.* **14**, 397–402.

Minchin, P.E.H. and Thorpe, M.R. (1989) Carbon partitioning to whole versus surgically modified ovules of pea: an application of the *in vivo* measurement of carbon flows over many hours using the short-lived isotope carbon-11. *J. Exp. Bot.* **40**, 781–787.

Minchin, P.E.H., Lang, A. and Thorpe, M.R. (1983) Dynamics of cold induced inhibition of phloem transport. *J. Exp. Bot.* **34**, 156–162.

Minchin, P.E.H., Ryan, K.G. and Thorpe, M.R. (1984) Further evidence of apoplastic unloading in the stem of bean: identification of the phloem buffering pool. *J. Exp. Bot.* **35**, 1744–1753.

Moorby, J. and Jarman, P.D. (1975) The use of compartmental analysis in the study of the movement of carbon through leaves. *Planta,* **122**, 155–168.

Moorby, J., Ebert, M. and Evans, N.T.S. (1963) The translocation of ^{11}C-labelled photosynthate in the soybean. *J. Exp. Bot.* **14**, 210–220.

More, R.D. and Troughton, J.H. (1973) Production of $^{11}CO_2$ for use in plant translocation studies. *Photosynthetica* **7**, 271–274.

Patrick, J.W. (1988) Assimilate partitioning in relation to crop productivity. *Hort. Sci.* **23**, 33–40.

Patrick, J.W., Jacobs, E., Offler, C.E. and Cram, W.J. (1986) Photosynthate unloading from seed coats of *Phaseolus vulgaris* L. — nature and cellular location of turgor-sensitive unloading. *J. Exp. Bot.* **37**, 1006–1019.

Pickard, W.F. and Minchin, P.E.H. (1990) The transient inhibition of phloem translocation in *Phaseolus vulgaris* by abrupt temperature drops, vibration, and electric shock. *J. Exp. Bot.* **41**, 1361–1369.

Pickard, W.F., Minchin, P.E.H. and Troughton, J.H. (1978a) Real time studies of carbon-11 translocation in moonflower. I. The effects of cold blocks. *J. Exp. Bot.* **29**, 993–1001.

Pickard, W.F., Minchin, P.E.H. and Troughton, J.H. (1978b) Real time studies of carbon-11 translocation in moonflower. II. The effects of metabolic and photosynthetic activity and water stress. *J. Exp. Bot.* **29**, 1003–1009.

Pickard, W.F., Minchin, P.E.H. and Troughton, J.H. (1978c) Transient inhibition of translocation in *Ipomoea alba* L. by small temperature reductions. *Aust. J. Plant Physiol.* **5**, 127–130.

Rocher, J.P. and Prioul, J.L. (1987) Compartmental analysis of assimilate export in a mature maize leaf. *Plant Physiol. Biochem.* 25, 531–540.

Roeb, G. and Britz, S.J. (1991) Short-term fluctuations in the transport of assimilates to the ear of wheat measured with steady-state ^{11}C-CO_2-labelling of the flag leaf. *J. Exp. Bot.* 42, 469–475.

Ruben, S., Hassid, W.Z. and Kamen, M.D. (1939) Radioactive carbon in the study of photosynthesis. *J. Am. Chem. Soc.* 61, 661–663.

Servaites, J.C. and Geiger, D.R. (1974) Effects of light intensity and oxygen on photosynthesis and translocation in sugar beet. *Plant Physiol.* 54, 575–578.

Spence, R.D., Rykiel, E.J. and Sharpe, P.J.H. (1990) Ozone alters carbon allocation in lobolly pine: assessment with carbon 11 labeling. *Environ. Pollut.* 64, 93–106.

Swanson, C.A. and Geiger, D.R. (1967) Time course of low temperature inhibition of sucrose translocation in sugar beet. *Plant Physiol.* 42, 751–756.

Thain, J.F., (1984) The analysis of radioisotope tracer flux experiments in plant tissues. *J. Exp. Bot.* 35, 444–453.

Thompson, R.G., Fensom, D.S., Anderson, R.R., Drouin, R. and Leiper, W. (1979) Translocation of ^{11}C from leaves of *Helianthus, Heracleum, Nymphoides, Ipomoea, Tropaeolum, Zea, Fraxinus, Ulmus, Picea,* and *Pinus*: comparative shapes and some fine structure profiles. *Can. J. Bot.* 57, 845–863.

Thorne, J.H. and Rainbird, R.M. (1983) An *in vivo* technique for the study of phloem unloading in seed coats of developing soybean seeds *Plant Physiol.* 72, 268–271.

Thorpe, M.R. and Minchin, P.E.H. (1987) Effects of anoxia on phloem loading in C_3 and C_4 species. *J. Exp. Bot.* 38, 221–232.

Thorpe, M.R. and Minchin, P.E.H. (1991) Continuous monitoring of fluxes of photoassimilate in leaves and whole plants. *J. Exp. Bot.* 42, 461–468.

Troughton, J.H., Currie, B.G. and Chang, F.H. (1977) Relations between light level, sucrose concentrations, and translocation of carbon-11 in *Zea mays* leaves. *Plant Physiol.* 59, 808–820.

Wang, G.M., Coleman, D.C., Freckman, D.W., Dyer, M.I., McNaughton, S.J. and Goeschl, J.D. (1989) Carbon partitioning patterns of mycorrhizal verus non-mycorrhizal plants: real-time dynamic measurements using $^{11}CO_2$. *New Phytol.* 112, 489–493.

Williams, E.J., Dale, J.E., Moorby, J. and Scobie, J. (1979) Variation in translocation during the photoperiod: Experiments feeding $^{11}CO_2$ to sunflower. *J. Exp. Bot.* 30, 727–738.

Williams, J.H.H., Minchin, P.E.H. and Farrar, J.F. (1991) Carbon partitioning in split root systems of barley: the effect of osmotica. *J. Exp. Bot.* , 212, 454–460.

Wolswinkel, P. and Ammerlaan, A. (1984) Turgor-sensitive sucrose and amino acid transport into developing seeds of *Pisum sativum*. Effect of a high sucrose or mannitol concentration in experiments with empty ovules. *Plant Physiol.* 61, 172–182.

Zierler, K. (1981) A critique of compartmental analysis. *Ann. Rev. Biophys. Bioeng.* 10, 531–562.

Pathways of phloem loading and unloading: a plea for uniform terminology

K.J. Oparka and A.J.E. van Bel

12.1 Introduction

Photosynthetic assimilates, manufactured in the green tissues of higher plants, are continuously mobilized from their sites of production ('sources') to sites of utilization or storage ('sinks'). The terms phloem loading and phloem unloading have been used to describe the entry of assimilates into, and exit from, the phloem and are now in extensive use in the plant physiology literature. Loomis (1955) first introduced the term 'vein loading' to refer to the movement of photosynthetic assimilates from the mesophyll to the minor veins of leaves. Eschrich (1970) later proposed that 'vein loading' should be more accurately replaced by 'phloem loading', since many organs lack distinct veins but still transport assimilates to the phloem.

In an early review of the subject of phloem loading, Geiger (1975), like many subsequent researchers in the field, carefully distinguished between assimilate transport into the sieve element (SE) from contacting cells (SE loading) and the cell-to-cell transport processes preceding this event. Initial studies of SE loading provided physiological evidence that assimilates were transported across the SE plasma membrane from the apoplast (reviewed by Geiger, 1975) giving rise to the concept of 'apoplastic SE loading'. It now seems very unlikely that in all species assimilates are loaded into the SE directly from the apoplast, and recent evidence would suggest that in several species symplastic SE loading predominates (van Bel and Gamalei, 1991). The picture has become equally complex in sinks, e.g. storage organs, where a variety of phloem unloading pathways may be operational (Oparka, 1990; Patrick, 1990; Thorne, 1985).

The complexity of available transport pathways leading to, or from, the sieve element–companion cell (SE–CC) complex has given rise to considerable ambiguity as to the precise nature of phloem (un)loading processes, particularly when the term 'phloem (un)loading' is preceded by descriptions of the pathway, such as 'apoplastic' or 'symplastic'. As an example, cereal caryopses are frequently referred to as 'apoplastic unloading sinks' because a symplastic discontinuity occurs at some point removed from the SE–CC complex (the nucellus/aleurone interface). However, the SE unloading step in many caryopses appears to be symplastic (Oparka, 1990).

We wish to propose a clarification of the current terminology relating to phloem (un)loading which we hope will lead to a more accurately defined use of existing terms. We see no need to introduce further (and possibly confusing) jargon into the literature and believe that the existing terminology is perfectly sufficient, provided that a simple set of descriptive definitions are adhered to. These definitions refer predominantly to the **pathways** utilized in phloem (un)loading and SE (un)loading, rather than the physiological mechanisms (e.g. active, passive, carrier-mediated transport) involved in these processes. The aim is to provide an accurate structural basis for physiological studies.

For the purposes of this article the SE–CC complex is regarded as a single functional unit.

12.2 Definitions

12.2.1 *Phloem (un)loading pathway*

We propose that the general term 'phloem loading pathway' should be used to refer to the pathway of solute movement from the mesophyll cell (or alternative source cell in the case of non-chlorophyllous tissues) to the SE–CC complex. Conversely, the term 'phloem unloading pathway' should refer to the pathway from the SE–CC complex to the assimilate-receiving cells (e.g. endosperm cells in the case of cereals).

12.2.2 *SE (un)loading*

This term should refer to the transfer of assimilates between the SE–CC complex and **contacting** cells only.

We propose that the vague term '(un)loading' be abandoned unless it is preceded by either 'phloem' or 'SE' since its use is likely to lead to confusion regarding the precise nature of the transport steps being studied. Furthermore, we recommend that '(un)loading' is not used to refer to transport processes in non-phloem tissues (e.g. mesophyll protoplasts or seed-coat cells). In such cases the term '(un)loading' should be replaced by more descriptive expressions relating to the transport process being stuided, for example, 'solute uptake' or 'assimilate release'.

12.2.3 *Symplastic phloem (un)loading*

When the entire functional pathway leading to and including the SE–CC complex is symplastic then the criterion for symplastic phloem (un)loading has been met (Figure 12.1A).

12.2.4 *Apoplastic phloem (un)loading*

This term should be used when symplastic continuity is interrupted on **any** part of the phloem (un)loading pathway, regardless of its distance from the SE–CC

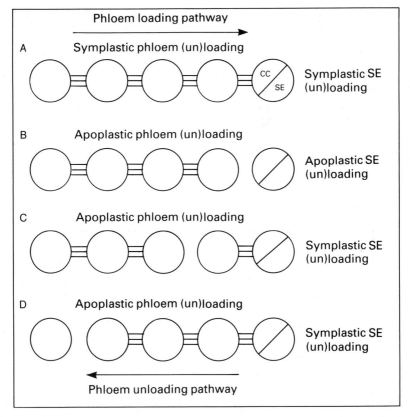

Figure 12.1. *Potential pathways available for phloem loading and phloem unloading. The SE-companion cell (SE–CC) complex is considered as a single functional unit. The presence of plasmodesmata (=) is used to indicate **functional** symplastic continuity between cells, as well as the dominant transport pathway. An absence of plasmodesmata between cells indicates a lack of functional symplastic continuity, i.e. an apoplastic transport step. Such steps may be absent along the entire pathway (A), present between the SE–CC and adjoining cells only (B) or present at some distance removed from the SE–CC complex (C and D).*

complex. In many tissues (e.g. stems; van Bel and Kempers, 1990) such a symplastic discontinuity occurs between the SE–CC complex and adjoining cells (Figure 12.1B). However, several tissues (particularly sinks) display a symplastic discontinuity at some point removed from the SE–CC complex. This therefore defines the entire phloem (un)loading pathway as apoplastic, even if the the SE (un)loading step is symplastic (Figure 12.1C and D).

12.2.5 *Symplastic SE (un)loading*

This term should be reserved for demonstrated functional symplastic continuity between the SE–CC complex and its adjoining cells (Figure 12.1A, C and D).

12.2.6 *Apoplastic SE (un)loading*

This term should be used when transfer of solutes occurs between the SE–CC complex and its contacting cells directly from the apoplast, that is, across the plasma membrane of the SE–CC complex (Figure 12.1B).

Examples of the types of phloem (un)loading and SE (un)loading shown in Figure 12.1 are given in Table 12.1. The list is not meant to be exhaustive but merely to provide examples of the above transport pathways. It is appreciated that the presence of plasmodesmata between adjacent cells does not necessarily invoke

Table 12.1. *Examples of the phloem (un)loading pathways defined in Figure 12.1*

Transport category	Genus	Organ	Phloem loading	Phloem unloading	Reference
A	*Cucurbita*	Mature leaf	√		Schmitz *et al.* (1987)
	Coleus	Mature leaf	√		Turgeon *et al.* (1975)
	Solanum	Tuber		√	Oparka (1986)
	Hordeum	Root apex		√	Warmbrodt (1985a)
	Zea	Root apex		√	Warmbrodt (1985b)
	Nicotiana	Expanding leaf		√	Ding *et al.* (1988)
B	*Zea*	Mature leaf	√		Evert *et al.* (1978)
	Commelina	Mature leaf	√		van Bel *et al.* (1988)
	Phaseolus	Stem	√	√	Hayes *et al.* (1985)
	Ricinus	Stem	√	√	van Bel and Kempers (1991)
C	*Phaseolus*	Seed coat		√	Offler and Patrick (1984)
D	*Vicia*	Leaf	√		Bourquin *et al.* (1990)
	Oryza	Caryopsis		√	Oparka and Gates (1981)

The examples were selected largely on the basis of quantitative ultrastructural evidence (rather than physiological evidence) purely to conform with the descriptive definitions given in Figure 12.1. It is appreciated that the most convincing evidence for a given phloem (un)loading pathway will be derived from studies which utilize ultrastructural and physiological approaches simultaneously. It is significant that physiological evidence has, on occasions, contradicted ultrastructural evidence (see Offler and Patrick, 1984; Oparka, 1990).

symplastic continuity. Similarly, cells containing abundant plasmodesmata might at some stage in their development exhibit apoplastic transport. However, for the purposes of this article the presence of plasmodesmata is used to indicate **functional** symplastic continuity between cells.

It is our hope that the definitions used in this brief article will provide a framework within which researchers can describe their findings accurately and unambiguously. We know that many anatomists and physiologists already use the terms 'phloem (un)loading' and 'SE (un)loading' broadly as they have been defined in this article. However, it is clear that frequent misuse of the term 'unloading' has given rise to considerable ambiguity in the literature (Oparka, 1990). This unfortunate situation has arisen as a natural consequence of the expansion of research which has occurred in the field of phloem transport. If the situation is not remedied in the near future then existing workers may find it increasingly difficult to relay their findings to a wider audience while newcomers may find it a minefield of semantics.

We therefore urge that the terms defined above be adopted where appropriate. We are in no doubt that future studies will precipitate a re-evaluation of this terminology. However, we feel it is time to provide a new framework for the further investigations of the transport events preceding (or subsequent to) long-distance transport in the phloem.

Acknowledgement

K.J.O. acknowledges the financial support of the Scottish Office Agriculture and Fisheries Department (SOAFD).

References

van Bel., A.J.E. and Kempers, R. (1991) Symplastic isolation of the sieve element-companion cell complex in the phloem of *Ricinus communis* and *Salix alba* stems. *Planta*, 183, 69–76.

van Bel., A.J.E. and Gamalei, Y.V. (1991) Multiprogrammed phloem loading. In: *Recent Advances in Phloem Transport and Assimilate Compartmentation* (eds J.-L. Bonnemain, S. Delrot, W.J. Lucas and J. Dainty). Ouest Editions, Nantes, pp. 128–140.

van Bel., A.J.E., van Kesteren, W.J.P. and Papenhuijzen, C. (1988) Ultrastructural indications for coexistence of symplastic and apoplastic phloem loading in *Commelina benghalensis* leaves. Differences in ontogenic development, spatial arrangement and symplastic connections of the two sieve tubes in the minor vein. *Planta*, 176, 159–172.

Bourquin, S., Bonnemain, J.-L. and Delrot, S. (1990) Inhibition of loading of ^{14}C-assimilates by p-chloromercuribenzenesulfonic acid. *Plant Physiol.* 92, 97–102.

Ding, B., Parthasarathy, M.V., Niklas, K. and Turgeon, R. (1988) A morphometric analysis of the phloem-unloading pathway in developing tobacco leaves. Planta, 176, 307–318.

Eschrich, W. (1970) Biochemistry and fine structure of phloem in relation to transport. *Ann. Rev. Plant Physiol.* 21, 193–214.

Evert, R.F., Eschrich, W. and Heyser, W. (1978) Leaf structure in relation to solute transport and phloem loading in *Zea mays* L. *Planta*, 138, 279–294.

Geiger, D.R. (1975) Phloem loading. In: *Transport in Plants I. Phloem Transport*, Volume 1, *Encyclopedia of Plant Physiology New Series* (eds M.H. Zimmermann and J.A. Milburn). Springer-Verlag, Berlin, pp. 395–431.

Hayes, P.M., Offler, C.E. and Patrick, J.W. (1985) Cellular structures, plasma membrane surface areas and plasmodesmatal frequencies of the stem of *Phaseolus vulgaris* L. in relation to radial photosynthate transfer. *Ann. Bot.* 56, 125–138.

Loomis, W.E. (1955). Resistance of plants to herbicides. In *Origins of Resistance to Toxic Agents* (eds M.G. Sevag and R.D. Reid). Academic Press, New York, pp. 99–121.

Offler, C.E. and Patrick, J.W. (1984) Cellular structures, plasma membrane surface areas and plasmodesmatal frequencies of seed coats of *Phaseolus vulgaris* L. in relation to photosynthate transfer. *Aust. J. Plant Physiol.* 11, 79–99.

Oparka, K.J. (1986) Phloem unloading in the potato tuber. Pathways and sites of ATPase. *Protoplasma*, 131, 201–210.

Oparka, K.J. (1990) What is phloem unloading? *Plant Physiol.* 94, 393–396.

Oparka, K.J. and Gates, P. (1981) Transport of assimilates in the developing caryopsis of Rice (*Oryza sativa* L.). *Planta*, 151, 561–573.

Patrick, J.W. (1990) Sieve element unloading: cellular pathway, mechanism and control. *Physiol. Plant.* 78, 298–308.

Schmitz, A., Cuypers, B. and Moll, M. (1987) Pathway of assimilate transfer between mesophyll cells and minor veins in leaves of *Cucumis melo* L. *Planta*, 171, 19–29.

Thorne, J.H. (1985) Phloem unloading of C and N assimilates in developing seeds. *Ann. Rev. Plant Physiol.* 36, 317–343.

Turgeon, R., Webb, J.A. and Evert, R.F. (1975) Ultrastructure of minor veins of *Cucurbita pepo* leaves. *Protoplasma*, 83, 217–232.

Warmbrodt, R.D. (1985a) Studies on the root of *Hordeum vulgare* L. — ultrastructure of the seminal root with special reference to the phloem. *Am. J. Bot.* 72, 414–432.

Warmbrodt, R.D. (1985b) Studies on the root of *Zea mays* L. — structure of the adventitious roots with respect to phloem unloading. *Bot. Gaz.* 176, 169–180.

Index